Multicultural Science Education

Mary M. Atwater • Melody L. Russell
Malcolm B. Butler
Editors

Multicultural Science Education

Preparing Teachers for Equity
and Social Justice

Editors
Mary M. Atwater
Department of Mathematics
and Science Education
University of Georgia
Athens, GA, USA

Melody L. Russell
Department of Curriculum and Teaching
Auburn University
Auburn, AL, USA

Malcolm B. Butler
University of Central Florida
Orlando, FL, USA

ISBN 978-94-007-7650-0   ISBN 978-94-007-7651-7 (eBook)
DOI 10.1007/978-94-007-7651-7
Springer Dordrecht Heidelberg New York London

Library of Congress Control Number: 2013952735

© Springer Science+Business Media Dordrecht 2014
This work is subject to copyright. All rights are reserved by the Publisher, whether the whole or part of the material is concerned, specifically the rights of translation, reprinting, reuse of illustrations, recitation, broadcasting, reproduction on microfilms or in any other physical way, and transmission or information storage and retrieval, electronic adaptation, computer software, or by similar or dissimilar methodology now known or hereafter developed. Exempted from this legal reservation are brief excerpts in connection with reviews or scholarly analysis or material supplied specifically for the purpose of being entered and executed on a computer system, for exclusive use by the purchaser of the work. Duplication of this publication or parts thereof is permitted only under the provisions of the Copyright Law of the Publisher's location, in its current version, and permission for use must always be obtained from Springer. Permissions for use may be obtained through RightsLink at the Copyright Clearance Center. Violations are liable to prosecution under the respective Copyright Law.
The use of general descriptive names, registered names, trademarks, service marks, etc. in this publication does not imply, even in the absence of a specific statement, that such names are exempt from the relevant protective laws and regulations and therefore free for general use.
While the advice and information in this book are believed to be true and accurate at the date of publication, neither the authors nor the editors nor the publisher can accept any legal responsibility for any errors or omissions that may be made. The publisher makes no warranty, express or implied, with respect to the material contained herein.

Printed on acid-free paper

Springer is part of Springer Science+Business Media (www.springer.com)

Foreword: Ruminations About a Half Century of Equity in Science Education

When contacted by the coeditors, I eagerly accepted their offer to write the Foreword for this most timely and important book. I envisioned an easy task of summarizing and analyzing the findings in its chapters, but those things have been thoroughly done in the "Introduction" and the final chapter, "Conclusion and Next Steps for Science Teacher Educators." As I read the chapters and thought about the possible impact of this book, I slowly came to realize that this volume can make concrete changes in the way future and current teachers teach—and their students learn—science by addressing the complexity of science teacher education, with questions like how do universities educate future teachers of science, who are their professors, what curricula are used, and where and with whom do prospective teachers get their practical experience; and, once science teachers enter classrooms, how are they supported in their efforts to achieve both equity and excellence in teaching science, how do they relate to students who differ from them, and how do they encourage diverse students to study science and, possibly, become scientists; and last, and key to this groups of book, who is encouraged to become a science teacher educator, what are their career trajectories, what are their scholarly interests, and how can they affect state and national education policies. Answers to these and other questions are found in this book.

I also reflected on the long journey that science education has taken toward equity, multicultural education, and social justice, the themes of this book. Fifty-five years ago, Margaret Mead assessed students' images of scientists by asking them to *Draw a Scientist*. Needless to say, in the 1950s, as well as today, drawings are consistently of middle-aged White males, dressed in white laboratory coats with pocket protectors and wearing glasses, a stereotype based on a Western/European perspective. Thirty-five years later, colleagues and I revisited Dr. Mead's premise, asking the same question of students in several countries, and the stereotype remains. At the same time in similar studies in mathematics, students drew both men and women mathematicians who were also younger and less eccentric. The impact of this stereotyping of scientists (and science) on students is explored in several chapters, while other authors provide examples of instruction and curricula that help to counter the stereotype.

For at least 60 years, we have been on a trajectory of trying to enhance science education to meet the interests and needs of all students. However, as delineated in this book, we have not met—or even come close to—that goal. Most of the efforts in the late twentieth and early twenty-first centuries have been focused on classrooms and students. Changing science education courses in order to educate science teachers differently has not been part of the many reforms. In fact, science teacher educators (those who educate potential and practicing teachers) have been marginalized in many of the recent reform movements. That is, principal investigators of federally funded projects had to be drawn from science, mathematics, or engineering faculty. This practice, plus the emphasis on changes in K-12 schools, ignored the critical area of educating the teachers themselves, an education that needed to address teacher attitudes and beliefs, as well as teaching practices, in order to counteract historically held biases and stereotypes.

The reason for hope is that this volume addresses cultural and historical prejudices and provides specific strategies for addressing them. It also makes us aware of the important role that educators of teachers of science play in creating and maintaining quality science education for all students. A science teacher educator may be the first instructor to challenge long-held (and often unacknowledged) views about the competencies of different groups of students in science, about the interest of parents from different backgrounds in a quality science education for their children, or about how social justice affects science education. They may be the first to introduce future teachers to culturally relevant pedagogy or to Critical Race Theory or to problem-based learning. All of those issues—and many more—are discussed in the following chapters. But, probably the most important reason why I think there is cause for hope is that the editors and chapter authors of this book constitute a critical mass of diverse science teacher educators. Their voices are heard, not only in this volume, but also in the pages of teacher education journals, in articles in research journals, and in opinion columns in local newspapers. They are bringing their voices to PTA meetings, to state department of education meetings, and to review panels at state and national funding agencies. The fields of medicine, biology, pharmacy, dentistry, chemistry, and engineering have changed as women have entered each of those areas. The emerging critical mass of diverse science teacher educators is sure to change and improve the way we educate and support science teachers in the future.

Condit Professor of Science Education   Jane Butler Kahle
Emerita, Miami University, Oxford, OH, USA
January 20, 2013

Preface

A major challenge in science education is how to encourage and support science teacher educators as they help preservice and practicing science teachers to encourage and support their adolescent students to participate at the highest level in science and ensure that all students learn quality science (Bryan & Atwater, 2002; Irvine, 1990; Russell & Atwater, 2005). Historically, science has been viewed by many as a culture-free, ethnicity-free, and gender-free discipline (Carter, Larke, Singleton-Taylor, & Santos, 2003). Fortunately, an increasing number of science education researchers acknowledge that there needs to be a significant change in the way science has been traditionally taught and instructional approach that emphasizes White, male, Western, and European perspectives (Atwater, 1994, 1995a, 1995b; Carter, Larke, Singleton-Taylor, & Santos, 2003). For example, a key component in this paradigm shift is the necessity to emphasize culturally relevant pedagogy in the classroom, a "pedagogy characterized by individual and collective empowerment" (Ladson-Billings, 1995a, 1995b, 2000).

With the aforementioned challenge in mind, it is imperative that science teacher educators understand how they can better prepare teachers to teach all students and be empowered with the best practices for promoting success. Research demonstrates that teachers tend to teach the way they were taught and subsequently they are more than likely to rely on methods that are heavily textbook-oriented and less on hands-on, inquiry-based learning (Carter, Larke, Singleton-Taylor, & Santos, 2003; Seung, Bryan, & Butler, 2009). Moreover, there are few based on research in best practices for equity in science teaching can serve as resources to science teacher educators that provide knowledge about multicultural science pedagogical strategies to better educate preservice and practicing teachers from multicultural and social justice perspectives in their college courses and professional development activities (Atwater & Butler, 2006; Butler, Lee, & Tippins, 2006; Eide & Heikkinen, 1998; Lawrence & Butler, 2010).

This book is designed as a resource for science teacher educators to ultimately emphasize the critical role that multicultural education, equity, and social justice have on the teaching and learning of science for adolescents from all backgrounds.

Moreover, science educators will find this book useful for professional development workshops and seminars for both novice and veteran science teachers. Many questions arise in discussions with teachers during professional development activities, as well as in the context of science teacher preparation programs that stem from a lack of commitment, knowledge, understanding, and obvious cultural dissonance between school professionals and the student populations that they teach on a daily basis. We thank the staff of Springer—Bernadette Ohmer, Publishing Editor Education, Marianna Pascale, Senior Editorial Assistant Education, and Shanthy Gounasegarane, Project Manager , SPi Content Solutions – SPi Global —for their support and assistance in the publication of the book. With this in mind, it is our hope to encourage science teachers to become actively involved in transforming their curriculum, pedagogy, and assessment to increase the scientific pool so that students from traditionally marginalized and underrepresented groups in STEM areas realize their full potential in science.

University of Georgia, Athens, GA, USA	Mary M. Atwater
Auburn University, Auburn, AL, USA	Melody L. Russell
University of Central Florida, Orlando, FL, USA	Malcolm B. Butler

Atwater, M. M. (1994). The multicultural science classroom part I: Meeting the needs of a diverse student population. *The Science Teacher, 62*(2), 20–23.

Atwater, M. M. (1995). The multicultural science classroom part II: Assisting all student with science acquisition. *The Science Teacher, 62*(4), 42–45.

Atwater, M. M., & Butler, M. B. (2006). Professional development of teachers of science in urban schools. In J. Kincheloe, K. Hayes, K. Rose, & P. M. Anderson (Eds.), *Urban education: An encyclopedia* (pp. 153–160). Wesport, CT: Greenwood Press.

Bryan, L. A., & Atwater, M. M. (2002). Teacher beliefs and cultural models: A challenge for science teacher preparation programs. *Science Education, 86*(6), 821–839.

Butler, M. B., Lee, S., & Tippins, D. J. (2006). Case-based pedagogy as an instructional strategy for understanding diversity: Preservice teachers' perceptions. *Multicultural Education, 13*(3), 20–26.

Carter, N. P., Larke, P. J., Singleton-Taylor, G., & Santos, E. (2003). Multicultural science education: Moving beyond tradition. In M. Hines (Ed.), *Multicultural science education: Theory, practice, and promise* (pp. 1–19). New York: Peter Lang.

Eide, K. Y., & Heikkinen, M. W. (1998). The inclusion of multicultural material in middle school science teachers' resource materials. *Science Education, 82*(2), 181–195.

Irvine, J. J. (1990). *Black students and school failure: Policies, practices, and prescriptions*. New York: Praeger.

Ladson-Billings, G. (1995a). Toward a culturally relevant pedagogy. *American Educational Research Journal, 32*(3), 465–491.

Ladson-Billings, G. (1995b). But that's just good teaching! The case for culturally relevant pedagogy. *Theory into Practice, 34*(3), 159–165.

Ladson-Billings, L. (2000). Fighting for our lives: Preparing teachers to teach African American students. *Journal of Teacher Education, 51*(3), 206–214.

Lawrence, M. N., & Butler, M. B. (2010). Becoming aware of the challenges of helping students learn: An examination of the nature of learning during a service-learning experience. *Teacher Education Quarterly, 37*(1), 155–175.

Russell, M. L., & Atwater, M. M. (2005). Traveling the road to success: A discourse on persistence throughout the science pipeline with African American students at a predominantly White institution. *Journal of Research in Science Teaching, 42*(6), 691–715.

Seung, E., Bryan, L., & Butler, M. B. (2009). Improving preservice middle grades science teachers' understanding of the nature of science using three instructional approaches. *Journal of Science Teacher Education, 20*(2), 157–177.

Acknowledgments

We are grateful to our colleagues who graciously agreed to offer their ideas and experiences in this book. Our thanks go to the following universities for providing the space and the time to think, talk, and write—Auburn University, the University of South Florida, the University of Central Florida and the Science Education Program at the University of Georgia; to those in the science education community who encouraged us to complete this tome; and to Valerie Kilpatrick for her time and talents in typing, editing, and getting the manuscript ready to send to the publisher. We thank the staff of Springer—Bernadette Ohmer, Publishing Editor Education, and Marianna Pascale, Senior Editorial Assistant Education—for their support and assistance in the publication of the book.

Finally, much gratitude is extended to our friends and families for sacrificing meals, games, and time away for us to write. Mary has been driven by her own experiences and those of her students and her colleagues to attempt to coedit a book for science teacher educators on multicultural science education, equity, and social justice. However, it is her grandchildren, Niké, Shimobi, and Ikenna, who remind her that K-12 students continue to be impacted by their educational passions, career goals, performance in science, and their science teachers. Melody is grateful to God for the blessings of her supportive husband Jared, and her son Jared II because they help make all the hard work possible and worthwhile. Malcolm is particularly thankful that Malcolm Lee and Vikki give him the motivation to attempt to "do good work." The three of us are indeed grateful.

Contents

Introduction: Culture, Equity, and Social Justice for Science Teacher Educators 1
Malcolm B. Butler, Mary M. Atwater, and Melody L. Russell

Part I Historical and Sociocultural Perspectives on Science Teacher Education

The Systematic Misuse of Science 11
André M. Green

Second-Class Citizens, First-Class Scientists: Using Sociocultural Perspectives to Highlight the Successes and Challenges of African American Scientists During the Jim Crow Era 29
Malcolm B. Butler

Science Education and Females of Color: The Play Within a Play 41
Leon Walls

Sociocultural Consciousness and Science Teacher Education 61
Brenda Brand

Part II Foundations of Science Teacher Education

The Impact of Beliefs and Actions on the Infusion of Culturally Relevant Pedagogy in Science Teacher Education 81
Natasha Hillsman Johnson and Mary M. Atwater

Motivation in the Science Classroom: Through a Lens of Equity and Social Justice 103
Melody L. Russell

Part III Pedagogical and Curricular Issues in Science Teacher Education

Negotiating Science Content: A Structural Barrier in Science Academic Performance 119
Barbara Rascoe

Internationally Inclusive Science Education: Addressing the Needs of Migrants and International Students in the Era of Globalization 137
Charles B. Hutchison

Using Problem-Based Learning to Contextualize the Science Experiences of Urban Teachers and Students 159
Neporcha Cone

Part IV Equity, Multiculturalism, and Social Justice: Diversity Issues in Science Teaching

African American and Other Traditionally Underrepresented Students in School STEM: The Historical Legacy and Strategies for Moving from Stigmatization to Motivation 175
Obed Norman

Preparing Science Teachers for Diversity: Integrating the Contributions of Scientists from Underrepresented Groups in the Middle School Science Curriculum 193
Rose M. Pringle and Cheryl A. McLaughlin

The Triangulation of the Science, English, and Spanish Languages and Cultures in the Classroom: Challenges for Science Teachers of English Language Learners 209
Regina L. Suriel

Part V Policy Reform for Science Teacher Education

STEM-Based Professional Development and Policy: Key Factors Worth Considering 233
Celestine H. Pea

Policy Issues, Equity, Multicultural Science Education, and Local School District Support of In-Service Science Teachers 253
Bongani D. Bantwini

Policy Issues in Science Education: The Importance of Science Teacher Education, Equity, and Social Justice 273
Sheneka M. Williams and Mary M. Atwater

Conclusion and Next Steps for Science Teacher Educators 285
Melody L. Russell, Malcolm B. Butler, and Mary M. Atwater

Index 293

About the Editors

Dr. Mary M. Atwater is presently a Professor of Science Education in the Department of Mathematics and Science Education at the University of Georgia, an American Educational Research Association Inaugural Fellow, and an American Association for the Advancement of Science Fellow. She has been the Principal Investigator or the Co-Principal Investigator of several federally funded and privately funded grants totally over $3.76 million dollars. Her publications include articles, book chapters, a coedited book on multicultural science education, and a coauthored K-8 science program. She uses both qualitative and quantitative methodologies in her research; her article, "Social Constructivism: Infusion into Multicultural Science Education Research," published in the *Journal of Research Science Teaching* in 1996 was recently selected as one of the article that started a systematic research movement in the area of multicultural science education. Recently, she served as guest editor and author of two articles in a special issue of *Science Activities* in 2010 that focused on multicultural science education. She now serves on several editorial boards and has held faculty positions in both education and the sciences and leadership positions in most of the national/international science education organizations such as the National Association for Research in Science Teaching, National Science Teachers Association, and the Association for Multicultural Science Education.

Dr. Malcolm B. Butler a former middle and high school mathematics and science teacher, is an Associate Professor of Science Education in the School of Teaching, Learning and Leadership, in the College of Education at the University of Central Florida in Orlando. His teaching and research interests include multicultural science education, science and underserved students, preservice and inservice science teacher education, environmental education, and physics education. His scholarship has been published in journals such as the *Journal of Research in Science Teaching*, the *Journal of Science Teacher Education*, *Science Activities*, the *International Journal of Environmental and Science Education*, and the *Journal of Multicultural Education*. His work has been generously supported by the National Science Foundation, the Environmental Protection Agency, and the US Department of

Education. Dr. Butler is also one of the authors of National Geographic Learning's "National Geographic Science," a national elementary science curriculum program. He is also coauthor of the book *Teaching Science to English Language Learners*.

Dr. Melody L. Russell from the University of Georgia in 2000. Dr. Russell's research interests include the investigation of strategies that enhance the recruitment and retention of people from traditionally underrepresented groups in the natural and physical sciences, and women at the precollege and college levels. She researches, publishes, and consults in the areas of equity in science education, mentoring, professional development for teachers, and motivation in the science classroom. Dr. Russell's teaching interests include preparing preservice teachers to teach through a lens of culturally relevant pedagogy and promoting equity and social justice through science teaching. She has published in the *Journal of Research in Science Teaching, The Negro Educational Review, Science Activities, The Professional Educator,* and *The Science Teacher*.

About the Contributors

Dr. Bongani D. Bantwini is Senior Research Specialist at the Human Sciences Research Council in Pretoria, South Africa. He works for the Research Use and Impact Assessment unit. He holds a Ph.D. in Elementary Education with a focus on Science Education from the University of Illinois at Urbana-Champaign. Formerly, he worked as an Assistant Professor of Science Education at Kennesaw State University in the Department of Elementary and Early Childhood Education, Georgia, USA. His area of research focuses on: professional development of inservice science teachers; science teacher learning and change; school district science officials; and science curriculum reforms. His recent publications include articles like: "Primary School Science Teachers' Perspectives Regarding Their Professional Development: Implications for School Districts in South Africa"; "Factors Affecting South African School District Officials' Capacity to Provide Effective Teacher Support"; and "How Teachers Perceive the New Curriculum Reform: Lessons from a School District in the Eastern Cape Province, South Africa."

Dr. Brenda Brand is an Associate Professor in Science Education at Virginia Tech. Her research interests include sociocultural factors influencing both teaching and learning in STEM disciplines with an emphasis on underrepresented groups. She has investigated these factors from the perspectives of preservice and inservice teachers, and also high school students. Currently, Dr. Brand is conducting a research study entitled "Why Women Stay: An Investigation of Two Successful Programs," which is an investigation of factors contributing to the retention of high school and undergraduate female students participating in two alternative engineering programs. The structures of both programs consist of activities that encourage students to take risks and contribute to their learning in a collaborative environment. Related publications include: "Sociocultural Factors Influencing Students' Learning in Science and Mathematics: An Analysis of the Perspectives of African American Students," "Using Critical Race Theory to Analyze Science Teachers Culturally Responsive Practices," "Crossing Cultural Borders into Science Teaching: Early Life Experiences, Racial and Ethnic Identities, and Beliefs About Diversity."

Dr. Neporcha Cone is an Assistant Professor of Education at Kennesaw State University. She received her Doctor of Philosophy at the University of South Florida, with an emphasis on Science Education. She has taught numerous science and science education courses, underpinned by a framework for social justice. Her publications include articles and book chapters that focus on equitable science education practices. Her most recent work, "Differentiating Through Problem-Based Learning: Learning to Explore More! with Gifted Students," delineates how Problem-Based Learning can be used to meet the needs of students in mixed-ability classrooms. Her professional and teaching experiences have led her to develop a research agenda that includes a focus on preparing teachers for the complexities that might emerge when teaching science for diversity.

Dr. André M. Green is an Associate Professor of Science Education at the University of South Alabama, and an experienced Principal Investigator with over $5 M in grants management from the National Science Foundation, Alabama State Department of Education through the AMSTI project, and various foundations and other governmental agencies. His research interests focus on underrepresented groups access to the STEM disciplines, the education of STEM teachers, STEM teacher leadership, mentoring, and the induction of educators into the profession. He has extensive experience in working with students from urban environments and has developed educational programs of community outreach to improve the academic achievement of underrepresented groups. Three publications of interest are: "A Dream Deferred: The Experience of an African American Student in a Doctoral Program in Science," "African Americans Majoring in Science at Predominantly White Universities," and "Issues Influencing Students' Learning in Science and Mathematics."

Dr. Charles B. Hutchison is an Associate Professor of Education at the University of North Carolina at Charlotte. His research on international teachers was one of the seminal works that systematically conceptualized international teachers' issues, an area that is being studied internationally. He is the recipient of Recognition and Key to the City of Boston, and has appeared on, or been featured by, local and international news media, including CBS Night Watch, Voice of America, Boston Globe, and Washington Post. His articles have appeared in several journals, including *Phi Delta Kappan*, *Intercultural Education*, *Cultural Studies of Science Education*, and *School Science and Mathematics*. He has lived, studied, and worked in Africa, Europe, and the USA and is the author of five books, including *What Happens When Students Are in the Minority* (Rowman and Littlefield, 2009) and *Understanding Diverse Learners* (Copley, 2011).

Natasha Hillsman Johnson earned her Bachelor of Science degree in Chemical Engineering from Cornell University in 1999. After working in industry, she completed her alternative certification through the Teach for America organization. Over the past 12 years, she has worked as a secondary science teacher in the DeKalb County, Atlanta Public and Rockdale County School systems. She is also employed as an adjunct science teacher with the Georgia Virtual School. She completed her Master of Education degree in Secondary Science Education from Mercer

University in 2007 and is currently a Ph.D. student in the Department of Science Education at the University of Georgia. Her research interests include multicultural science teacher education, STEM undergraduate education, science teaching and learning, and online instruction.

Cheryl A. McLaughlin is a Ph.D. candidate in Science Education in the School of Teaching and Learning at the University of Florida. She has 10 years of experience in teaching science at the K-20 level and has been involved in the development of science and elementary education curricula. She has also conducted research on student motivation and engagement in urban science classrooms. Currently, her area of focus is teacher professional learning and development within the context of curriculum change. She is particularly interested in ways in which practicing science teachers develop their professional knowledge about inquiry teaching and how this knowledge enables their practice.

Dr. Obed Norman is Senior Research Associate at the CAPSTONE Institute at Howard University. He is also Associate Director of the Baltimore Education Research Consortium (BERC) based at the Center for the Social Organization of Schools at Johns Hopkins University. Norman holds a Ph.D. in Science Education (University of Iowa) and an M.S. in Biophysics (Penn State). His research focus is the identification of instructional, interactional, and structural interventions that will enhance the achievement profiles and trajectories of students from communities underrepresented in STEM. Norman has been the recipient of a number of NSF research and professional development grants. He was awarded the prestigious Early Career Award by the NSF in 1999. Norman's work has been published in the *Journal of Research in Science Teaching* and other outlets. He has also served on the editorial boards of the *Journal of Research in Science Teaching* and the *Journal of Science Teacher Education*.

Dr. Celestine (Celeste) H. Pea is a Program Director in the Division of Research on Learning (DRL), Education and Human Resource Directorate, National Science Foundation (NSF). In DRL, Celeste works primarily with the Research on Education and Learning (REAL), Innovative Technology Experiences for Students and Teachers program, the CAREER program, and the Albert Einstein Distinguished Educator Program. Her areas of interest include research on professional development, teacher beliefs about science teaching, stereotype and identity threats, and student achievement. She is coauthoring two books and has several other projects underway. Examples of her work include "Inquiry-Based Instruction: Does School Environmental Context Matter?" published in 2012 in the Science Educator and her coauthored book chapter "Why Is Understanding Urban Ecosystems an Important Frontier for Education and Educators" in 2002 in Understanding Urban Ecosystems: A New Frontier for Science Education edited by Berkowitz, Nilon, and Hollweg and published by Springer-Verlag. Before coming to NSF, Dr. Pea was the Science Coordinator for a Louisiana statewide reform initiative for four years and a middle school science teacher for East Baton Rouge Parish Schools for 24 years. In 1991, Dr. Pea won the Presidential Award for Excellence in Mathematics and Science Teaching.

Dr. Rose M. Pringle is Associate Professor in the School of Teaching and Learning at the University of Florida. As a science educator, her research focuses on interrelated themes of science teacher education: preservice teachers' positionality as science learners, science specific pedagogy for both prospective and practicing science teachers, and the translation of these practices into equitable inquiry-based science experiences for *all* learners. Currently, she is exploring pedagogical content knowledge as a framework for charting teachers' practices during the enactment of their curriculum including their use of formative assessment. In addition, she also examines how both preservice and inservice teachers develop knowledge about teaching and learning science and how they adjust their practices to accommodate the learning needs of underrepresented learners in their science classrooms.

Dr. Barbara Rascoe is an Associate Professor in Tift College of Education at Mercer University in Macon, Georgia; has a B.A. in Biology from the University of North Carolina at Greensboro; an M.A. in Science Education from East Carolina University; and a Ph.D. in Science Education from the University of Georgia. She has taught science in K-12 rural and urban schools. Her research trajectory involves examining performance dynamics that affect Black males in science learning environments, issues relative to the ethnic/gender performance gap, the language of science, the history and nature of science, and cultural capital variables that impact science performance. She has published articles and made conference presentations that reference Black males' science performance, scientific literacy, science knowledge transformations, myths of science, shaping science curriculum reform, and the products and processes of science.

Dr. Regina L. Suriel is a Latina from the Dominican Republic who learned English as a second language in New York City schools. She earned both her B.S. and M.S. in Science Education from the City University of New York and her Ph.D. in Science Education from the University of Georgia. Her dissertation work focused on the linguistic demands of Latino/as in secondary-level science. Prior to her doctoral work, Dr. Suriel taught science to bilingual and monolingual students at the secondary school level in New York City. Dr. Suriel has published works and the field of multiculturalism. Samples of her published works include a book chapter coauthored with M. M. Atwater and titled "Science Curricular Materials through the Lens of Social Justice: Research Findings" in 2008 in "The Practice of Freedom: Social Justice Pedagogy in the United States" edited by Chapman and Hobbel. She also coauthored an article titled "Using Educative Assessments to Support Science Teaching for Middle School English Language Learners" (2013) in the Journal of Science Teacher Education 24(2). Dr. Suriel has conducted a number of presentations and workshops for educational organizations in science education and in the field of English as a second language intended at developing pedagogical skills for effectively educating English language learners. Prior to Dr. Suriel current position as an assistant professor at Valdosta State University, she served as a Postdoctoral Fellow in the Department of Curriculum and Instruction at the University of Connecticut. Her current research at Valdosta State University focuses on the effects of dual language assessments on science achievement.

Dr. Leon Walls is an Assistant Professor of Elementary Science Education in the College of Education and Social Services at the University of Vermont. For the past four years, he has taught inquiry-based science methods to preservice teachers at the undergraduate level. He is lead author of the book chapter "Race, Culture, Gender, and Nature of Science in Elementary Settings" in the Cultural Studies of Science Education series and author of the research study "Third Grade African American Students' Views of the Nature of Science" published in the *Journal of Research in Science Teaching*. His scholarly and curricular interests include the nature of science, multicultural science education, and sustainability education. He holds a doctorate in Geoenvironmental Science Education from Purdue University in West Lafayette, Indiana; a master's degree in Instructional Leadership from Marquette University in Milwaukee, Wisconsin; and a degree in Electrical Engineering from St. Mary's University in San Antonio, Texas.

Dr. Sheneka M. Williams is an Assistant Professor in the Department of Lifelong Education, Administration, and Policy within the College of Education at the University of Georgia. Her research examines aspects of the policy process, including policy formulation, implementation, and evaluation. Dr. Williams recently published two policy implementation studies: "Maintaining Balanced Schools: An Examination of Three Districts" and "Micro-politics and Rural School Consolidation: The Quest for Equal Educational Opportunity in Webster Parish." These works were published in Richard Kahlenberg's *The Future of School Integration* and the *Peabody Journal of Education*, respectively. Dr. Williams has also published her policy research in *Education and Urban Society* and the *Journal of Public Management and Policy*. Additionally, Dr. Williams' research has been presented at state, national, and international conferences and forums, including the National Press Club in Washington, DC.

Introduction: Culture, Equity, and Social Justice for Science Teacher Educators

Malcolm B. Butler, Mary M. Atwater, and Melody L. Russell

> Voices sometimes reveal the great challenges and even the deep pain young people feel when schools are unresponsive, cold places. (Nieto, 1996, p. 106)

This book is a systematic attempt to address science teacher education issues related to preparing and working with science teachers to successfully instruct middle and high school students taking science courses in different school settings. School populations in the United States are very diverse in some settings and very monocultural in others. Many teachers, community leaders, and policy makers acknowledge there is a crisis in the education of Black students, especially Science, Technology, Engineering, and Mathematics (STEM) education in the United States. In the past and even today, a few still presume that the deficiencies in Black students' culture, attitudes, and their families and communities explain this crisis. If one accepts this explanation, then little can be done in schools until students' cultures, families, and communities are fixed. Many believe that fixing these things means the culture, families, and communities of all students become more like those of middle-class Whites so that cultural alienation does not occur but cultural subjugation and cultural annihilation do occur. For this crisis to continue after more than 100 years of abolition of slavery does not bode well for the United States as a global player in the twenty-first century. The US population in 2006 had 40 million Latino/a

---

M.B. Butler, Ph.D. (✉)
University of Central Florida, PO Box 161250, Orlando, FL 32816-1250, USA
e-mail: malcolm.butler@ucf.edu

M.M. Atwater, Ph.D.
Department of Mathematics and Science Education, University of Georgia,
376 Aderhold Hall, Athens, GA 30602-7126, USA
e-mail: Atwater@uga.edu

M.L. Russell, Ph.D.
Department of Curriculum and Teaching, College of Education,
Auburn University, 5004 Haley Center, Auburn, AL 36849, USA
e-mail: russeml@auburn.edu

residents fueled by past waves of immigrants from Latin America and the Caribbean (National Research Council, 2006). While immigration is still pushing these numbers, the children and grandchildren of past immigrants are also increasing these numbers. The potential costs of underinvesting in the education, especially science education, of this young adolescent Latino/a population as well as the perils of allowing a large and growing "undocumented population to live on the fringes of society" (National Research Council, p. 14) will be catastrophic to the United States in the twenty-first century. Science learning in middle school and high schools is a challenging experience for most students; however, it is more difficult for Black and Latino/a adolescents destined for oversized, resource-poor schools staffed with experienced and inexperienced teachers who may or may not be committed to high-quality science teaching and learning and lack the knowledge and skills for doing so with these student populations. Most science teacher educators will probably have the opportunity to prepare teachers or work with teachers who teach students whose cultures are very different from theirs. This book focuses on the issues science teacher educators should grapple with related to culture, equity, and social justice for preparing and working with middle and high school science teachers.

Pertinent to these issues is the onset of new science standards. With a framework for these standards having already been published (National Research Council, 2012), science teachers will have the dimensions of scientific and engineering practices, crosscutting concepts, and disciplinary core ideas on the forefront of their minds. Not to be ignored in this publication, a chapter of the framework is devoted to equity and diversity in science and engineering, which further highlights the need for science teachers to address these issues as they focus on the aforementioned dimensions. The *Next Generation Science Standards* (Council of Chief State School Officers [CCSSO] & the National Governors Association [NGA], 2010) are being written based on the frameworks. Not to be excluded, the *Common Core State Standards* (Achieve, 2013) also highlight science as an integral part of English Language Arts and Literacy, as students "come to understand other perspectives and cultures" (p. 7).

The readers of this book are probably asking themselves why write such a book and especially one for middle and high school science teacher educators. The editors of this book as science teacher educators embarked upon this project as a way to facilitate preservice science teacher educators in developing a framework for teaching through a lens of both equity and social justice. Currently, the student population in many areas of the United States (both rural and urban) is growing increasingly diverse, and the teacher population in many areas is still overwhelmingly White and female. This causes a concern for educators and particularly science teacher educators because though it goes without saying that science and culture are closely intertwined, rarely do teachers understand this context and present it in their teaching. Consequently, science is often viewed as a content area that is "for White males, with glasses, beakers and lab coats" which marginalizes the majority of the population which in many areas is made up of Blacks, Latinos/as, Native Americans, and Pacific Islanders. Students of color, primarily from underrepresented and traditionally underrepresented groups, are rarely able to see themselves in science or see how science is relevant to their daily lives in science classrooms. Moreover, since many science teachers are often from different backgrounds than their students, because of "cultural dissonance" these teachers often do not understand the critical role that equity and social justice in science teaching play in a student's success and persistence in science.

Equity can be most simply defined as fairness and justice and with respect to science teaching; our major concern is science learning and teaching regardless of the student background. This book was written in hopes to generate more discourse and promote change relative to the role that science teaching plays in the participation of students from traditionally underrepresented and marginalized students in the science. More importantly, this book was written through a lens of equity and social justice because these two factors underscore primary issues that science teachers face in teaching in their pedagogy. The chapters presented in this book put forth a call to action for science teacher educators, science teachers, and administrators to start from the "ground up" so to speak and examine how science is taught not only to students in the secondary science classroom but how we teach to preservice teachers in teacher preparation programs. Moreover, this book also addresses the important role that professional development for inservice teachers plays in promoting equity and social justice in the increasingly diverse classrooms of today. Challenging the status quo in how science has been traditionally taught is the first step in changing the outcomes of "who does science" and ensuring that the STEM pipeline is more inclusive for all students both on the secondary level and beyond:

> Those of us who work in teacher education may see incremental changes in the field, but we are unlikely to see and participate in a 'promised land' of teaching and learning. The joy is reserved for those new to the profession. Like the ancient Hebrew leader Moses, we are charged with the responsibility of liberating the field from the enslavement of narrow thinking about curriculum and human capacity. However, we will not enjoy the direct benefits of the struggle. Those benefits are to accrue to a new generation of leadership – a generation that will cross over to Canaan. (Ladson Billings, 2001, p. 142)

The contributing authors with expertise in different areas of science teacher education bring together a collection of ideas about US science teacher education that did not exist until the publication of this book. Some of the book authors are young to the field, while a few have been in the field of multicultural science teacher education for a long time. Some will cross over Canaan as Ladson-Billings has predicted, but a few like Mary M. Atwater are "charged with the responsibility of liberating the field from the enslavement of narrow thinking about curriculum and human capacity" (Ladson Billings, 2001, p. 142) but will not enjoy the direct benefits of the struggle.

> I have never encountered any children in any group who are not geniuses. There is no mystery on how to teach them. The first thing you do is treat them like human beings and the second thing you do is love them. (Hilliard as cited in Fallon, 2006)

Science education unless science teacher education is transformed so that science teachers understand the magnitude of culture, equity, and social justice in their own professional work. The publication of this book is possible because presently there are a substantial number of critical thinking science teacher educators of color who are in the field of science teacher education. The publication of this edited book demonstrates that when a considerable collection of science teacher educators come together who are committed to the communities that they *labor in, labor for, write with and about*, and sometimes *dwell in,* then significant insights can be added to the field of science teacher education. These book chapter authors not only seek to impact the theoretical, ideological, and methodological aspects of science teacher education but also have a positive impact on the educational quality of the lives of their college and university students and students and teachers in the community and in the United States.

This book in illuminates historically persistent, yet unresolved issues in science teacher education from the perspectives of a remarkably groups of science teacher educators and presents research that has been done to address these issues. Second, it centers on research findings on underserved and underrepresented groups of students and presents frameworks, perspectives, and paradigms that have implications for transforming science teacher education. Finally, the chapters provide an analysis of the sociocultural-political consequences in the ways in which science teacher education is theoretically conceptualized and operationalized in the United States.

Although this book is divided into five unique parts, it has a common thread that allows the reader to examine culture, equity, and social justice through various lenses. The first part entitled "Historical and Sociocultural Perspectives on Science Teacher Education" contains four chapters in which the authors seek to shed light on how historical and sociocultural issues have impacted the involvement of students from underserved and underrepresented group in science and how these should influence what happens in science teacher education programs. In the first chapter, André Green discusses ways to effectively confront the negative stereotypes that distribute assertions of inadequacy and also declarations of entitlement. His ultimate goal is to shed light on the infrastructure that currently exists and demonstrates that this infrastructure is a major contributor to the lack of participation by African Americans in scientific fields. Next, Malcolm Butler highlights Black scientists who succeeded during the Jim Crow period in the United States in order to create a knowledge base for secondary science teacher educators and candidates. He also proposes several activities and strategies as starting points for expanding and enhancing the dialogue of critical race theory and inclusion of "Others" in who can do science. Following Butler's chapter, Leon Walls writes about females of color in science education using a Shakespearean "play within a play" device as the backdrop for allowing the reader to experience to science through the lenses of race and gender at every turn. He closes the chapter by offering a model of both equity and social justice particularly for females of color. The last chapter in Part I is written by Brenda Brand, where she discusses how marginalized groups have been left with the challenge of sustaining their lives in an environment where declared norms isolate and devalue them and their cultures. She goes on to focus on understanding this system in terms of its history and impact on society and schooling and how it translates into sociocultural awareness and social consciousness of students, teachers, and science teacher educators. This guides science teacher educators toward better understanding how sociocultural awareness is the crux of culturally responsive pedagogy.

The second section of this book focuses on the foundations of science teacher education. The two chapters in this part of the book examine the role science teachers' beliefs and culturally relevant pedagogy plays in promoting equity and social justice in the science classroom. The first chapter, written by Natasha Johnson and Mary Atwater, offers insights on how science teachers' beliefs and actions impact if and how culturally relevant pedagogy is implemented in today's classrooms. Effective strategies for doing so are offered to the reader. In the second chapter, Melody L. Russell considers the impact of science teaching on the motivation and student achievement and strategies for increasing the participation of students from Traditionally underrepresented groups in the STEM fields.

The reader will find in the third part of this book three chapters that view equity and social justice through a lens of classroom challenges both pedagogical and curricular relative to preparing secondary science teachers. Barbara Rascoe proposes strategies designed to promote and enhance science educators' effectiveness to address the capabilities of preservice science teachers to negotiate science content and use different perspectives to engage science learners in critical thinking, creativity, and collaborative problem solving for equity and social justice. Charles B. Hutchison offers us a global idea—*internationally inclusive teaching*—as a way of thinking about preparing our teachers and students to live and succeed in a constantly changing world, as well as a country that is more diverse than ever before. Finally, Neporcha Cone highlights problem-based learning (PBL) as a powerful instructional model that can be used to contextualize the science experiences of urban teachers and students in real-world problems relevant to themselves and their community while concurrently developing the critical thinking skills necessary to participate in a global society. She advocates that science teacher educators must immerse their preservice and inservice teachers in authentic PBL contexts so that they may see value of inquiry, even as it relates to their own learning.

The authors of the three chapters in the penultimate part of the book entitled "Diversity Issues in Science Teaching" look at various aspects of the classroom environment. Obed Norman, Rose M. Pringle and Cheryl McLaughlin, and Regina L. Suriel discuss the importance of culturally relevant teaching and high expectations toward enhancing the participation of traditionally underrepresented groups in the STEM pipeline. These authors also posit that there are numerous inequities and cultural barriers that impact the level of participation of traditionally underrepresented and marginalized groups (e.g., females, African American, Latinos/as/Latinas, and Native Americans) in the STEM pipeline. Obed Norman advocates that science teacher educators must focus their efforts on science teachers creating learning environments that are welcoming and nurturing for all students but particularly for students of color in STEM education. The chapter deals mainly with African American students (and the history cited is that of African American teachers) and discusses issues of marginalization, stigmatization, and the cultivation of positive possible selves for all students. Rose Pringle and Cheryl McLaughlin explicate their efforts to provide pedagogical opportunities for preservice teachers to broaden their concept of multicultural science education and ways to engage the personal and cultural identities of their learners into their science lessons. They understand that preservice science teachers and middle school students must develop images of scientists beyond the monoculture of White male dominance in order to effectively implement science curriculum that acknowledges the contributions made to science by scientists from underrepresented groups. Regina L. Suriel provides a teaching scenario that serves as the platform for understanding the linguistic barriers and challenges experienced by Latino/a students while in science classrooms and specific approaches addressing science learning and language development in Latino/a learners and English language learners. She also delineates the cultural barriers that may impede Latinos/as/Latinas from participating in science and offers a synopsis of challenges science educators face relative to how Latino/a students experience science. They put forth a call to action to provide equity and multicultural education curricula in preservice teacher preparation programs to change perceptions of "who can do science" and promote high expectations to enhance the participation of traditionally underrepresented groups in the STEM pipeline.

In the fifth and final part of this book, "Policy Reform for Science Teacher Education," are three chapters that take the reader beyond the classroom and consider some factors that directly impact science teaching and learning. Sheneka Williams and Mary M. Atwater, Celeste Pea, and Bongani D. Bantwini go beyond the classroom and consider how various policy issues on both the national and local level directly impact what goes on in the science classroom. These authors discuss the role that providing science teachers more support, smaller class sizes, and STEM professional development could play in better preparing science teachers to teach in culturally diverse classrooms. Sheneka Williams and Mary M. Atwater review science teacher education policy in conjunction with standards to which teachers teach and set forth a new policy agenda to improve science teacher practices and science performance among low-income rural and urban students of color by redefining the policy problem in science teacher education. Celeste Pea discusses STEM-based professional development in the context of research across different domains and uses evidence from research to highlight effective STEM-based professional development that now includes both teachers and principals. She advocates that policy must become more supportive of STEM-based approaches in order to fully embrace evidence from research and encourages science teacher educators to pay special attention to STEM-based professional development, policy, curriculum, and cultural pedagogy as critical factors aimed at improving the performance of teachers and leaders, who, in turn, will be better positioned for improving the performance of all students. Bongani D. Bantwini argues that science teachers on their own can hardly achieve the desired goals to help students perform well or achieve the desirable outcomes and discusses policy issues and why it is imperative that school districts and their officials should support science teachers, specifically for students from diverse and low socioeconomic backgrounds. He emphasizes that it is imperative that school districts increase their science teacher support and develop science education policies that are not just political symbols. Moreover, these authors emphasize the importance of enhancing preservice teachers' knowledge of culturally relevant teaching and pedagogical content knowledge relative to student achievement outcomes in science education. A concluding chapter brings the five previous parts together and identifies themes that are most salient to science teacher educators working with teachers to resolving science education problems that are pervasive in dealing with cultural, equity, and social justice issues. In this final chapter, the coeditors offer some suggestions for where we are and should be going with science teacher education as we move along in the twenty-first century to transform the learning and teaching experiences of middle and high school science teachers.

References

Achieve. (2013). *Next generation science standards*, January 2013 draft. Accessed at: www.nextgenscience.org

Council of Chief State School Officers (CCSSO) and the National Governors Association (NGA). (2010). *Common core state standards for English Language Arts & Literacy in History/Social Studies, Science, and Technical Subjects*. Accessed at: www.corestandards.org

Fallon, K. (2006, February 22). Education symposium focuses on cultural gap. *Marietta Daily Journal*. Retrieved from http://www.newspaperclips.com/npcapp/readArticle.aspx?from=publisher&OrgID=19986&ArticleID=2004343915

Ladson Billings, G. (2001). *Crossing over to Canaan*. San Francisco: Jossey-Bass.

National Research Council. (2006). *Multiple origins, uncertain destines: Hispanics and the America future*. Washington, DC: National Academies Press.

National Research Council. (2012). *A framework for K-12 science education: Practices, crosscutting concepts and core ideas*. Washington, DC: National Academies Press.

Nieto, S. (1996). *Affirming diversity: The sociopolitical context of multicultural education*. White Plains, NY: Longman.

Part I
Historical and Sociocultural Perspectives on Science Teacher Education

The Systematic Misuse of Science

André M. Green

Opportunities continue to grow in the United States for those with specific education in STEM areas; however, there should be great concern among citizens, educators, and experts that African Americans and other underrepresented groups are not pursuing careers in STEM fields, the key to US long-term global involvement (Smyth & McArdle, 2002). Although the demand for science and engineering backgrounds is on the rise, it is troublesome to note that there are fewer individuals seeking these careers (National Science Foundation, 2004b). According to Weiss (2009), a manpower survey indicates that US engineering jobs are difficult to fill by qualified employees. Research confirms that careers necessitating advanced science and mathematics education are not attracting African American interest (e.g., Lewis, 2003; National Science Foundation, 2004a). Atwater, Wiggins, and Gardner (1995) document that many urban students who plan to engage in a science-related career do not take high school science courses in preparation for advanced educational achievements. Researchers and educators are greatly concern at this profound underrepresentation of African American students in science and mathematics vocations.

The AAAS (1998) reports that over the course of our nation's history, science and science-related careers have been regarded as a privilege of the upper class; as a result, only a small number of African American students achieve success in science (Russell & Atwater, 2005). Although equity, equal opportunity, and fairness are supposedly foundational factors in U.S. culture, that foundation is not consistently the case for African Americans pursuing science credentials. Russell and Atwater (p. 692) write that "although, in the last few decades, African Americans have made significant strides in science and mathematics (Oakes, 1990a, 1990b), their increased participation in the sciences has been miniscule compared with

---

A.M. Green (✉)
College of Education, University of South Alabama,
UCOM 3100, Mobile, AL 36688-0002, USA
e-mail: green@usouthal.edu

Whites." Twenty years ago, it was observed that White males were becoming less interested in STEM occupations (Johnson, 1992), and for that reason, the United States is now forced to attend to the problem of declining numbers of the majority population as well as to the absence of other groups of people such as African Americans in STEM careers to remain a viable leader in the twenty-first century. With the majority group losing interest in STEM areas and with the lack of African Americans pursuing STEM areas, this trend could hinder the scientific and technological advancements of the country.

To understand why so few African Americans pursue careers in STEM areas, the history of how we got to this point must first be understood. This chapter traces historically the idea of perceived racial inferiority in regard to African Americans and how that label has inhibited the full participation or inclusion of African Americans in science and other human endeavors. Implicit throughout this chapter is the theme of Social Darwinism, because Darwinism was the casing that gave shelter to racialist ideologies that provided the validity, the credence, and the power to convince a nation that the idea of natural selection should be applied to humanity. Most importantly, it gave a scientific foundation for the belief that the structure of society was the way that nature intended. The question that should be asked after reading this chapter is as follows: why would African Americans want to be a part of something that has continuously tried to disenfranchise them and has tried to prove since its inception that they could not think on a higher order?

Science teacher educators should be interested in this chapter as it attempts to explore the discriminatory ideologies that framed science with regard to African Americans, the conceptions that resulted from these foundational ideologies, and the subtleties embedded within the present infrastructure of society that are the residuals of these ideologies in an attempt to show that where we are with regard to African American participation in STEM fields is no accident but are the fruits of the seeds that were planted many years ago. Understanding the history of African American experiences with science has the potential to equip science teacher educators with the ammunition needed to tackle the problem of African American underrepresentation in STEM fields. Knowing plausible reasons as to why a problem exists is the first step in attempting to solve it.

Social Darwinism and a General Overview

Long before Social Darwinism was established, the relationship between race and intelligence had been a subject of conversation among numerous European intellectuals (Dennis, 1995). Social Darwinism provided a foundation that allowed ideas of European supremacy to manifest because it provided a framework that allowed these ideas to rationally function (Dennis). According to Dennis, individuals such as Buffon (1797) and Gobineau (1853/1915, 1995) used this framework to establish a trend in racialist ideology by connecting the pigmentation of a person's skin to their conduct and human capabilities.

Darwin himself used his theories of evolution to explain occurrences within the animal species. He never applied his theories to human beings. It was others like Herbert Spencer that applied Darwin's evolutionary theories to those of the human race. In fact, it was Spencer (1874) who coined the phrase "survival of the fittest," not Darwin. It was also Spencer who believed that the rules of natural selection applied to the human species as well as to those biological species (Dennis, 1995). Spencer believed that humans are guided by rules of opposition and power and that they progress from an uncouth and antiquated condition to one of separation and advancement. According to Spencer, those not able to adjust should by nature's law perish or be beneath those who have adjusted (Dennis).

This doctrine of Social Darwinism promoted racial conflict because the key to social advancement required "a continuous over-running of the less powerful or less adapted by the more powerful, a driving of inferior varieties into undesirable habitats, and occasionally, an extermination of inferior varieties" (Greene, 1963 as cited in Dennis, 1995, p. 244). Darwinism, explained in simpler terms, can be construed as the battle for survival in which competitions between the races occur. In this competition, the fittest or superior will replace the weakest or inferior (Montagu, 1965). Put into these terms, the conflict among the races is justifiable because it supplies a biologically impartial resolution that is neat but most of all natural (Montagu).

The idea of Social Darwinism most notably presented itself in the United States during the antebellum period by the nations' leading Social Darwinist, William Graham Sumner (Dennis, 1995). Sumner situated the ideal of slavery into Social Darwinism and reasoned within this framework that because "slavery permitted superior groups the leisure to construct and develop more refined cultures, it actually advanced the cause of humanity" (Bierstedt, 1981; Dennis, 1995, p. 244). Sumner also believed that the current status of certain groups of people was a result of the natural selection of nature.

Scholars such as Spencer and Sumner helped to create the atmosphere and disposition towards race relations in the United States. In their assessment of society, aptitude and merit were characteristics only identifiable within the European community. Their view, which was housed in the framework of Social Darwinism, also supported the reality of institutional structures that already existed in society.

The Nature of Science in Science Education

Throughout history, humanity has found and developed many interesting theories about the order of the world and about the people who live in it. Some theories have been proven legitimate based on the evidence provided, while other theories have not fared so well. The interesting occurrence, however, is that these theories, legitimate or not, have provided road maps of processes to future generations. The procedure in which these processes are formulated is the foundation that gives science the credence that renders it unique from other disciplines. The processes of examining, reasoning, testing, and authenticating are all pivotal components in the construction of that foundation, and those components are at the core of the nature of science (AAAS, 1989).

According to Lederman, Khalick, Bell, and Schwartz (2002), "typically, the nature of science refers to the epistemology and sociology of science, science as a way of knowing, or the values and beliefs inherent to scientific knowledge and its development" (p. 498). Glasson and Bentley (1999) write, "the most influential current curriculum documents in science education consider the nature of science as basic content for the K-12 curriculum for all students" (p. 470). Project 2061's AAAS (1989) and AAAS (1993) are both major contributors to the establishment of the current *National Science Education Standards* (National Research Council [NRC], 1996). These documents establish the nature of science to include three categories: the scientific worldview, scientific inquiry, and the scientific enterprise.

The scientific worldview relays that those who practice science have specified fundamental standards that guide their way of thinking about how they practice and regard science. This line of thought is concerning the nature of the world and what knowledge can be obtained from it. This scientific worldview is supported by four tenets: the world is understandable, scientific ideas are subject to change, scientific knowledge is durable, and science cannot provide complete answers to all questions (AAAS, 1989).

Scientific inquiry implies that every discipline of science, from chemistry to physics to biology, etc., requires evidence to substantiate claims. Although scientists may differ in the process in which their research is conducted, the basic premise of how they conducted that research should be similar. It is that premise which makes research scientifically legitimate. This characteristic is what makes science inquiry based, and everyone, regardless of whether they practice science, could employ these skills on a daily basis on issues of importance to them if they so choose. Scientific inquiry is supported by five tenets: science demands evidence, science is a blend of logic and imagination, science explains and predicts, scientists try to identify and avoid bias, and science is not authoritarian (AAAS, 1989).

The scientific enterprise recognizes that science has individual, societal, and foundational facets. The activity or practice of science, presently, is what separates it from the practices of other disciplines. The scientific enterprise consists of four tenets: science is a complex social activity, science is organized into content disciplines, science is conducted in various institutions, and there are generally ethical principles in the conduct of science (AAAS, 1989).

Even with those three principles established and with those principles being the foundation and framework of sciences' curriculum within the K-12 system, theorists, philosophers, academics, sociologists, and educators of science are prompt to dispute on particular matters concerning the nature of science (Lederman et al., 2002). Perhaps the reason for this is that it is impossible, or at least very difficult, to define specifically an ideal such as the nature of science, because that nature can take on so many meanings. The nature of science has many sides to it; it is very complicated and has many layers. Also, the views about the nature of science, like scientific knowledge, are provisional and tentative. Throughout the history of the nature of science, views about it have changed (Lederman et al., 2002 & see Abd-El-Khalick & Lederman, 2000, for a broad survey of these changes).

It may be argued, for example, that science has been used throughout history as a means to separate, classify, and rank things according to some type of order. Science in the past as well as in the present separates everything, the good from the bad, trees from other trees, trees from insects, people from animals, good methods from bad methods, and so on. The very methods provided by Science for All Americans are given to separate good science from bad science to legitimize the scientific process. The question that this raises is: Is the nature of science inherently good or bad given what it has been used for? It could be reasonably debated that science really has no nature at all, because how can something that is inanimate have a nature?

Many talk of science as if it is an entity that lives, breathes, and operates separately from the rest of the world. In that aspect and that aspect alone, science can be pure and objective, but science does not operate in this manner. The science that society has come to know cannot exist independently from the world because science is a tool that takes on the very nature of whoever controls it. Science in essence is a set of principles, established by man, which help to guide man to "pure and objective" science, an ideal that he will never come to know. Science with the involvement of man cannot have one true nature. Given this parameter, the nature of science can be good or bad depending on whose hands control it.

Working on the assumption, for example, that guns were created to kill, it may be asked if the nature of a gun is inherently good or evil. In some hands, a gun kills, and in other hands a gun may serve to protect from evil. The point is that the gun takes on the characteristics or the intent of the person using it. Science can act in much the same way; it can be used for good or evil. Both guns and science have no say in how they are used because both are only tools.

The nature of science, like scientific knowledge, is a concept that is comprised of educated conjectures made by those who practice and study science. Since science is an entity that cannot exist separately from society, a scientist's opinions, prior experiences, preparation, and viewpoints may have some bearing on their practice (Lederman et al., 2002), and "all these background factors form a mindset that affects the problems scientists investigate and how they conduct their investigations, what they observe (and do not observe), and how they interpret their observations" (p. 501). This is important because many people believe that scientists and their observations are always impartial (Lederman et al., 2002; Popper, 1992), but in reality it may not consistently happen in this manner. This is because science and the practice of it is a societal construct, and practitioners of science are members of this society and can be as given to presuppositions as anyone else (Grant, 1992). "Observations and investigations are always motivated and guided by, and acquire meaning in reference to questions or problems, which are derived from certain theoretical perspectives" (Lederman et al., 2002, p. 501).

The question becomes: can one separate the nature of an individual from the nature of science? Stanfield (1995) argues that science cannot be separated from its creators. He contends that:

> Social realists argue that for far too long there has been reluctance to view scientists as human beings with biases derived from their historical and cultural contexts, politics, and idiosyncrasies. They claim that the traditions, institutions, communities, and networks scientists, as cultural baggage carriers, create, stabilize, and transform are sociological and anthological phenomena. (p. 223)

Stanfield also states that:

> One cannot divorce the history of the human sciences from the sociology, politics, and economics of capitol formation. It is this sense that the human sciences, by their very nature are social, cultural, and political and therefore intrinsically biased. (p. 223)

The nature of science has been throughout the course of history both good and bad. It could be reasonably debated that in the case of African Americans, science or the misuse of science has been used to hinder the full inclusion of them into society.

The very nature of science, in the hands of certain persons, excludes and separates, systematically using information to project certain images or beliefs. As a result, general laws are implied within society, not laws that are recorded or spoken, but invisible or implicit laws of social practices, a kind of hidden curriculum (Apple, 1986). Those unwritten, unspoken, and invisible laws imply that African Americans have no worth in this society, are mentally inferior, are second-class citizens, and deserve their lot in life. The nature of science when misused has made those of African descent appear less than they really are.

The History of African Americans and the Misuse of Science

It may be argued that science has been represented as something that is free of personal beliefs and values; something that is uncorrupted, without fault; and something that is above all else, objective. Since the eighteenth century, science has on many occasions been used as a rationalization to recommend, develop, and endorse bigoted social practices in this society (Dennis, 1995). Science does and has always had great authority in society. It is because of the authority given to science that it has had great effect on the attitudes towards the idea of race in society as known in present time. Science, since its inception, has had a reputation of being exclusionary. Science was, and remains, an institution in which not everyone can participate because it was designed to be such. Norman (1998) describes the institution of science in this way:

> The institution of science by way of the Royal Society in England and other academies in Europe rendered science a powerful force in the hegemonic projects of Europe. It was the scientific establishment that reinforced the widely held notions that the bodies of women, the lower class, and the colonized were mere "signs" that were to be interpreted and

incorporated into narratives aimed at consolidating as natural and legitimate the position of privilege occupied by European males at the top of the gender, class, and race hierarchy. The almost unassailable position of prestige and influence attained by science through its institution was used to legitimize the tendencies of exclusion and dominance manifested in the wider society. (p. 366)

In order to preserve this institution of privilege, much pure and objective science was conducted in an effort to keep the European male in control of society.

Since the seventeenth century, science has been so esteemed and powerful that it prevailed over all other thoughts that opposed tactics of supremacy and separation. Scientists, because of the esteem and power that science encompassed, were held in high regard in society. During the seventeenth and most of the eighteenth centuries, the works of scientists were deemed indisputable; so without dispute, the scientists' findings about racial inequality were basically unchallenged. Since science was viewed by humanity as a discipline that was incontestable, their assertions about race were accepted by the mainstream (Norman, 1998; Schiebinger, 1989; Steppan & Gilman, 1993). Three reasons can account for the acceptance of these social theories of the time:

1. Science has done a spectacular job in its persuasive declaration to absolute impartiality.
2. Institutional science has been successful in positioning itself outside the grasp of ethical, political, and spiritual examination (Norman, 1998).
3. Science provided clear and precise evidence that showed the natural inferiority of African Americans, as well as women and those from different socioeconomic classes.

It was not until the late eighteenth and early nineteenth centuries that opposition started to emerge, but by this time the damage had been done. The doctrine of racial inferiority had already been allowed to infiltrate the fabric of society. Perhaps through the use of two scientific methodologies in particular, craniometry and IQ testing, scientists managed to use science to really embed the notion of racial superiority of European Americans and the racial inferiority of African Americans in US society.

Craniometry

In the 1800s, scientists such as Carleton S. Coon, Samuel G. Morton, and Paul Broca measured and weighed the human brain to document unequal intelligence between races, and all came to the conclusion that African Americans were inferior to European Americans and that women were inferior to men. Social Darwinism would give theoretical sophistication to the methodology these scientists used that claimed that people of African descent, because of the size of their skull in relation to those of European descent, were not on the same level intellectually as European Americans, and for this reason, their less significant status among society was merited (Stanfield, 1995). This message was allowed to penetrate society even though these scientists found substantial amounts of evidence that contradicted their original hypothesis.

For example, the average European American's brain that was measured during this time had a volume of about 1,400 cm^3, while those of African descent had some 50 cm^3 less. What the scientists using this methodology failed to communicate was that the Neanderthal man, Mongols, and Eskimos all had brain volumes that exceeded those of European descent by at least 150 cm^3 (Montagu, 1965). Also added to the list of those with greater brain size would be Native Americans, as well as some entire African nations (Montagu) which are an indication that scientist selected what they wanted to report to advance their line of thought.

The fact that none of these findings were discussed in regard to African Americans when such claims were made about them being inferior because of their brain size is not shocking. This is further evidence that indicates that science or scientists are influenced by the social constructs in which they live. This is evident in that scientists neglected to discuss their entire findings because it went against their worldviews about race in this society. The truth is that no one in the past or present has been able to make a correlation between brain size and intellectual ability (Grant, 1992). This is because brain size, skull size, weight, volume, cell number, etc., have no relation at all to intelligence (Montagu, 1965).

Intelligence Testing

In the latter part of the nineteenth century and well into the twentieth century, this ideology of inferiority continued with intelligence testing (better known as IQ testing). This methodology again was used to show that African Americans were not as intelligent as White counterparts and that their position in society was therefore deserved. The tests were used as an extension from craniometry in that scientists wanted to relate smaller skull size, as well as the volume of the brain, to low performance on intelligence tests that were designed. Again Social Darwinism gave theoretical and scientific validity to these methodologies. The IQ tests were used to exclude African Americans from certain fields of work requiring a higher level of thought.

For example, the US Army developed tests to place soldiers in particular lines of duty in World War I. These tests showed that on average, White Americans outscored African Americans, but ironically those African Americans from the North in many cases outscored their White counterparts from the South. One possible explanation for this outcome could be attributed to the conditions in which African Americans of the North lived. The racial climate that African Americans from the North lived in was not as harsh as the environment for those African Americans who lived in the South. In the North, segregation was less prominent, and this allowed African Americans to attend school alongside those of European descent. The results of these tests indicated that environment and opportunity to learn had more to do with the results on the intelligence tests than did genetics (Hines, 2002).

Racialists did not agree with the finding concerning environment related to African Americans. Scientists such as Professor Richard Lynn of the University of Ulster believed that those with European blood would continually outscore those of African descent. The thought was that the differences in scores were too large to be explained by the environmental conditions in which African Americans lived; therefore, the reason must be genetic makeup (Grant, 1992). The argument was made that those African Americans with higher scores had more European ancestry than those that scored lower and those European Americans who scored lower had significantly more African ancestry. However, regardless of their scores, African Americans were still placed in subservient roles because of their race.

Social Darwinism

To give an idea of how much Social Darwinism was, and to some extent still is, entrenched in this society, Henry E. Garrett, a visiting professor at the University of Virginia, published in 1961 "The Equalitarian Dogma" in *Perspectives in Biology and Medicine,* in which he asserted that holding African Americans to the mental equals of European Americans was the scientific hoax of the century (Synder, 1962). The article received national attention because of Garret's reputation within the scientific community. Garrett believed that the idea that all men were born of equal endowments was ludicrous as well as deceptive because he believed those of African descent have never accomplished anything of significance. He contended that the environment in which African Americans lived had little to do with their intelligence and that their scores on intelligence tests were mostly a sign of their genetic composition (Synder). According to Synder, Garrett was under the belief that some in society suppressed evidence of African American mental and social immaturity in an effort to help them. Garret believed that their efforts were sincere, but unfortunately erroneous, and he referred to these actions as the equalitarian dogma.

Although the findings by these scientists in craniometry and intelligence testing may have been filled with racial prejudices, for many White Americans, these scientific methods only confirmed what they already believed about African Americans: "that there was White ethnic hierarchy, and that this hierarchy, despite differences, stood atop all other races, especially the African American race (Dennis, 1995, p. 247)." Even without the backing of science, there was a real need for White Americans to believe that African Americans were inferior to them, and due to the validation that science provided, even those with low economic status could take solace in knowing that those of African descent were beneath them (Dennis).

Science provided the objective confirmation needed for those of lower economic status to believe, without a shadow of doubt, that at the very minimum they were made better than the "Negroes." A professor from the University of Virginia was quoted in 1900 that "the Negro race is essentially a race of peasant farmers and laborers. As a source of cheap labor for a warm climate he is beyond competition; everywhere else he is a foreordained failure" (Perkinson, 1991, p. 42). The misuse of science led persons to think in this manner and promoted a "natural bias toward analysis that glorifies one's own status groups and deprecates those of others" (Jorgensen, 1995, p. 236).

Slavery

According to Dennis (1995), the science methodologies of that day accomplished two things: "they confirmed White Superiority and they strengthened the idea that Blacks should be excluded from the core culture of American society" (p. 247). These thoughts are still prevalent today because science laid the foundation for these thoughts to manifest through the years, regardless of the fact that science has been recognized to be imperfect and not beyond letting personal biases or agendas into its absolute objectivity.

For example, the South in the 1840s and the 1850s received tremendous pressure from the North to abolish slavery. In order to ease some of this pressure or tension about slavery, the South badly needed a way to justify its position on slavery to the North because at this time, other countries in the world had completely eradicated slavery. Science served as the South's justification. It was around this time that much of the literature that was discussed earlier in regard to African inferiority and White superiority started to emerge (Dennis, 1995; Oakes, 1982).

In order to justify slavery, those of the South used the Declaration of Independence for its foundation. In *The Idea of Race*, Ashley Montagu articulates how the signatories of the Declaration of Independence did not mean in the biological sense that "all men are created equal" and that "they are endowed by their creator with certain unalienable rights; that among these are life, liberty, and the pursuit of happiness." It was believed the signatories were speaking from a political sense only, in part because the US Declaration of Independence was by no means intended to represent equal aptitude; its intent was to establish the position that a person within set parameters is entitled to live without restrictions and to realize himself/herself to his/her fullest potential and that a person has the unequivocal right to develop without repercussion.

This view caused those who had the most interest in slavery to prove that African Americans were not biologically, or in any other way equal to them. Because of this, they could not enjoy the rights and privileges granted by the Declaration of Independence to the level of European Americans. This doctrine of racism gained strength through the various scientific experiments like craniometry and IQ testing that were previously mentioned and the various interpretations of those experiments. Science convinced society that it was justified in enslaving African Americans because they were by nature beneath those with White ancestry, were in a sense not human, and were "scarcely capable of mental endowment" (Jorgensen, 1995, p. 234).

According to Montagu (1965), three things were to be accomplished by this doctrine: (1) "To prevent homogenization or magnetization and thus deterioration of the superior race; (2) to keep the races segregated so that each has the opportunity to pursue life, liberty, and happiness within the prescribed limits; and (3) to provide educational and social opportunities for the members of each race according to the limits of their assigned capacities, the superior race, of course, enjoying superior opportunities to those of which the inferior race is held to be capable of taking advantage (p. 45)."

Today's Messages Regarding Racial Inferiority

Science as a whole still enjoys that cloak of irrefutable exactness that it enjoyed when it was making the claims of the past. The scientists of the past have always claimed objectivity when questioned, and to an extent the scientific community of today invokes that same claim to objectivity, with essentially the same effect when questioned. In 1923 Carl Brigham published *A Study of American Intelligence*, and in 1994 Richard Herrnstein and Charles Murray published *The Bell Curve*. Both publications claimed to be scientifically reliable and completely objective when reporting the findings that European Americans were superior to African Americans and other races of people (Vera & Feagin, 1995).

These two books would probably be of little importance if they went unnoticed, but the fact remains that both sets of authors had an audience. *A Study of American Intelligence* was offensive but understandable due to the time in which it was written and published; however, *The Bell Curve* was totally shocking due to the fact that it was published in 1994, a time when supposedly the use of science was used to unite instead of separate. The message about the inferiority of different ethnic groups was again allowed to permeate through society. That message was the same message that has been with this country for generations and that message is "groups of people should learn to appreciate what they do well and not aspire to other things outside their natural capabilities" (Zappardino, 1995, p. 6). This view is the offspring of the misuse of science, and this perspective has had great effect on African Americans.

The Effects of the Misuse of Science in the African American Community

The status of future generations of African Americans could easily be predicted by some due to the oppression experienced by African Americans of the past. Four tenets given by Jorgensen (1995) depict a realistic synopsis of African Americans, and those realities can and have created a climate of racism in the field of science and in the society as a whole.

First and Second Tenet Discussion

The first and second tenets state, (1) "racial oppression creates negative social facts such as the low economic, political, and social status of the oppressed and its harmful effects on the character of a portion of the oppressed population," and (2) "the negative social facts that are the consequences of the oppression are used as justification of oppression" (Jorgensen, 1995, p. 235). These tenets are communicated daily about African Americans. In today's society the suppressors try to hide their continuous study of the suppressed by masking their experiments in social problem approaches. Through these approaches, the misuse of science has validated and reconfirmed the notion that African Americans are mentally inferior by creating an undertone that leads people to draw these conclusions (Stanfield, 1995).

Take, for instance, the "sociological studies of dysfunctional African American families and gender categories, the educational psychological studies of poor African American performance on standardized tests, and the identity pathologies of children of mixed descent to the neurological explanation of inner city African American violence" (Stanfield, 1995, p. 226). In addition, according to Power, Murphy, and Coover (1996), in a content analysis of prime-time fictional programming from 1955 to 1986, Lichter, Lichter, Rothman, and Amundson (1987) found a strong association between crimes, drug trafficking, and African American characters.

Similarly, in a series of studies on reality-based news reports, Entman (1994) suggests that the television news "paints a picture of Blacks as violent and threatening towards Whites" (p. 29). Entman (1994) also notes a "dearth of positive portrayals of African Americans as contributors to American Society." The negative images place African Americans at a disadvantage. They are not only faced with the challenge of overcoming the expected hurdles for achievement, but their hurdles are further compounded by their struggles to prove their self-worth.

These studies and perceptions of African Americans cannot help but validate the notions of White superiority, White normality, and above all else African American inferiority (Stanfield, 1995). "The historical origins, institutionalization, and transformation of science as sources of racially and ethnically bounded knowledge reaffirms its legitimacy" (Stanfield, p. 224). Society in general is fine with these results because it legitimizes the dominant group position in this society. It gives privileges and advantages in which everyone cannot participate.

The system is a very complicated entity that has maintained its advantage and privilege by destroying the self-efficacy of an ethnicity's hopes and dreams, causing African Americans to question their value in this society. Those questions guide the journey that African Americans travel in their quest to define themselves. They must wade through images perpetuated in today's society that are in most cases not positive. In these circumstances, African Americans are forced to maintain vigilance and thus must devote major energy to discerning, preventing, and ameliorating such negative presumptions.

Many African Americans live life, confronting stereotypes that affect their existence. "In effect, stereotyped assumptions greatly determine the salience of African Americans physical and psychological presence in many contexts" (Franklin & Franklin, 2000, p. 45). Their experience, the history of African Americans, and those representations of their race in which they see in the media all have an effect on the psyche of African Americans.

Due to this stigma that science has established and validated about race, many African Americans live their entire lives trying to "refute the degrading, humiliating and offensive racial images and stereotypes" (Yeakey & Bennett, 1990, p. 12) that have plagued their race. The images that are perpetuated have caused frustration as well as aggression in African Americans. To take an entire race on their shoulders truly has an effect on the consciousness of African Americans, especially when "the drive towards achievement and accomplishment that the African American professional inspires is overwhelmed and distorted by the social reality it conceals" (Yeakey & Johnston, 1979, p. 12).

Almost every problem that plagues African Americans can be traced back to the roots of perceived racial inferiority and how the misuse of science helped establish those roots. The roots that were validated by science have developed into what is formally known as racism. The residue of this misuse of science has manifested in the lives of many African Americans.

Racism can be seen, according to Harrell (2000), as:

> A system of dominance, power, and privilege based on racial group designations; rooted in the historical oppression of a group defined or perceived by dominant group members as inferior, deviant, or undesirable; and occurring in circumstances where members of the dominant group create or accept their societal privilege by maintaining structures, ideology, values, and behavior that have the intent or effect of leaving non-dominant group members relatively excluded from power, esteem status, and/or equal access to societal resources. (p. 43)

Another author, Tatum (1997, p. 7), believed that racism was "not only a personal ideology based on racial prejudice, but a system involving cultural messages and institutional policies and practices as well as the beliefs and actions of individuals." She further notes "in the context of the United States, this system clearly operates to the advantage of European Americans and to the disadvantage of people of color" (p. 7). The system of privilege that European Americans enjoy oppresses and denies African Americans and other ethnic groups of those unalienable rights that are dictated in the Declaration of Independence.

Studies report the connections between the impact of racism on African Americans and their social and physical conditions (Franklin & Franklin, 2000; Gordon, Gordon, & Nembhard, 1994; Leary, 1996). These studies analyze the degree to which the complicated and frequently pathological state of affairs unconstructively affects the development, self-identity, and self-esteem of African Americans (Gordon et al., 1994). African Americans are psychologically injured by their demoralized standings and treatment (Kardiner & Ovessey, 1951).

In the system of advantage, those in power set the parameters in which those without power operate, meaning that the individuals in power have a large amount of control in shaping the structure of society. The structure of society places African Americans at a disadvantage because their predetermined positions have devalued significance due to implications such as African Americans being considered throughout history as less intelligent than European Americans, or incapable of performing in high cultural capital professions such as science. This line of thought is embedded so deeply in this society that African Americans may internalize the representations or images that the dominant group holds about them, making it challenging for them to have faith in their own ability (Tatum, 1997).

Each generation of African Americans throughout history has experienced obstacles that they had to overcome. Those generations that follow have the history of those that came before them and the present circumstances in which they now live. The effects of having experiences that include overcoming racial obstacles and operating in a system that was designed to keep them in place leave a people feeling invisible or not of worth because the cycle of injustice repeats itself.

In this society many messages are conveyed about African Americans. Images and information that have been made popular by the media can easily be interpreted to mean that African Americans are lazy and unintelligent. From these depictions, many believe that African Americans deserve the secondary status that they hold in this society. Society, for the most part, has absolved itself of the responsibility for the negative state that many African Americans may find themselves.

Third and Fourth Tenet Discussion

The third and fourth tenets relay that (3) "in addition to oppression justifying itself by blaming the negative social consequences on the nature of the oppressed, oppression justifies itself by ignoring positive social facts about the oppressed" and (4) "oppressors must always find a way to scientifically and morally justify their oppression" (Jorgensen, 1995, p. 235). An argument can be made that no matter how much advancement is made by people of color, oppressors will always find ways to hinder their progress. The misuse of science has validated and laid underpinnings for the justification of racial oppression.

Although many African Americans have been successful at doing and using science, science and African Americans have been at opposite sides of the spectrum for many years. While the misuse of science has been used throughout history to attempt to show the mental inferiority of African Americans their survival through over 400 years of oppression counter ideas of mental inferiority with very little success at changing beliefs. If any of the assertions made by science were true, African Americans would not have made any advancement since slavery. Thus the use of science has essentially violated and misrepresented the identity of African Americans in an effort to maintain and sustain a system of privilege for White Americans. It can be debated that African Americans are in a no-win situation because even when they operate by the rules that White Americans establish, their accomplishments essentially are belittled and twisted into other evidence of their insufficiency.

What if the social structure of society actually dictated that African Americans could not participate fully within this society? If society was arranged in a way that primarily benefited those that possessed certain characteristics, could society blame African Americans for their current situation? Turner (1984) believes that oppression is the result of the following conditions:

1. "When a social system reveals populations that are biologically, culturally, and/or socially distinguishable
2. When one population perceives another as a threat to its well-being, particularly when (a) There is competition over scarce resources, and (b) Political leaders need to unify a population by focusing on a common enemy
3. When populations possess vastly unequal degrees of power; and
4. When discriminatory actions can become institutionalized in specific social structures and in cultural beliefs that legitimate these structures. (p.7)"

If these conditions are met, which they are in this country, then according to Turner (1984), oppression will take place.

The need to suppress certain groups of people makes it clear that privilege can be obtained by the suppressors, so much so that in order to maintain this sense of entitlement, the suppressors must condemn the aptitude of the suppressed even though the actions contradict what the suppressors believe is ethically correct (Dennis, 1995). The actions and practices of the suppressors are in complete contradiction to the "fundamental yet abstract antiracist moral principles embodied in the U.S. Constitution and Declaration of Independence and the virulent racism evident in American social practice" (Jorgensen, 1995, p. 234). Knowing these actions were and are ethically and morally wrong, the assumption can be drawn that power (political, social, and economic) can be gained by maintaining the suppressed in substandard places in society (Dennis, 1995).

Conclusion

The misuse of science has created a system of privilege that has over the years guaranteed, as a whole, White Americans control and success in all aspects of life in the United States. Science has, through its systematic arrangements of truths, managed to create a system that separates and oppresses those who do not possess the same skin color as White America. "A science that is the reflection of a White ethnic-dominated, race-centered society that creates and nurtures it cannot help but view non-White others in a lesser light than those who are given, by virtue of skin-color privileges, divine qualities of superiority" (Stanfield, 1995, p. 229). All underrepresented groups are affected by this system of privilege, but African

Americans are the primary beneficiaries of all the hate and bigotry that exist in society. With regard to African Americans, the misuse of science has allowed a system that lacks parity and equity for all to be established and maintained.

This chapter will end by asking the same question with which it started: why would African Americans want to be a part of something that has continuously tried to disenfranchise them and has tried to prove since its inception that they could not think on a higher order? We know that African American students are just as competent as any group of students in achieving in STEM fields, but society has told them repeatedly that they are not capable and has even used "science" to prove that they are not. These actions and these messages have through the years led many African Americans and others to believe that STEM is beyond their intellectual capabilities and just not for them. In order for change to occur, science teacher educators must first answer the question as to why African Americans should pursue STEM professions for themselves. They must then recognize the critical role they play in assisting more African American students to answer the question in a manner that encourages them to pursue and persist in STEM fields. Science teacher educators at every level must recognize that an increase in African American participation in STEM is dependent on them as they are the gatekeepers to the profession. Again, recognizing why a problem exists is the first step in solving it.

References

Abd-El-Khalick, F., & Lederman, N. G. (2000). Improving science teachers' conceptions of the nature of science: A critical review of the literature. *International Journal of Science Education, 22*, 665–701.
American Association for the Advancement of Science. (1989). *Project 2061. Science for all Americans*. Washington, DC: AAAS.
American Association for the Advancement of Science. (1993). *Project 2061. Benchmarks for scientific literacy*. Washington, DC: AAAS.
American Association for the Advancement of Science. (1998). *Project 2061. Blueprints for reform in science, mathematics, and technology education*. Washington, DC: AAAS.
Apple, M. W. (1986). *Ideology and curriculum* (2nd ed.). New York/London: Routledge.
Atwater, M., Wiggins, J., & Gardner, C. (1995). A study of urban middle school students with high and low attitudes towards science. *Journal of Research in Science Teaching, 32*(6), 665–677.
Bierstede, R. (1981). *American sociological theory*. New York: Academic Press.
Brigham, C. (1923). *A study of American intelligence*. Princeton, NJ: Princeton University Press.
Buffon, G. L. L. (1797). *Barr's Buffon: Buffon's natural history-Containing a theory of the earth, a general history of man, of the brute creation, and of vegetables, minerals, etc*. (J. S. Barr, Trans). London: H. D. Symonds.
Dennis, R. M. (1995). Social Darwinism, scientific racism, and the metaphysics of race. *Journal of Negro Education, 64*(3), 243–252.
Entman, R. (1994). Representation and reality in the portrayal of blacks in network television news. *Journalism Quarterly, 71*(3), 509–520.
Franklin, A., & Franklin, N. (2000). Invisibility syndrome: A clinical model of the effects of racism on African American males. *American Journal of Orthopsychiatry Incorporating, 70*(1), 33–41.
Glasson, G., & Bentley, M. (1999). Epistemological undercurrents in scientists' reporting of research to teachers. *Science Education, 84*, 469–485.

Gobineau, A. (1915). *The inequality of human race*, Vol. I (A. Collins, Trans.). New York: G. P. Putnam's Son's. (Original work published 1853)

Gordon, E. T., Gordon, E. W., & Nembhard, J. G. (1994). Social science literature concerning African American men. *Journal of Negro Education, 63*, 508–531.

Grant, N. (1992). "Scientific" racism: What price objectivity. *Scottish Educational Review, 24*(1), 24–31.

Greene, J. C. (1963). *Imperialism*. London: Allen & Unwin.

Harrell, S. P. (2000). A multidimensional conceptualization of racism-related stress: implications for the well-being of people of color. *American Journal of Orthopsychiatric Incorporating, 70*(1), 42–57.

Hernstein, R., & Murray, C. (1994). *The bell curve intelligence and class structure in American life*. New York: Free Press.

Hines, M. (2002). A science educator, a sociologist, a historian, and a social studies educator on race in US history. *Race, Ethnicity and Education, 5*(1), 107–123.

Johnson, R. C. (1992). *Providing African American students access to science and mathematics* (Research report #2). Cleveland, OH: Urban Child Research Center.

Jorgensen, C. (1995). The African American critique of white supremacist science. *Journal of Negro Education, 64*(3), 232–242.

Kardiner, A., & Ovessey, L. (1951). *The mark of oppression: A psychosocial study of the American Negro*. New York: Norton.

Leary, W. E. (1996, October 24). Discrimination may affect risk of high blood pressure in blacks. *New York Times*, p. A20.

Lederman, N. G., Khalick, F. A., Bell, R. L., & Schwartz, R. S. (2002). Views of nature of science questionnaire: Towards valid and meaningful assessment of learners' conceptions of nature of science. *Journal of Research in Science Teaching, 39*, 497–521.

Lewis, B. (2003). *The under-representation of African Americans in science: A review of the literature*. Paper presented at the National Association of Research in Science Teaching, Philadelphia.

Lichter, S. R., Lichter, L. S., Rothman, S., & Amundson, D. (1987, July/August). Prime-time prejudice: TV's images of Blacks and Hispanics. *Public Opinion*, pp. 13–16.

Montagu, A. (1965). *The idea of race*. Lincoln: University of Nebraska Press.

National Research Council. (1996). *National science education standards*. Washington, DC: National Academy Press.

National Science Foundation. (2004a). Division of Science Resources Statistics, special tabulations of U.S. Department of education, National Center for Education Statistics, Integrated Postsecondary Education Data System, Completions Survey, 1997–2001.

National Science Foundation. (2004b). An emerging and critical problem of the science and engineering labor force: A companion to science and engineering indicators. www.nsf.goi>/she/srs/nsb04O7/start.htm. Last accessed Apr 2011.

Norman, O. (1998). Marginalized discourses and scientific literacy. *Journal of Research in Science Education, 35*(4), 365–374.

Oakes, J. (1982). *The ruling race*. New York: Random House.

Oakes, J. (1990a). *Lost talent: The underparticipation of women, minorities, and disabled students in science*. Santa Monica, CA: The Rand Corporation.

Oakes, J. (1990b). *Multiplying inequalities: The effects of race, social class, and tracking on opportunities to learn mathematics and science*. Santa Monica, CA: The Rand Corporation.

Perkinson, H. J. (1991). *The imperfect panacea: American faith in education 1865–1990* (3rd ed., pp. 40–61). New York: McGraw-Hill.

Popper, K. R. (1992). *The logic of scientific discovery*. London: Routledge. (Original work published 1934)

Power, J. G., Murphy, S. T., & Coover, G. (1996). Priming prejudice: How stereotypes and counter-stereotypes influence attribution of responsibility and credibility among in-groups and out-groups. *Human Communication Research, 23*(1), 36–58.

Russell, M. L., & Atwater, M. M. (2005). Traveling the road to success: A discourse on persistence throughout the science pipeline with African American students at a predominantly white institution. *Journal of Research in Science Teaching, 42*(6), 691–715.

Schiebinger, L. (1989). *The mind has no sex? Women in the origins of modern science*. Cambridge, MA: Harvard University Press.

Smyth, F., & McArdle, J. (2002). *Ethnic and gender differences in science graduation at selective colleges with implications for admission policy and college choice*. Paper presented at the 2002 American Educational Research Association (AERA) annual meeting, New Orleans, LA.

Spencer, H. (1874). *The study of sociology*. New York: Appleton.

Stanfield, J. H. (1995). The myth of race and the human sciences. *Journal of Negro Education, 64*(3), 218–231.

Steppan, N. L., & Gilman, S. L. (1993). Appropriating the idioms of science: The rejection of scientific racism. In S. Harding (Ed.), *The racial economy of science*. Bloomington: Indiana University Press.

Synder, L. L. (1962). *The idea of racialism*. Princeton, New Jersey: D. Van Nostrand.

Tatum, B. D. (1997). *Why are all the black kids sitting together in the cafeteria?* Basic Books. (1944). *The life and selected writings of Thomas Jefferson* (p. 262). New York: Modern Library.

Turner, J. H. (1984). *Oppression: A socio-history of black-white relations in America*. Chicago: Nelson-Hall Publishers.

Vera, H., & Feagin, J. R. (1995). Superior intellect?: Sincere fictions of the white self. *Journal of Negro Education, 64*(3), 295–306.

Weiss, T. (2009). The 10 hardest jobs to fill in America. http://www.forbes.com/2009/06/03/hard-jobs-fill-leadership-careersemployment.html. Last accessed Apr 2011.

Yeakey, C. C., & Bennett, C. T. (1990). Race, schooling, and class in American society. *Journal of Negro Education, 59*(1), 3–18.

Yeakey, C. C., & Johnston, G. S. (1979). A review essay. *American Journal of Orthopsychiatry, 49*(2), 353–359.

Zappardino, P. (1995). *Science, intelligence, and educational policy: The mismeasure of Frankenstein*. Presented at the AERA/NCME Annual Conference, 2–10. CA: San Francisco.

Second-Class Citizens, First-Class Scientists: Using Sociocultural Perspectives to Highlight the Successes and Challenges of African American Scientists During the Jim Crow Era

Malcolm B. Butler

Introduction

> This is a strange time to be alive in America, in that regard. Close one eye, and we can seem to be moving toward a one-race society; close the other and we seem as racially conflicted and stratified as ever. Racism is still our madness. (Sullivan, 2012, para. 24)

The aforementioned quote may not be well received by many science teacher educators, yet it is difficult to argue that for a large number of students in our schools, it is still a challenge to move beyond the common myths and concomitant deleterious behaviors associated with the social construct of race. To move forward, we must understand our current status and our past. Within this context, educators can use anthropological skills to better understand science education in the United States. Indeed, these understandings can further develop in the science teacher educator the will to effect change through one's science teaching and scholarship.

The National Science Education Standards emphasize the need to focus on science, technology, and society (STS) as a part of science instruction (National Resource Council [NRC], 1996). For example, the NRC (1996) states that "middle school students are generally aware of science-technology-society from the media, but their awareness is fraught with misunderstandings. Teachers should begin developing student understanding with concrete and personal examples…" (pp. 167–168). In addition, the standards also highlight the need for high school students to understand the important role social issues play and have played in scientific and technological advances (NRC, 1996, p. 199). In the *Benchmarks for Science Literacy*, the American Association for the Advancement of Science (AAAS, 2009) devotes an entire chapter to "the scientific enterprise," where it is

---

M.B. Butler (✉)
School of Teaching, Learning and Leadership, College of Education,
University of Central Florida, PO Box 161250, Orlando, FL 32816-1250, USA
e-mail: Malcolm.Butler@ucf.edu

stressed that middle and high school students must come away from their science classrooms with knowledge and understanding about the many different kinds of people who have contributed to scientific and technological developments, some in spite of the restrictions placed on them within their particular society. Even in *A Framework for K-12 Science Education* (2012), it is acknowledged that science understanding is a cultural accomplishment and science instruction should include the contributions of people from diverse cultures and ethnicities. No matter where advances are made or who makes them, the world has the potential to benefit from them. Thus, science teacher educators have a strong and solid rationale for pursuing societal matters in science education.

Within societal matters rests the need to explicate the roles of African American scientists in the making of America, especially in the United States. Pragmatically, we need more Black scientists and students who are interested and prepared to succeed as science majors. Yet reports indicate that there continues to be a relative dearth of Black students who graduate with degrees in science (Czujko, Ivie, & Stith, 2008). However, it must be noted here that historically Black colleges and universities (HBCUs) are the places where most Black science majors are found (Czujko et al., 2008). These schools have some unique characteristics (e.g., identifying, nurturing, and recruiting future science majors, mentoring future scientists, and developing synergistic collaborations and partnerships with governmental agencies, private industries, and traditionally White colleges and universities) that contribute to their success in the production of scientists of color. These characteristics, which are beyond the scope of this chapter, merit further exploration by the reader.

A Story

Having recently earned a bachelor's degree in physics from a historically Black university, Lee was prepared to live the life of a physicist as he began work on his Ph.D. in physics at a historically White university. A year later, a confluence of events and experiences led Lee to reconsider his decision and think more about influencing the teachers who would educate the next generation of scientists. Thus he changed his graduate school aspirations to becoming a high school physics teacher. It was during this preparation that one of Lee's professors told him and his classmates that they needed to be aware and able to work with diverse groups of students. So, Lee walked into his first high school physics class with the awareness that the racially and ethnically diverse group of students in front of him deserved his best efforts to meet their needs. But what was deficit was his very limited set of knowledge and skills in how to teach this great group of students. Thus began a year of learning from his students in ways that could have at least been addressed in his teacher preparation program. Thank goodness for teachable and teaching students!

Questions to Consider

1. Is it possible for Lee to bring his prior experiences and background as a Black physicist to bear on his classroom's curriculum and instruction? If so, how should he go about doing it?
2. How could Lee's university professor have better prepared him for the realities of a diverse high school science classroom?

3. What might Lee learn from his students and their lived experiences that he can use to effectively teach them science?

The aforementioned story is an all too common experience for beginning secondary science teachers. Trying to address this dearth of knowledge about who does science has been the work of many noted scholars. Recently, science educators have attempted to use a psychological construct—sociocultural theory—to better explain what happens and what should happen in the science classroom. Sociocultural theory is typically associated with the work of the Russian psychologist, Lev Vygotsky. This theory posits that children's thoughts and behaviors are inextricably linked to the social context in which they find themselves (Vygotsky, 1979). In essence, what we know and come to know is intimately connected with our lived experiences, as well as the lives of others.

Recently, Verma (2009) focused on sociocultural perspectives as she examined curricular approaches linked to sociocultural perspectives for urban students. Verma's work builds on the research of numerous science education scholars (e.g., Lemke, 2001). This chapter will attempt to highlight the best of what we know about science teaching and learning, sociocultural theory (SCT), and Black scientists to create a knowledge base for secondary science teacher educators and candidates. For this chapter, I will attempt to address the following questions:

- Why Black scientists?
- Why the Jim Crow era?
- Who are some first-class Black scientists?
- What are some curricular connections and pedagogical strategies we can use to help secondary science teacher candidates prepare to teach multicultural science in their classrooms?

Why Black Scientists?

Let us clarify a few matters before proceeding further. First, the terms *African American* and *Black* will be used interchangeably throughout this chapter. Justifications can be found for using either or neither term (e.g., see Newport, 2007). While not trying to simplify a complex social issue, we just do not want to lose sight of the key issues to be addressed here. Secondly, we will focus on a short two-decade period in US history. However, the racial caste system known as Jim Crow was actually in effect in the United States from the 1870s to the mid-1960s. This legalized and institutionalized way of life, was widely accepted as the norm across the United States, being more acute in southern United States (Jim Crow Museum, 2012).

Thirdly, we will use several aspects of Critical Race Theory (CRT) to connect our exploration of Black scientists to the science education we hope to share with future science teachers. Critical Race Theory has beginnings in the legal profession in the 1970s (Delgado, 1995) and began to take root in education with the work of noted scholars such as Gloria Ladson-Billings and William Tate, who began to articulate a connection between CRT and education (Ladson-Billings, 1998; Ladson-Billings & Tate, 1995). Four underpinnings are commonly associated with CRT:

- Racism is embedded in the fabric of the culture of the United States.
- Stories and narratives are essential to providing context.
- The concept of and actions based on liberalism merit constructive critique.
- Whites have benefited from civil rights legislation more than Blacks (Ladson-Billings, 1998, pp. 11–12).

Why the Jim Crow Era?

The first two aforementioned components of CRT (e.g., the permanence of racism and the effective use of stories and narratives) will serve as a backdrop for this chapter, as we will seek to use the period of Jim Crow to tell the story of African American scientists. Indeed, renowned historian Dr. Kenneth Manning (1999) made the case as follows:

> Although science purports to be objective and supposedly has imbedded in it a kind of democratic core, scientists are not science, they are not the thing itself—they are people who live in the world with other people and have many of the same social views and behaviors of society at large. Their institutions are hardly any different than institutions of other professions. The pursuit of science education conforms to the structure of that for any other kind of education. A segregated educational system has had the same effect, if not greater, on science in this country with regard to blacks as is has had on other provinces of learning. Even though the 1954 Brown vs. The Board of Education landmark decision was intended to eliminate segregation in education, we know that in many parts of the country segregation persisted. Not until the 1964 Civil Rights Bill was a minor milestone in the direction of eliminating segregation achieved. Then, opportunities for African Americans opened up at both the undergraduate and graduate levels at many white colleges and universities, and as a result, careers in the field of science became a firmer reality for many African American students. (Mickens, 1999, p. 3)

Based on the cogent points made by Manning, this chapter will take a closer look at Black scientists during the 20-year time period of 1945–1965. Strategies for using the challenges and successes of several scientists with secondary students will be shared, along with resources that can serve as primers for the science teacher educator and secondary science teacher.

Who Are Some First-Class Black Scientists?

We will consider the stories of some African American scientists whose careers included at least a portion of the Jim Crow Era, in particular 1945–1965 (see Table 1). There are many others who could be discussed, and resources for identifying them can be found at the end of this chapter.

Table 1 African American scientists who worked during the Jim Crow era

Scientist	Gender	Science area	Profession
Archie Alexander	Male	Physical science	Chemist
Austin Curtis	Male	Life science	Biologist
Charles Drew	Male	Life science	Biologist
Katherine Johnson	Female	Earth/space science	Aeronautics mathematician
Carl Rouse	Male	Physical science	Physicist
Marie Maynard Daly	Female	Physical science	Chemist

Scientific Highlights

Before looking at incorporating these should-be-famous scientists into our teacher education courses, we need to know something about them. A brief synopsis of each of the following scientists will be offered here. Included also will be some insights as to how each person overcame numerous societal obstacles and succeeded in her/his chosen profession.

Archie Alexander (1888–1958)

After earning an engineering degree from Iowa University (then called the State University of Iowa) in 1912, Alexander opened his own design firm. This firm designed and built structures encompassing the entire United States, including sewage treatment plants, freeways, and airfields. Most prominently, Alexander was responsible for the Tidal Basin Bridge and Seawall in the United States' capital city, Washington, DC. In his later years, Alexander was appointed Territorial Governor of the Virgin Islands by US President Dwight Eisenhower in 1954. During Alexander's time at Iowa, he faced numerous obstacles because of his race. In fact, he was warned by many of his professors that he would face such challenges as a "Negro engineer," concomitant with the racial prejudices of the 1950s. Alexander was also the first Black football player at his alma mater. Alexander's very successful design firm was a partnership between him and one of his White classmates, Maurice Repass.

Austin Curtis, Jr. (1911–2003)

Curtis is best known as the protégé of Dr. George Washington Carver. Dr. Curtis earned a Ph.D. in chemistry from Cornell University. After this achievement, he completed a fellowship in the laboratory of Dr. Carver in Tuskegee, Alabama, where he collaborated on several projects, including research on peanuts and sweet potatoes. He was also instrumental in the establishment of the George Washington Carver Cabin in Detroit, Michigan, the G.W. Carver Museum in Tuskegee, and the Carter Research Foundation. Later in life, after Dr. Carter's death, Dr. Curtis moved to Michigan and founded his own company, Curtis Laboratories, and created over 50 natural and organic-based products, several of which were made from peanuts. Although he was Black and living in the southern United States during his most productive years, Curtis was afforded numerous advantages because of his unique affiliation with Dr. Carver. Indeed, in some circles Curtis was referred to as "Baby Carver", carrying with it a certain amount of cachet. This sobriquet afforded Curtis many opportunities to be quite successful in his later years after Carter's death. Thus two Black scientists used their association to benefit themselves and humanity in spite of the racially sensitive times in which they lived.

Charles Drew (1904–1950)

Dr. Drew is well known for his work with the American Red Cross and is known as the "father of blood banks". He earned his medical degree from McGill University in Montreal, Canada. During his lifetime he worked at several hospitals in the United States, even serving as chief surgeon at one. Dr. Drew's short life ended when he and three other doctors were involved in a car accident in North Carolina. Legend has it that needing a blood transfusion, Dr. Drew was denied one at the nearest hospital because he was Black. Lacking the necessary blood to sustain life, Dr. Drew subsequently passed away. While the details of his death are difficult to corroborate because of conflicting reports and accounts, what is clear is because of segregation in the South and other negative activities in the country during the Jim Crow era, these sorts of stories tended to take on mythical status when it came to successful and well-known Blacks. And no one, especially Blacks who were fighting against racist treatment, was going to pass up an opportunity like this story to support their cause.

Katherine Johnson (1918–)

Johnson earned a bachelor's degree in mathematics and French at West Virginia State College. Johnson was the first African American woman to work at the National Aeronautics and Space Administration (NASA) as a research mathematician and physicist. She was based at the Langley Research Center in Hampton, Virginia. Among her numerous accomplishments, Johnson's most remarkable achievement was her contribution to the development of the mathematical method used to keep track of space ships while in orbit. While Johnson's race and gender could have served as major deterrents to her success as a scientist during the turbulent Civil Rights Era, her knowledge, talent, resourcefulness, and determination proved to be enough to overcome those hurdles. Dr. Johnson continues to serve as a role model for many Black women scientists today.

Carl Rouse (1926–)

Dr. Rouse earned a Ph.D. in physics from the California Institute of Technology (Caltech). Most of Rouse's accomplishments were in the field of astronomy, where he was the *first person* (not first African American!) to solve the Saha equation, a mathematical equation associated with the interior structure of the sun. In 1969, some of Dr. Rouse's solar work was published in the prestigious journal, *Nature*. This was no small achievement, as Black scientists typically worked within communities that associated the value of their work with their skin color. However Rouse was able to accomplish so much because his colleagues in the astrophysics community respected him for his knowledge and scientific acumen.

Marie Maynard Daly (1921–2003)

Dr. Daly was the first Black woman to earn a Ph.D. in chemistry, accomplishing this achievement in 1947 at Columbia University in New York. A native of New York City, Dr. Daly was intimately involved in the early work associated with the organizational structure of DNA. The research she conducted with her colleagues was so well received and regarded that James Watson, Francis Crick, and Maurice Wilkins won a Nobel Prize in 1962, using some of Daly's work to further their understanding of the double-helix structure of the DNA molecule. Dr. Daly was fortunate to colloborate with several White scientists during her illustrious professional career, including her doctoral mentor, Mary L. Caldwell and Dr. A. E. Mirsky, her partner in the study of the cells' nucleus. While it is evident that part of Dr. Daly's success can be attributed to the teamwork that is fairly common in science, it is equally important to highlight that Daly was successful because of her strong working relationships with prominent White scientists. However, Dr. Daly's scientific prowess played just as important of a role in her success.

While the aforementioned biosketches give some sense of the accomplishments of these six amazing scientists, the reader is encouraged to seek more information about each, as their lives were much richer than space allows to be mentioned in this chapter. So now that we know some scientists and their scientific contributions, let us look more closely at the role race played in their lives and how we can situate their accomplishments within the sociocultural times of a portion of the Jim Crow era, 1945–1965.

What Are Some Curricular Connections and Pedagogical Strategies?

The six scientists in the table accomplished much with limited support from the communities in which they lived. Indeed, they succeeded in spite of and not because of such support. Family and key individuals were the nucleus for them, spurring them on to higher heights in their chosen professions.

Much of their adult years were spent during significant civil unrest in the United States. Although citizens, Blacks were subjected to rules and regulations that were not applied to Whites. De jure segregation (i.e., legalized separation) was a natural part of the country's landscape. De facto segregation (e.g., Black people living in a particular neighborhood) also existed, which was just as influential in the scientists' work. Considering the times, we could pose the following questions:

1. How did the Jim Crow laws impact the scientists' recognition in their respective communities and the country?
2. If any one of the six scientists was interviewed today, what would that person highlight as the key factors to her/his scientific success?
3. How might these scientists' lives (both personally and professionally) have been different had Jim Crow laws not been in place during the most productive periods in their careers.

Science teacher candidates should be encouraged (and sometimes forced!) to think about *who* did science, *what* they did in science, *how* they did science, and *why* they did science. Black scientists should be an integral part of the thought process, and it must be an explicit part of teachers' preparation.

Another strategy that could prove fruitful is to role play one of the six scientists. There are many facets of the scientists' lives that are not in view when we focus on their many scientific contributions. For example, focusing on the life of Archie Alexander will cause one to find out more about his relationship with George Washington Carver. Indeed the two scientists spent a significant amount of time together both inside and outside the lab, in a very symbiotic partnership.

A Revisit of the Story of Lee

At the beginning of this chapter, a brief narrative was shared about an experience of a new high school physics teacher facing the challenge of teaching a diverse group of students. Three questions were posed at the end of the story. Questions 1 and 3 should serve as excellent opportunities for preservice science teachers to engage in dialogue about race and its impact on science learning and teaching. If racism is embedded in the US culture, then it truly influenced Lee's major—physics—and his reasons for becoming a physics teacher and not a physicist. Since Lee learned much from his students, how can his students lived experiences be used in his teaching to motivate them to learn science? Are there any racist challenges his students overcame just to be present in a physics class? These ideas and others can be part of a dialogue in a science teacher education class in which racism and culture are the foci. This part of that discussion should push the preservice teachers to consider the why, what, and how of teaching in culturally diverse science classrooms. Thus, these two questions and their class-generated responses should serve as antecedents to what actually happens in the preservice science classrooms.

Subsequently, Question 2, which queries how university professors could have better prepared Lee for teaching in classrooms with culturally diverse groups of students, now should become of interest to science teacher educators. While prescription is not the objective, Question 2 is most germane to the previous section, for if science teacher educators incorporate these and/or similar ideas that weave sociocultural issues (tied to race) into their teacher preparation programs, their graduates will be at least a modicum better prepared than Lee for working with their students. The success of the teacher and the students necessitates such preparation so that issues that connect race and science are on the forefront of teachers' minds as they seek to prepare students to succeed in a race-conscious society.

Final Thoughts

In this chapter, I set out to enlighten us about African American scientists from the Jim Crow era whose contributions are nationally and internationally significant. In addition, attempts were made to connect the scientists' professional successes with the societal challenges, particularly due to their race, and how they overcame to achieve recognition in science. These and many others like them may have been treated like second-class citizens because of the color of their skin, but they were definitely first-class scientists because of their impact in their respective fields. Their accomplishments and lived experiences can serve as fertile ground for helping teacher candidates understand and appreciate the role history can play in motivating students to want to learn science. Such motivation can serve as the genesis for students to learn the important science knowledge, concepts, and skills they need to succeed in science and in life.

Hopefully, the thoughts and ideas shared in this chapter will whet one's appetite to learn more about these particular scientists and share more strategies for infusing our future middle and high school science teachers with understanding and appreciation for the contributions of these often-overlooked yet phenomenal scientists. They are certainly worthy of celebration.

A Sociocultural Exercise for the Secondary Science Classroom

Over the years, in preparing teacher candidates to teach science, I have used the following assignment to help the preservice teachers think about how they can and should include sociocultural perspectives in their future science classrooms. It has been modified to address the issue of race in the United States.

Cultural Adaptation of a Science Lesson

Using the activities we have done in class as resources and examples, locate a science lesson that you would use in your classroom. It could even be one that you have used with students already. Be sure that the lesson you locate has at least the following components clearly identified:

- Grade level
- Goals and/or objectives
- Materials
- Procedures
- Assessment(s)

If the located lesson does not have the appropriate national and state science standards identified, please include them.

Now, take the identified science lesson and describe how you could modify and/or adapt this lesson to be more socioculturally relevant. In your description, show how you could include issues related to race that are germane to the lesson plan. The maximum length of the description should be three pages. Be sure to include any resources that you would use with the lesson.

Turn in the original lesson as found and the three-page description of your suggested modification(s)/adaptation(s).

References

American Association for the Advancement of Science. (2009). *Benchmarks for science literacy*. New York: Oxford University Press.
Czujko, R., Ivie, R., & Stith, J. (2008). *Untapped talent: The African American presence in physics & the geosciences*. Retrieved August 2, 2012, from http://www.aip.org/statistics/trends/reports/minority.pdf
Delgado, R. (Ed.). (1995). *Critical race theory. The cutting edge*. Philadelphia: Temple University Press.
Jim Crow Museum. (2012). *What was Jim Crow?* Retrieved August, 9, 2012, from http://www.ferris.edu/htmls/news/jimcrow/what.htm
Ladson-Billings, G. (1998). Just what is critical race theory and what's it doing in a nice field like education? *International Journal of Qualitative Studies in Education, 11*(1), 7–24.
Ladson-Billings, G., & Tate, W. F. (1995). Toward a critical race theory. *Teachers College Record, 97*(1), 47–68.
Lemke, J. L. (2001). Articulating communities: Sociocultural perspectives on science education. *Journal of Research in Science Teaching, 38*(3), 296–316.
Manning, K. R. (1999). Can history predict the future? In R. E. Mickens (Ed.), *The African American presence in physics*. Atlanta: Author.
National Research Council. (2012). *A framework for K-12 science education: Practices, crosscutting concepts, and core ideas*. Washington, DC: The National Academies Press.
National Resource Council. (1996). *National science education standards*. Washington, DC: National Academy Press.

Newport, F. (2007, September). *Black or African American? "African American" slightly preferred among those who have a preference.* Retrieved August 9, 2012, from http://www.gallup.com/poll/28816/black-african-american.aspx

Sullivan, J. J. (2012, June). How William Faulkner tackled race- and freed the South from itself. *New York Times.* Retrieved from http://www.nytimes.com

Verma, G. (2009). *Science and society: Using sociocultural perspectives to develop science education.* Amherst, NY: Cambria Press.

Vygotsky, L. S. (1979). Consciousness as a problem in the psychology of behaviour. *Soviet Psychology, 17*(4), 3–35.

Resources

With the advent of the Internet, there are many sources of information on Black scientists, many of which can provide the context for analyzing science during the integration of the Jim Crow era and the Civil Rights Movement. Be sure to check sources carefully, as legends, myths, and truths can coexist in some of the stories of life.

In addition to the resources cited in the chapter, here are a few more sources of information that could prove useful.

Brown, J. E. (2011). *African American women chemists.* New York: Oxford University Press.

Detroit Area Pre-College Engineering Program. (1997, April). *Minority Contributors Unit.* Presented at the annual convention of the National Science Teachers Association, Philadelphia, PA.

Fouché, R. (2003). *Black inventors in the age of segregation.* Baltimore, MD: The Johns Hopkins University Press.

Jones, L. (2000). *Great Black heroes: Five brilliant scientists.* East Orange, NJ: Just Us Books.

Mickens, R. E. (1999). *The African American presence in physics.* Atlanta: Author.

Mid-Atlantic Equity Center. (1992). *Introducing African American role models into mathematics and science.* Washington, DC: Author.

Science Education and Females of Color: The Play Within a Play

Leon Walls

The Tragedy of Hamlet: Prince of Denmark by William Shakespeare is noteworthy for its dramatic use of the "play within a play" device. As the play opens, what we the viewing audience know is that the king has just died; his brother has ascended to the throne; the new king has wooed and married his brother's former wife, the queen; and finally, the deceased king's son, Hamlet, is considered mentally unstable and delusional. This of course is only what we *think* we know. Soon Hamlet reveals himself to be quite lucid and aware. By way of the "play within a play" technique, the events are now cast in a new light. What the audience "within" the play and *we* in the viewing audience simultaneously discover is that all is not as it appears. So what does Shakespeare's masterful tragedy have to do with females of color in science education as the title of this chapter proclaims? The answer surprisingly is quite a lot. Like Hamlet's Denmark, all is not well in the state of science education either. This "double bind" is particularly troublesome for females of color who *simultaneously* experience sexism and racism throughout their science education and into their STEM careers (Ong, Wright, Espinosa, & Orfield, 2011). In this chapter, I take a slightly Shakespearean perspective on the experiences of females of color as they progress along the science education–science career continuum. This chapter will hopefully prove particularly useful to science teacher educators who are interested in infusing equity into their instructional framework. To that end, I close the chapter by addressing how science teacher educators can work to mitigate and ultimately work to transform the experiences and outcomes for females of color. For the sake of clarity, I use the capitalized version of the term "Science" throughout this chapter in an attempt to encompass and more accurately reflect the multiple

---

L. Walls, Ph.D. (✉)
University of Vermont, Waterman Building 532,
85 South Prospect Street, Burlington, VT 05405, USA
e-mail: lwalls@uvm.edu

definitions embedded in this single term. I have divided this Science/play dialectic into three major parts or *acts*. Each act further discusses "Science" from the perspectives and experiences of females of color.

Imagine for a moment the tale that is *Hamlet,* rewritten and cast as a one-man show featuring only the title character. Ophelia, the King, Rosencrantz, and Guildenstern all are relegated to peripheral and insignificant roles, virtually airbrushed away. Just such a feat has marked the evolution of Science and in particular, *science education*. Much like Hamlet himself, the lead role in Science coincidentally is also embodied by a "White non-Hispanic" male (Leggon, 2006). Those airbrushed out include, well, everyone else not fitting that demographic, most notably females and people of color. To provide some context and perspective on the role race has played in Science, James McKeen Cattell, past Vice President of the American Association for the Advancement of Science (AAAS) offered the following summation concerning people of color and Science: "There is not a single mulatto who has done credible scientific work" (Catell, 1914). Likewise, with gender working against them, women too fared no better. Conner (2005) highlights similar hostile treatment at the hands of "men of Science" when he notes, "The subordination of women was an essential component of their worldview, which was entirely committed to maintaining male dominance in a patriarchal society" (p. 364). These are but two examples of the overall environment of contempt and subjugation faced by those who failed to gain access into the exclusively White- and decidedly male-controlled institution called Science. If race and gender were disregarded individually, what chance would anyone embodying both of these traits have? As Catell and Conner's remarks highlight, females and people of color have both been ill-served by Science.

Though much time and effort has been committed to researching and documenting the many barriers that females *and* members of underrepresented groups face in science, less has been researched or written about females who dually *are* members of an underrepresented group in science. The term *underrepresented* as its used here has generally meant to include all individuals not racially or ethnically described as White or of European descent. Although the definition of what it means to be "White" has shifted throughout US history, the power and privilege inherent with that title has nevertheless remained consistent. Just as consistently, anyone labeled as *underrepresented* in the United States has certainly suffered discrimination, prejudice, and bias based upon that descriptor. I will first discuss Science or the "play" we think we know. I then revisit those acts and present them from the perspective of females of color. Thus, the "play within" then critically looks at the actors as well as their actions through a lens of equity.

Science: The Play

Western Science or, more precisely, the scientific enterprise dominated by western civilizations, is really a sum of many parts. Much like the acts and scenes of a play, each part in some way connects to the next to give a fuller picture of the entity itself.

For instance, when each of us thinks of the word *Science*, we may conjure up entirely different images, conceptualization, and definitions in our heads. That is okay and, in fact, is to be expected. A general concept that all users of Science can agree upon, however, is that one of its main purposes is to help us humans understand the natural phenomena around us. Ultimately, the more we learn and discover about our world, the more predictable it becomes. In that respect, Science is uniquely equipped to reveal patterns and areas of consistency in observed phenomena. One unquestionable benefit from being able to understand and predict how the world around us operates is that we subsequently learn how to better improve our viability or prospect of survival as a species. At the very least, Science has thus far allowed us to avoid going the way of the dinosaurs. The three acts I have chosen to discuss are as follows: Act I, *Science as Knowledge to Be Taught and Learned*; Act II, *Science as Process*; and Act III, *Science as Fields of Study*. It is from these three perspectives that I selected to refer to the term "Science" monolithically to highlight its multiple meanings.

Act I: Science as Knowledge To Be Taught and Learned

If you are a product of the US system of public education, chances are you have a science story to tell. That story may be a happy one or perhaps the worst nightmare of your schooling experience. The charts, graphs, formulas, and assorted experiments we encounter in school all make up that discipline some derisively at times refer to as the *subject* of Science. Beginning with learning about plants and animals in elementary school and continuing through physics, chemistry, and biology in high school and college, Science in essence teaches us about, well... Science. Or in other words, in school we learn the *knowledge* that is Science, which is used in the *processes* and activities of Science, so that someday we may be employed and earn a living in one of the *fields* of Science.

Act II: Science as Process

Science is also a process or a series of actions directed towards the aim of increasing our knowledge about natural phenomena in our world. Within this context it is important to understand that Science operates by a clearly established set of rules. These rules help to differentiate Science from other ways of knowing such as myths, folklore, and mysticism. For example, although basketball and golf are both considered sports, the rules and procedures governing each are drastically different and serve to distinguish one from the other. Specifically, Science as process likewise distinguishes itself from all other "ways of knowing" by its systematic and unwavering adherence to evidence. Or as a leading science organization states:

> Sooner or later, the validity of scientific claims is settled by referring to observations of phenomena. Hence, scientists concentrate on getting accurate data. Such evidence is obtained by observations and measurements taken in situations that range from natural

settings (such as a forest) to completely contrived ones (such as the laboratory). (American Association for the Advancement of Science, 1989, p. 4)

One of the several ways that this process is carried out is through applying the scientific method (SM). McComas (1998) states that the SM generally includes (a) defining a problem, (b) gathering information, (c) forming a hypothesis, (d) making relevant observations, (e) testing the hypothesis, (f) forming a conclusion, and finally (g) reporting the results. Certainly, scientists frequently use the SM and the steps outlined above to answer questions and curiosities about the natural world, but not always. Sometimes scientific knowledge is gained purely by accident or via an entirely different path altogether. There is, as I have alluded to, no single way that this process of discovery is carried out. Research on scientists at work in fact has confirmed that no research method is applied universally (Carey, 1994; Chalmers, 1990; Gibbs & Lawson, 1992; Gjertsen, 1989).

Science as process also includes the use of specific tools and instruments designed to ensure accuracy and reduce the effects of human error, for humans do not make especially good interpreters of the observations we gather through our five senses. Technologically unaided, our ability to provide truly objective accounts of observations is overshadowed by the subjectivity imposed by each of our uniquely individual experiences and backgrounds. Take, for example, five different people being asked to estimate the weight of a large stone simply by handling it. Let us assume that the five estimates are 5, 8, 2, 10, and 15 pounds. Obviously the stone has a *single* discreet weight, not five. Despite making the same observation on the same object, each individual provides a different estimated quantity. Once the stone is placed on a scale, the value read actually may show up as nine pounds. Specially constructed instruments and mutually agreed upon tools such as scales and balances are designed to remove as much human subjectivity from scientific observations as possible. However, at no time and under any circumstances can *all* of our subjective impulses be removed from scientific experimentation, data collection, and analytical interpretations.

Act III: Science as Defined Fields of Study

Increasing knowledge about the natural world is not just knowledge to be learned or processes and functions to be undertaken but also an industry serving as a source of full-time employment. Science as an institution including research efforts therefore can be divided into several major fields or disciplines: physical sciences, life sciences, earth sciences, and social sciences. Viewed individually, each is quite different from the other, and in most cases, they bear scant resemblance when closely inspected. Yet the tie that binds them is their common pursuit of knowledge through discovery, creativity, invention, and imagination. It is important to note that the following categorization is meant to be a representative sampling rather than an exhaustive list. Each field was selected because of its well-established and durable history. One of the very central tenets of the nature of science or specifically one of its characteristics is that it is tentative and, therefore, subject to changing over time. That holds true for the methods used as well as the areas or fields studied. Mathematics and technology have also been included due to the close relationship that Science has with each. *Physical sciences*, including physics and chemistry, are the study of relationships between matter, energy, force, and time; *life sciences*, including biology, are all fields of study that deal with living organisms; *earth sciences*, including geology, are concerned with the structure and composition of our planet and the physical processes that have helped to shape it; *social sciences*, including sociology and psychology, all explore human society past and present, specifically in the way human beings interact and behave; *mathematics* investigates the relationships between things that can be measured or quantified in either a real or an abstract sense; and *technology* can be thought of as science put to practical use.

Science: The Play *Within*

All the world's a stage, and all the men and women merely players; they have their exits and their entrances, and one man in his time plays many parts. (Shakespeare, trans. 1980a, 2.7.139–142)

What is a play without its players? Though rhetorical in nature, this question is really at the heart of Science itself. In the lines above, from Shakespeare's pastoral comedy *As You Like It,* Jaques succeeds in accurately summing up not just the world as he sees it but the world of Science as we have come to experience it. As I have previously shown, Science is indeed much like a play, and like a play it is made up of several parts; each viewed separately can only reveal part of the story. In a play these individual parts are referred to as *acts*. Yet sometimes even the most straightforward telling of a tale requires some reading between the lines. For instance, in my example of *Hamlet*, there operates a subtext, one in which the title character reveals to us in a very clever and unique way. Let us now revisit Science the play and its three acts, where I will likewise reveal just what the "play within" is really about.

Act I: Science as Knowledge To Be Learned or Who Teaches and Who Learns?

In many ways Act I is the most crucial and telling of all because it deals with epistemology or scientific knowledge itself. Science has up to now followed a simple maxim: He who possesses the knowledge of Science determines who teaches Science; he who teaches Science determines who learns Science; he who learns Science determines who earns a living in Science; he who earns a living in science determines what methods and processes will govern Science; he who uses and understands those methods and processes determines who possesses the knowledge to teach Science. This ouroboric process has continued virtually unabated since at least the late sixteenth century, when Francis Bacon proclaimed, *scientia potentia est*—for also knowledge itself is power! It was therefore not by accident that in the previously outlined gradation, the pronoun "he" was used exclusively. How ingenious is such a human construct, one in which gender, race, teacher, learner, and creator are all in one the same.

In the United States, one institution in particular has, more than any other, wielded the greatest power of gatekeeper; and that is our educational system. This system, mirroring and evolving in tandem with the norms of US society itself and with similar aims, was incidentally never meant to serve *all* of its citizens equitably. In fact, so powerful was the fear of educating African slaves and their descendants that laws making it illegal to do so were deemed necessary. These laws, taking the form of a racial caste system known as "Jim Crow," most liberally and violently wielded in Southern states, assisted in maintaining this unequal lifestyle especially between Blacks and Whites. Though by no means limited to just the South, these suffocating and controlling edicts affected nearly every aspect of a Black person's public and private life. Under this system it would have been unthinkable to have persons of color ascend to the highly respected level of scientist or engineer in any great numbers. Taking into account even the modest gains towards racial inclusion and parity in Science professions made to date, it is clear how overwhelmingly successful this system of exclusion by race was and still is today. In addition, as the theme of this writing highlights, females and females of color in particular, too, have been marginalized via science education.

The notion that females have suffered tremendously in K-12 science education (American Association of University Women, 2010; Baker, 2002; Elgar, 2004; Kahle & Meece, 1994), resulting in a dearth of representation in science and engineering careers (American Association of University Women, 1992, 2004; Aud et al., 2012; Beede et al., 2011; National Science Foundation, 1999), is not a matter of debate. Assisted by people and policies that coerce them to do so, females' exclusion from science education begins as early as kindergarten. Research has shown that males and females learn socially appropriate behavior by age two to two and a half. By this time male and female stereotypes are set, and boys, more than girls, define what they will and will not do (Kahle, 1998). Physical science education and professions have been particularly difficult for females of color to join. By as early as fourth and fifth grade, African American girls are more positive about their ability to do physical science than White girls. However, by middle-grade levels, all racial/ethnic groups, including females, held less positive attitudes towards science. African American girls have fewer interactions with teachers than do White girls, despite evidence that they attempt to initiate interactions more frequently (American Association of University Women, 1992). As the predicament of women of color in Science points out, a "blaming of the victim" mentality often prevails. Yet research itself presents a clear refutation of this false culpability.

Beyond the early and middle grades, the representation of all US women and girls in science, technology, engineering, and mathematics (STEM) fields has increased dramatically in recent decades (National Science Foundation, 2007). Girls now take as many high school science course as boys and perform as well (American Association of University Women, 2004; U.S. Department of Education & National Center for Education Statistics, 2007), but despite taking advanced science courses in high school, these do not continue with science in college. Teaching, at least at the elementary and secondary levels, remains a predominantly female profession.

Approximately 75 % of full-time teachers were women in 2007–2008. At the elementary level, 84 % of public school and 87 % of private school teachers were female. At the secondary level, 59 % of public school teachers were female, up from 57 % in 2003–2004. Females represented 53 % of private secondary school teachers in 2007–2008. Eighty-three percent of full-time teachers were White, 7 % were Black, 7 % were Hispanic, and 1 % were Asian in 2007–2008. The racial/ethnic distribution of full-time teachers was similar at both the elementary and secondary level (U.S. Department of Education & National Center for Education Statistics, 2012). Where it counts the most, women of color begin at the bottom rung, even within a marginalized group characterized by gender. As you will see in Act II, this push towards invisibility continues.

Act II: Science as Process or Who Decides What Counts?

Modern or Western science is decidedly the product of European influence. In using the term "influence," I do not, however, mean to suggest that this outcome was brought about through democratic egalitarianism. On the contrary, as a result of the Europeans' well-documented history of colonization of conquered peoples, a more appropriate word would be *dominance*. Through the domination of civilizations, Western scientific methods and processes have also successfully subsumed and, in many cases, obliterated most other forms of competing indigenous and cultural ways of deciphering the natural world. Iaccarino (2003) is correct when he states that, "…science is part of culture, and how science is done largely depends on the culture in which it is practiced" (p. 221). However, what he failed to include was that there is a hierarchical tier upon which the European tradition firmly rests. With this position at the pinnacle comes also the power to literally dictate whose Science is worthy of being called thus and whose is not. For example, if the public dissemination of scientific research via publications is any indicator, Europe and North America have produced the most by far. According to Science and Technology Indicators (Observatoire des Sciences et des Techniques [OST], 2004), nearly three quarters (73.9 %) of all scientific publications in 2001came from Europe (46.1 %) and North America (36.2 %). By comparison Latin America produced 2.6 %, sub-Saharan Africa .7 %, and North Africa .2 %. Furthermore, if we take the awarding of the Nobel Prize for Science as an indicator of scientific excellence, not much is different from who publishes. More than 90 % of the laureates in the natural Sciences are also from Western countries, despite being home to only 10 % of the world's population (Iaccarino, 2003, p. 221).

Two other foundational components of the modern scientific process that bear mention here are its patriarchal orientation and its claim of objectivity. Certainly, within the scientific enterprise historic jerry rigging towards male superiority has been the key in producing the skewed, racial, and gender-biased playing field we have today. Shiva (1993) noted that an outcome of just such patriarchy necessitated the subjugation of both nature and women. However, it is my opinion that it was the label, or as I like to describe it, the *shroud*, of objectivity placed upon Science that created the fertile ground upon which seeds of division and exclusion were sown. When I speak of Science being objective, some clarification is warranted. Clearly it is not Science that bears this purported characteristic but the humans who use it, particularly those we call "scientists." However, McComas (1998) leaves no doubt in dispelling the illusion of scientists' objective nature with the following statement:

> Scientists are no different in their level of objectivity than any other professionals. ... [they], like all observers, hold myriad preconceptions and biases about the way the world operates. These notions, held in the subconscious, affect the ability of everyone to make observations. It is impossible to collect and interpret facts without bias. (p. 10)

Not just women but all people of color have known since the initial claim of objectivity was made that it bore scant resemblance to their reality. For instance, how particularly objective was the conducting of the 40-year-long "Tuskegee experiment" on 399 Black men simply to "scientifically" document their slow deterioration and death due to the ravages of syphilis? How objective were the motives that lay behind eugenics creator Francis Galton to produce "a highly gifted race of men?" Is it merely coincidence that this so-called objective science so thoroughly distorted by White males like *The Bell Curve* authors Richard Herrnstein and Charles Murray has so often been used to *scientifically prove* their superiority over people of color?

Shiva (1993) adds a final characteristic of Science that distinguishes it from all other knowledge systems which it has subjugated and replaced, one she describes as *reductionist*. She describes it in the following manner: "Primarily, the ontological and epistemological assumptions of reductionism are based on uniformity, perceiving all systems as comprising the same basic constituents, discrete, and atomistic, and assuming all basic processes to be mechanical" (p. 23). The human as well as environmental toll resultant from this simplistic formulation is today all too evident. Merchant (1980) sums it up this way:

> In investigating the roots of our current environmental dilemma and its connections to science, technology and the economy, we must re-examine the formation of a world-view and a science that, reconceptualizing reality as a machine, rather than a living organism, sanctioned the domination of both nature and women. (p. xxi)

Act III: Science as Professional Fields of Study or Who Is a Scientist?

So who are the biologists, chemists, physicists, ecologists, geologists, and other "ists" that we collectively call scientist? A long-established and oft-used activity known as the Draw-A-Scientist Test (DAST) developed by Chambers (1983) could offer some clues. Historically this simple drawing activity has helped to uncover how children conceptualize their idea of a scientist. Overwhelmingly, when students of varying grade levels are asked to "draw a scientist," they produce a consistently stereotypical figure. Most often it is a White, Einstein-like male, with crazy hair, wearing a laboratory coat. Seldom do the drawings produced represent females and even less of females of color. This is true even when the students producing the drawing are themselves female (Walls, 2012). Coincidently, in his hit song *It's a Man's, Man's, Man's World* (Brown & Newsome, 1966), the proclaimed "Godfather of Soul," James Brown, while echoing Jaques' earlier perspective of the world, was just as accurate in his assessment of the players in Science who we call scientists. He could have more precisely described the essence of Science had he said, *It's a (White) Man's, Man's, Man's World*. A look at some US statistical evidence could explain why.

Women make up approximately 49 % of the US college educated workforce and are approximately 24 % of the science and engineering (S&E) workforce (Beede et al., 2011). However, closer inspection shines a different light even on this modestly impressive statistic. Less than 3 % of this total are employed as computer and information scientists, 1 % are engineers, and even fewer, <1 %, are physical scientists (National Science Foundation, 2009). This last statistic can be explained by the fact that out of roughly 40,000 members of the American Physical Society, only about 2,400 or 6 % are women (Coles, 2007). While women are more likely than men to graduate from high school and enroll in college and are equally likely to graduate from college, they are significantly less likely to major in S&E fields (National Science Foundation, 2002). The sad reality for females of color is that things can always be worse as indicated by the following statistics: Although Blacks and Latino/as are about as likely to major in S&E fields, they are less likely than Whites or Asians to graduate from high school or to enroll in or graduate from college (National Science Foundation, 2002). Of the 27 % that constitute women in S&E employment, White females accounted for 74 %, Asian females accounted for 9.7 %, Black females accounted for 7.5 %, Hispanic females accounted for 5.9 %, and <1 % were accounted for by American Indian/Alaskan Native females (National Science Foundation, 2006). The number of women with science and engineering (S&E) doctorates employed in colleges and universities rose continuously between 1973 and 2006. In 1973 women constituted 33 % of all academic S&E doctoral employment and 30 % of full-time faculty in 2006 up from 9 % to 7 %, respectively (National Science Board, 2008). Again, racial breakdowns highlight just how misleading these numbers can be for females of color. African American and Hispanic girls have high interest in STEM, high confidence, and a strong work ethic but have fewer supports, less exposure, and lower academic achievement than White females (Modi, Schoenberg, & Salmond, 2012). Of all the S&E doctorates awarded between 1997 and 2006, 77 % went to White females, 7.8 % to Asian females, 5.9 % to Black females, and 5.8 % to Hispanic females. The aforementioned statistics hold true even though African American females, for instance, express more interest in STEM fields than do young White females (Fouad & Walker, 2005; Hanson, 2004).

Changing the Script

As Act III closes and the curtain falls a final time, it is clear from the "play within" that the plight of females of color in Science, as debilitating as it may be, is merely a symptom of a much larger affliction. That larger problem can be summed up in one word, *humanity*. Science is ultimately about the humans that use it, and in a perfect world, none would be barred by race or gender. Yet in almost every facet of the scientific enterprise highlighted in the preceding text, it was the human factor that was ultimately the culprit. Another way to describe this systemic preoccupation is simply that it is an undervaluing of human capital. Freire (1988) stated it best when he said, "Concern for humanization leads at once to the recognition of humanization, not only as an ontological possibility but as an historical reality" (p. 27). The opposite of humanization is what has sustained Science to this point in history—that being dehumanization. Yet however incremental and snail-like this march towards equity continues to be, I believe as did Freire that only humanization is the normal state or "man's vocation" (p. 28). Plainly put, it is about how we utilize and treat people in society in general that will ultimately determine when the homeostatic condition of humanization is reached in Science. The good news is that unlike the Bard's classics whose endings were cast in stone centuries ago, we have the power and, increasingly, the will to write a different outcome for all females, especially females of color in science education. In closing I have listed some action steps that science teacher educators can take to address and hopefully alter the "play within."

Language Carries with It Power

> What's in a name? That which we call a rose by any other name would smell as sweet. (Shakespeare, trans. 1980b, 2.2.43–44)

The meaning of Juliet's words as she spoke so eloquently of her beloved Romeo is clear; it is not what you call someone or, even for that matter, some*thing* that defines its essence. Harry and Klingner (2007) agree when they point out that "Language in itself is not the problem. What is problematic is the belief system that this language represents" (p. 16). I partly agree with both the researchers and Juliet in this classic "chicken and egg" scenario. However, I contend that the language we use and the beliefs we hold cannot be separated neatly by saying one precedes the other. Haberman (2000) concurs by stressing that "language is not an innocent reflection of how we think. The terms we use control our perceptions, shape our understanding, and lead us to particular proposals for improvement" (p. 203). While we may use invented terms such as *underserved, disadvantaged, underrepresented, at risk, underprivileged, or excluded* to describe those that are harmed by injustice, we often lose track of the fact that they simply are just children. Preservice teachers therefore should be challenged by teacher educators to consider their own language and belief systems as future teachers of children, including females of color. For no matter what euphemistic language we choose to dress it up in, inequity retains its power to marginalize and oppress. In short, that which we call discrimination, by any other name, is equally devastating to those being affected.

How Science Is Taught Is as Important as Who We Teach

Preparing future teachers in ways to adequately instruct females of color requires that they be well versed in the latest research-based theories and methods of pedagogy. Therefore, in order to achieve this goal, the following must be standard curriculum in any teacher education program: constructivist learning theory, inquiry-based instruction, the nature of science (NOS), and multiculturalism.

Constructivist Learning Theory

Constructivist learning theory simply put is based upon two principles. The first principle states that students do not come to the learning process as empty vessels waiting to be filled or *tabula rasa* (blank slates). Instead, they bring with them prior knowledge, experiences, and backgrounds that influence how they will therefore receive and interpret new information. The second principle states that the learner must take an active role in their learning and are not passively awaiting knowledge to be transmitted via teacher to student. Learning, therefore, is an active endeavor undertaken by the learner in which new knowledge is scaffolded, built, or *constructed* upon. Piaget (1971) explained it this way, "the essential functions of the mind consist in understanding and in inventing, in other words, in building up structures by structuring reality" (p. 27). This view of constructivism concerns only the individual and what they themselves are capable of learning based upon their age or developmental stage they reached. However, social constructivists believe that factors and interactions external to the learner also played a key role in their learning. Vygotsky (1962), for example, theorized that language development simultaneous with working alongside either an adult or more knowledgeable "other" could move the learner beyond predetermined stages bound by age. This process was what he referred to as the "zone of proximal development" or ZPD. In his study involving young African American third-grade students, Walls (2012) found that they clearly distinguished learning Science from other school learning by its social and active components. The following comment expressed by one young African American female speaks volumes:

> In science, we do like projects and we mix stuff, and in math and all that other stuff we get to do at school we got to use a piece of paper and write down the stuff, and in science we got to mix stuff together to see what it makes, and in math all you have to do is just have to write stuff down. Like in science we partner up and do activities with your friends and talk and in math we got to be quiet and do our work. (p. 20)

It would appear that they not only learned from each other but valued conversing and "doing stuff" in the process. Therefore, as part of their training, future teachers should consider the possibility that quiet non-inquiry-structured science classrooms may not be the most conducive learning environments for females of color.

Inquiry-Based Instruction

Based upon constructivist principles, inquiry-based instruction is first and foremost a student-centered process where active involvement and the learner's prior knowledge are essential components. In inquiry it is also necessary for science instructors to view themselves and their function in a new light as well. Instead of being the dispenser of knowledge, the teacher's role becomes that of *facilitator* of their students' learning by relinquishing some level of control over the class. This is often a difficult transition for new teachers to make and a source of much anxiety in learning to teach in an inquiry-based fashion. The National Science Education Standards (NSES) defines inquiry as:

> Scientific inquiry refers to the diverse ways in which scientists study the natural world and propose explanations based on the evidence derived from their work. Inquiry also refers to the activities of students in which they develop knowledge and understanding of scientific ideas, as well as an understanding of how scientists study the natural world. (NRC, 1996, p. 23)

Coburn (2000) adds that inquiry-based instruction is "the creation of a classroom where students are engaged in essentially open-ended, student centered hands-on activities" (p. 42). He further describes the inquiry hierarchical continuum in the following ways:

1. Structured inquiry—The teacher provides students with a hands-on problem to investigate, as well as the procedures and materials, but does not inform them of expected outcomes.
2. Guided inquiry—The teacher provides only the materials and problem to investigate. Students devise their own procedure to solve the problem.
3. Open inquiry—This approach is similar to guided inquiry, with the addition that students also formulate their own problem to investigate. Open inquiry, in many ways, is analogous to doing science.

The age-old tale of the three little pigs provides a clear illustration of how the strength of a house is only as good as the foundation it is built upon. Likewise, preservice teachers often find it difficult to teach effective inquiry-based science lessons because they too so often begin from poorly developed foundations as well, namely, their lesson plans. To assist them in this endeavor and to insure that their lessons do indeed follow an inquiry/constructivist tract, many science teacher educators make available a method known as the 5E Instructional Model (Bybee et al.,

2006). The 5Es consist of engage, explore, explain, elaborate, and evaluate. A brief description of each phase of the 5Es is listed below:

1. Engage—The focus of the engage phase is to essentially do two things, gain the learner's interest or "hook'em," and to access their prior knowledge.
2. Explore—In the explore phase, the students are provided with a common base of activities or hands-on experiences. The main focus of this phase is to allow them to tactilely interact with concepts central to the lesson and to formulate their own questions.
3. Explain—During the explain phase, students are encouraged through questioning and prompts by the teacher to essentially "explain" what conceptual understanding they took away from the explore activity. Effective questioning during this phase is critical to getting the students to verbalize their explore phase experience. This is where it is important for the teacher to listen carefully and to take note of any potential misconceptions the students may have at this point. This is also the opportunity for teachers to do some further clarifying and explanations of their own. However, teachers must not lose sight of the fact that it is the students who should be doing most of the talking and explaining.
4. Elaborate—The focus of the elaborate phase is for the teacher to challenge and extend students' conceptual understanding. One of the essential goals here is for the students to be able to take the conceptual understanding gained in the narrow context of the classroom and apply that understanding to a larger real-world application.
5. Evaluate—The evaluate phase is designed to assess the students' conceptual understanding at the culmination of the lesson. This is also an opportunity for teachers to evaluate their own success by assessing whether the student learning objectives (SLO) they set for the lesson were obtained.

Finally, an additional benefit of utilizing the 5E method is the confidence it instills in preservice teachers. With consistent use, over time its flexibility ultimately allows each user to tailor the method to his or her own needs. However, the fact that it initially works as a template enables easy verification that the essential pieces of the inquiry lesson are built into the foundational lesson plan.

Nature of Science (NOS)

Just as constructivist principles support inquiry-based science instruction, inquiry-based instruction is the foundation for the most effective way to teach the NOS. What is the NOS? NOS can be generally defined as the epistemology of science or the values and beliefs inherent in the development of scientific knowledge (Lederman, 1992).

Walls (2012) added the following definition of NOS from the perspective of the learner:

> Operationally this includes, an individual's beliefs about, how scientific knowledge is constructed; where scientific knowledge originates; who uses science (including scientists); who produces scientific knowledge; and most importantly, where the individual places themselves within the community of producers and users of science. (p. 1)

What this means is that in order for females of color to be successful in learning Science in K-12 classrooms, they must first understand how it *really* operates, where it *really* comes from, and what it *really* can and cannot do. More importantly however, they must understand that they themselves are *really* integral to Science no matter what the world at large has depicted. The desired outcome of teaching K-12 Science of course is to prepare students to be scientifically literate. Though there has been intense debate over the exact definition of science literacy (Hodson, 1999), the American Association for the Advancement of Science (AAAS, 1989) offers this portrait of a scientifically literate person:

> One who is aware that science, mathematics, and technology are interdependent human enterprises with strengths and limitations; understands key concepts and principles of science; is familiar with the natural world and recognizes both its diversity and unity; and uses scientific knowledge and scientific ways of thinking for individual and social purposes. (p. 4)

While not specifically referring to the phrase "science literacy," *A Framework for K-12 Science Education* (NRC, 2012) states that:

> America's children face a complex world in which participation in the spheres of life—personal, social, civic, economic, and political—require deeper knowledge of science and engineering among all members of society. Such issues as human health, environmental conservation, transportation, food production and safety, and energy production and consumption require fluency with the core concepts and practices of science and engineering. (p. 278)

In other words, we want all students to be able to understand, logically analyze, and make informed decisions based on the scientific knowledge they learn in schools. In order to do this, students must first understand not only the content or facts that make up Science but also the characteristics, rules, and boundaries of Science as well. These characteristics, rules, and boundaries are embedded in what we refer to, and as I've previously defined, as the NOS. Preservice teachers of course must also have the proper understanding of NOS before they can then attempt to assist all K-12 students to become scientifically literate.

Multicultural Science Instruction

A close inspection of each of the preceding pedagogical theories and methods reveals a consistent constructivist thread connecting each to the other. Specifically, it is the idea that the females of color are not simply *receivers* of information but instead are active *contributors* to their own learning. Equally inherent to these pedagogical practices and greatly impacting how they learn are the prior experiences and backgrounds each brings with them into the classroom. Taking this logic a step further, it is clear that the overarching influence shaping these experiences and backgrounds are the *multiple cultures* females of color are immersed in prior to encountering school Science. In her seminal work, Atwater (1994) defined *multicultural science education* as "a recognized field of disciplined inquiry devoted to research using quantitative and qualitative approaches and the development of educational policies and practices so that all students can learn" (p. 1). This is echoed very clearly in *A Framework for K-12 Science Education* (NRC, 2012):

> There is increasing recognition that the diverse customs and orientations that members of different cultural communities bring both to formal and to informal science learning contexts are assets on which to build—both for the benefit of the student and ultimately of science itself. (p. 28)

Including the cultural impacts of language, Lee and Luykx (2006) concur:

> All students come to school with knowledge constructed within their home and community environments, including their home language (s) as well as cultural beliefs and practices. Learning is enhanced – indeed, made possible – when it occurs in contexts that are culturally, linguistically, and cognitively meaningful and relevant to students. Effective science instruction must consider students' home cultures and languages in relation to the pedagogical aims of science instruction. (p. 72)

The rationale behind preparing teachers of science to be cognizant of the various cultural experiences children bring to the learning environment is that the backgrounds and cultural norms of the communities are often at odds or incongruent with those of school (Lee & Luykx, 2006). For instance, the cultural importance of language and the natural environment to Native Americans learning science is of utmost importance to teaching these females (Gilbert, 2011). Elmesky and Seiler (2007) also highlight how African American cultural norms of music and movement, when expressed in urban science classrooms, are often negatively interpreted by educators unaccustomed and unprepared to capitalizing on them. Ladson-Billings (1995) advocates for the use of culturally relevant pedagogy (CRP) to counter the effects of the incongruence just described. She outlines three central goals for teaching females of color using CRP:

1. An ability to develop students academically. This means effectively helping students read, write, speak, compute, pose, and solve higher-order problems and engage in peer review of problem solutions.
2. A willingness to nurture and support cultural competence in both home and school cultures. The key for teachers is to value and build on skills that students bring from the home culture.
3. The development of a sociopolitical or critical consciousness. Teachers help students recognize, critique, and change social inequities.

While the above are general practices, methods, and curriculum necessary for the effective teaching of females of color, the following are examples of what an actual science classroom environment would look like. Utilizing recommendations from Halpern et al. (2007), for instance, the following suggestions can be used to encourage and improve the participation of females of color in K-12 classrooms:

1. Teach females of color that academic abilities are expandable and improvable. It is important that these learners understand that they are as capable of learning

scientific concepts as anyone and that neither gender nor race predetermines who can or cannot learn.
2. Provide prescriptive, informational feedback. Try to avoid general terms when assisting females of color. Feedback should be specific to the task they are trying to achieve.
3. Expose females of color to role models who have succeeded in mathematics and science. When possible invite other females of color from science, technology, engineering, and math (STEM) fields into the classroom. Arrange for field trips that will highlight females of color in STEM-related professions.
4. Create a classroom environment that sparks initial curiosity and fosters long-term interest in math and science. Provide opportunities for females of color to interact with and successfully master fun and creative science lessons. Frequent access and exposure to science equipment and scientific experimentation will benefit them by instilling confidence.

Provide spatial skills training. Success in solving science problems often relies on the ability to visualize and think about objects from multiple perspectives. Activities requiring rotating of objects or puzzle solving are examples where spatial skills are used. This ability like most can be improved with practice. It is therefore important that females of color be provided these opportunities as part of their science instruction.

Conclusion

To be, or not to be, that is the question. (Shakespeare, trans. 1980c, 3.1.56)

In closing, it is fitting that Hamlet poses the query that is clearly most central to this writing. He appears to be wondering aloud the very thing that those interested in equity and justice in science education too are mulling over. Will we be the egalitarian institution based in equity that we aspire to be or not? Will we create welcoming learning environments for all children, especially those who have been marginalized throughout U.S. history as females of color clearly have been, or won't we? However, as compelling a rationale as equity may be for why recruiting and educating females of color in Science is important, there are other factors at work as well. In the United States roughly 51 % of the population is female, and according to projected demographic trends, females of color will occupy an even larger percentage of the female population going forward. An obvious reality of course is that society will depend heavily upon this demographic to fill vitally important STEM roles that they have been excluded from in the past. It is therefore no longer possible to simply say that we have a problem educating females and in particular females of color to be scientists and engineers. We must either solve the problem as a nation or face falling farther and farther behind other developed and developing nations in the world. It is no secret that part of the solution begins in K-12 classrooms, and we most certainly cannot continue to operate the *play within* as we always have. For as Hamlet so aptly put it, "the play is the thing."

References

American Association for the Advancement of Science. (1989). *Project 2061: Science for All Americans*. Washington DC: Author; also published by Oxford University Press (1990).

American Association of University Women. (1992). *How schools shortchange girls, executive summary*. Washington, DC: Author.

American Association of University Women. (2004). *Under the microscope: A decade of gender equity projects in the sciences*. Washington, DC: Author.

American Association of University Women. (2010). *Why so few: Women in science, technology, engineering and mathematics*. Washington, DC: Author.

Atwater, M. M. (1994). Introduction: Invitations of the past and inclusion of the future in science and mathematics. In M. M. Atwater, K. Radzik-Marsh, & M. Strutchens (Eds.), *Multicultural education: Inclusion of all* (pp. 1–3). Athens, GA: The University of Georgia.

Aud, S., Hussar, W., Johnson, F., Kena, G., Roth, E., Manning, E., Wang, X., & Zhang, J. (2012). *The Condition of Education 2011* (NCES 2012-045). U.S. Department of Education, National Center for Education Statistics. Washington, DC: U.S. Government Printing Office.

Baker, D. R. (2002). Good intentions: An experiment in middle school single-sex science and mathematics classrooms with high minority enrollment. *Journal of Women and Minorities in Science and Engineering, 8*, 1–23.

Beede, D., Julian, T., Langdon, D., McKittrick, G., Khan, B., & Doms, M. (2011, August). *Women in STEM: A gender gap to innovation*. U.S. Department of Commerce Economics and Statistics Administration. Washington, DC: U. S. Government Printing Office.

Brown, J., & Newsome, B. J. (1966). It's a man's, man's, man's world [Recorded by J. Brown]. On *It's a man's, man's, man's world* [record]. New York: King (February 16, 1966).

Bybee, R. W., Taylor, J. A., Gardner, A., Van Scotter, P., Powell, J. C., Westbrook, A., et al. (2006). *The BSCS 5E instructional model: Origins, effectiveness, and applications*. Colorado Springs: BSCS.

Carey, S. S. (1994). *A beginner's guide to scientific method*. Belmont, CA: Wadsworth Publishing Company.

Catell, J. M. (1914). Correspondence. *Science, 39*, 154–164.

Chalmers, A. (1990). *Science and its fabrication*. Minneapolis, MN: University of Minnesota Press.

Chambers, D. W. (1983). Stereotypic images of the scientists: The draw-a-scientist test. *Science Education, 67*, 255–265.

Coburn, A. (2000, March). An inquiry primer. *Science Scope*, 42–44.

Coles, P. (2007). The trailing spouse syndrome. *The UNESCO Courier, 2*, 8–9.

Conner, C. D. (2005). *A people's history of science: Miners, midwives and low mechanicks*. New York: Nation Books.

Elgar, A. G. (2004). Science textbooks for lower secondary schools in Brunei: Issues of gender equity. *International Journal of Science Education, 26*, 875–894.

Elmesky, R., & Seiler, G. (2007). Movement expressiveness, solidarity and the (re)shaping of African American students' scientific identities. *Cultural Studies of Science Education, 2*(1), 73–103.

Fouad, N. A., & Walker, C. M. (2005). Cultural influences on responses to items on the Strong Interest Inventory. *Journal of Vocational Behavior, 66*, 104–23.

Freire, P. (1988). *Pedagogy of the oppressed*. New York: The Continuum Publishing Corporation.

Gibbs, A., & Lawson, A. E. (1992). The nature of scientific thinking as reflected by the work of biologists and by biology textbooks. *The American Biology Teacher, 54*, 137–152.

Gilbert, W. S. (2011). Developing culturally based science curriculum for Native American classrooms. In J. Reyhner, W. S. Gilbert, & L. Lockard (Eds.), *Honoring our heritage: Culturally appropriate approaches to indigenous education* (pp. 43–55). Flagstaff, AZ: Northern Arizona University.

Gjertsen, D. (1989). *Science and philosophy past and present*. New York: Penguin.

Haberman, M. (2000, November). Urban schools: Day camps or custodial centers? *Phi Delta Kappan, 82*(3), 203–208.

Halpern, D., Aronson, J., Reimer, N., Simpkins, S., Star, J., & Wentzel, K. (2007). *Encouraging girls in math and science.* Washington, DC: National Center for Education Research, Institute of Education Sciences, U.S. Department of Education.

Hanson, S. L. (2004). African American women in science: Experiences from high school through the post-secondary years and beyond. *NWSA Journal, 16,* 96–115.

Harry, B., & Klingner, J. (2007). Discarding the deficit model. *Educational Leadership, 64*(5), 16–21.

Hodson, D. (1999). Going beyond cultural pluralism: Science education for sociopolitical action. *Science Education, 83,* 775–796.

Iaccarino, M. (2003). Science and culture. *European Molecular Biology Organization Reports, 4,* 220–223.

Kahle, J. B. (1998). Gender equity in science classrooms. *Teacher Education Materials Project.* Retrieved October 29, 2009, from http://www.te-at.org/Essays/kahle_pf.aspx

Kahle, J. B., & Meece, J. (1994). Research on gender issues in the classroom. In D. L. Gabel (Ed.), *Handbook of research on science teaching and learning.* New York: Macmillan.

Ladson-Billings, G. (1995). But that's just good teaching! The case for culturally relevant pedagogy. *Theory into Practice, 34,* 159–165.

Lederman, N. G. (1992). Students' and teachers' conceptions about the nature of science: A review of the research. *Journal of Research in Science Teaching, 29,* 331–359.

Lee, O., & Luykx, A. (2006). *Science education and student diversity: Synthesis and research agenda.* Cambridge, UK: Cambridge University Press.

Leggon, C. B. (2006). Women in science: Racial and ethnic differences and the differences they make. *Journal of Technology Transfer, 31,* 325–333.

McComas, W. F. (1998). The principle elements of the nature of science. In W. F. McComas (Ed.), *The nature of science in science education* (pp. 53–70). The Netherlands: Kluwer Academic.

Merchant, C. (1980). *The death of nature.* New York: Harper & Row.

Modi, K., Schoenberg, J., & Salmond, K. (2012). *Generation STEM: What girls say about science, technology, engineering, and math.* New York, NY: Girl Scout Research Institute.

National Research Council. (1996). *National Science Education Standards.* Washington, DC: National Academy Press.

National Research Council. (2012). *A Framework for K-12 science education: Practices, crosscutting concepts, and core ideas. Committee on a conceptual framework for new K-12 science education standards.* Board on Science Education, Division of Behavioral and Social Sciences and Education. Washington, DC: National Academies Press.

National Science Board. (2008). *Science and engineering indicators: 2008.* Arlington, VA: National Science Foundation.

National Science Foundation. (1999). Division of science resources statistics, *Women, minorities, and persons with disabilities in science and engineering: 1998* (NSF 99-338). Arlington, VA. Available from http://www.nsf.gov/statistics/wmpd

National Science Foundation. (2002). Division of science resources statistics, *Women, minorities, and persons with disabilities in science and engineering: 2002* (NSF 02-336). Arlington, VA. Available from http://www.nsf.gov/statistics/wmpd

National Science Foundation. (2007). Division of science resources statistics, *Women, minorities, and persons with disabilities in science and engineering: 2007* (NSF 07-315). Arlington, VA. Available from http://www.nsf.gov/statistics/wmpd

National Science Foundation. (2009). Division of science resources statistics *Women, minorities, and persons with disabilities in science and engineering: 2006* (NSF 09-305). Arlington, VA. Available from http://www.nsf.gov/statistics/wmpd

Observatoire des Sciences et des Techniques [OST]. (2004). *Science and technology indicators.* Paris: Economca. 576 pp.

Ong, M., Wright, C., Espinosa, L. L., & Orfield, G. (2011). Inside the double bind: A synthesis of empirical research on undergraduate and graduate women of color in science, technology, engineering, and mathematics. *Harvard Educational Review, 81,* 172–209.

Piaget, J. (1971). *Science of education and the psychology of the child*. New York: Viking.
Shakespeare, W. (1980a). As you like it. In John Dover Wilson (Ed.), *The complete works Of William Shakespeare* (pp. 243–266). Cambridge, UK: Cambridge University Press.
Shakespeare, W. (1980b). Romeo and Juliet. In John Dover Wilson (Ed.), *The complete works of William Shakespeare* (pp. 781–808). Cambridge, UK: Cambridge University Press.
Shakespeare, W. (1980c). The Tragedy of Hamlet: Prince of Denmark. In John Dover Wilson (Ed.), *The complete works of William Shakespeare* (pp. 885–919). Cambridge, UK: Cambridge University Press.
Shiva, V. (1993). Reductionism and regeneration: A crisis in science. In M. Mies & V. Shiva (Eds.), *Ecofeminism* (pp. 22–35). Halifax, NS: Fernwood Publications.
U.S. Department of Education, National Center for Education Statistics. (2007). *The nation's report card: America's high school graduates: Results from the 2005NAEP high school transcript study*, by C. Shettle, S. Roey, J. Mordica, R. Perkins, C. Nord, J. Teodorovic, J. Brown, M. Lyons, C. Averett, & D. Kastberg. (NCES 2007–467). Washington, DC: U. S. Government Printing Office.
U.S. Department of Education, National Center for Education Statistics. (2012). *The condition of education 2012* (NCES 2012–045). Washington, DC: U.S. Government Printing Office.
Vvgotsky, L. S. (1962). *Thought and language*. Cambridge, MA: MIT Press.
Walls, L. (2012). Third grade African American students' views of the nature of science. *Journal of Research in Science Teaching, 49*, 1–37.

Sociocultural Consciousness and Science Teacher Education

Brenda Brand

Villegas and Lucas (2002) describe the task of preparing teachers to effectively teach students from culturally diverse backgrounds as a pressing issue that will continue for years to come. This concern is compounded by the fact that US classrooms are becoming increasingly diverse. Perspectives among researchers differ in terms of the knowledge and experiences necessary for preparing teachers to manage the complexities associated with teaching students who are quite different from them, ethnically and culturally. Of a surety, instructional practices should be responsive to students' needs as influenced by their cultural differences. Culturally responsive teaching by its name implies instruction that takes into account the needs of students as influenced by the cultural diversity existing in the classroom. At the surface, the term culture distinguishes individuals by their ethnic groups. However, culture is more expressly defined as a set of shared values and beliefs that belong to a particular group (Banks, 2010). So, how does culture relate to teaching students? For some teachers, culturally responsive teaching primarily consists of including representations of the contributions of individuals from the students' ethnic backgrounds into the curriculum or integrating relevant customs and traditions. While these strategies may be considered beneficial, in that the contributions of individuals from underrepresented groups are represented in a positive light, they do not account for the cultural distinctions resulting from the purposeful singling out of groups of people as deviant or inferior, cultural marginalization (Ferguson, Gever, Minh-ha, & West, 1990). The messages conveyed through these acts exist at the core of the challenges confronted by students from marginalized groups and are intuited differently dependent upon the individual, which determines the nature and extent of the impact:

> I think that it's important that they [students] see strong minority people in strong leadership roles, whether they be women or men… You wonder why all these minority students want to be rap stars, singers, television stars, and football, basketball, and baseball players. Okay,

---

B. Brand, Ph.D. (✉)
Virginia Tech, 321 War Memorial Hall, Blacksburg, VA 24061, USA
e-mail: bbrand@vt.edu

> let's look at this, how many doctors and lawyers do you see? How many baseball and football players and singers do you see? That's all we've grown up with. Okay the Black person sings. The Black person plays football. The Black person plays basketball. The White person is the lawyer. The White person is the teacher. That's all that's seen. I think that kids start to believe or start to think they are inferior, which is part of self-esteem, and also part of a bad learning environment. Because if you think you're inferior, you're not going to do as best as you could have if you thought you were just up there with everybody else. (p. 232)

This statement was expressed by an African American high school student (Lezly) in a research study investigating sociocultural factors influencing the achievement of students from underrepresented groups in science and mathematics (Brand, Glasson, & Green, 2006). It conveys the inner workings of marginalization in classifying groups, having the potential to negatively impact individuals' self-esteem and identity. These ideas are transmitted subtly and are all-encompassing, influencing beliefs and actions without being realized, which lies at the root of the problem. These are realities that science teachers must understand to effectively design instruction that is aligned with the needs of marginalized students. Providing educational experiences that foster preservice and inservice teachers' understanding of the problems confronted by these students and their needs is the responsibility of science teacher education programs.

Understanding the Problem

The misrepresentation or lack of representation of marginalized groups in the curriculum is not a mere oversight, neither are biased portrayals of them in the media. Historically, societal constructions insinuating messages of inferiority have been a means for preserving and protecting the mainstream culture's way of life. According to Giroux (1983), capitalistic principles advocated by the dominant culture secured the capitalist way of life and the marginalization of select groups. These principles resulted in policies and infrastructures leading to widespread unemployment, segregated and under resourced schools, racist violence, and low-income housing, conditions that plague the communities of marginalized groups, particularly African Americans. Likewise, schools function as agents of social and cultural reproduction. The hegemonic power of the mainstream curriculum is actualized through what it includes, Westernized accounts and ideologies, and excludes, ideas of interest to the working class and other "subordinate" groups. Giroux (1992) contends that the indifference of the dominant discourse to the plight of marginalized people is too often the response, dissuading individuals from embracing their identities and actualizing their voice and sense of agency. Michael Apple (1978) also substantiates this notion of the reproductive elements of society and schooling as catalysts for cultural and economic reproduction, emphasizing the tendency of education to relegate individuals from select groups to specified positions in society. Michael Apple discussed the formal and hidden curriculum, with the formal curriculum consisting of that information represented in the core disciplines, and the hidden curriculum which consists of subtleties asserting and reinforcing the social order. According to Apple, both have been influenced by the dominant systems in society. The curricula convey meanings that socialize groups of individuals to accept as legitimate certain roles and lifestyles. According to Apple (1986), curriculum and pedagogy are not neutral but instead are by-products of interactions between class, economic, and cultural power. He emphasizes that an evaluation of the social and educational outcomes of schooling should take into account the unequal cultural and economic power that produced them.

Pierre Bourdieu (1986) discussed these power dynamics in terms of the presence or lack of capital. He explained capital as having the potential to be profitable. In other words, the amount of capital individuals possess could advantage them, making their participation in society less influenced by chance. The process of accumulating capital takes time, and an individual's position in the social order influences the ease at which capital is accrued. While capital can be acquired, it is primarily obtained through hereditary transmission. Bourdieu discusses three forms of capital: economic capital which is associated with financial wealth, cultural capital which could be considered educational or intellectual qualifications, and social capital which are networks or connections. Bourdieu explains that the amount of capital individuals possess influences their capacity to meet the demands of the educational system, and the lack of cultural capital is directly linked to the social inequities yielding different student outcomes in schools. The presence or lack of capital influences students' dispositions and attitudes toward themselves in terms of their abilities and schooling. The amount of capital an individual possesses is influenced by society's positioning of groups, which is perpetuated through social media. According to Bourdieu (1979), these are acts of symbolic violence, socially publicized images signifying normalcy and worth throughout society in accordance with the habits and interests of the dominant class.

Thus, schools, like society, are undergirded by an infrastructure that is dictated by the mores of the dominant culture. Students who relate to, or identify with, these codes of existence are considered as having more economic, cultural, and social capital. Alternatively, students who do not relate are considered as having less capital and in order to be successful must acquire capital through other means. The state of having less capital subliminally signifies a lesser state, resulting in barriers that restrict, discourage, or challenge the engagement of students who are not part of the dominant group. Consequently, traditional classrooms consist of an explicit curriculum of facts and concepts, as well as a hidden curriculum affirming social norms, roles, social class, and work (Anyon, 1981; Apple, 1986). Since the hidden curriculum is taught implicitly and powerfully through social media in out-of-school contexts, it is not alarming that messages about social class and social roles are processed more deeply by students than the content within the explicit curriculum (Anyon, 1981). Those students who lack capital with the behavioral patterns, values, and viewpoints of the dominant culture are at a disadvantage, hence the term socioculturally disadvantaged.

Socioculturally Disadvantaged

> You know like my family is struggling now, but I still come to school. I come to school, make my grades and work. And you know, I can do all of this. I'd like to be praised for doing all of this... So, I feel that people like me, Black males who are trying so hard should be recognized for it, instead of everybody paying attention to the ones who drop out. (p. 232)

This statement was expressed by an African American male high school student (Alfred) in the research study investigating sociocultural factors influencing students from underrepresented groups' achievement in science and mathematics (Brand et al., 2006). Evidence of this student's sociocultural disadvantages can be identified and interpreted from his statement that describes his challenges with the negative implications that characterize his identity. In it, he expressed pride in the fact that in spite of his family's struggles, he is able to work and achieve academically. His mention of the need to be praised for his achievements is based upon his frustration with where the primary focus of attention is placed in society, on the Black males who drop out. Clearly, he shouldered the weight of what it means to be a Black male in society, and somehow he had acquired enough capital to resist the negativity and achieve. His use of the word "everybody" signified the vastness of this challenge for him and his awareness of how his ethnicity and gender is viewed by many in society. The nature of his struggle does not impact all of his peers in the same manner; however, the implications of what it means to be a Black male applies to all. Rather than succumb to the pressures of accepting and conforming to the implicit expectations for his peer group, he chose the alternative. It is important to note that while this student's response to his life's challenges allowed him to reap favorably, his success also invites the judgment of those to use his success to blame the students who fail, which is unjustifiable. When applying the principle of cultural capital, it can be concluded that Alfred had more capital than other students who fail in their attempts to navigate these psychological barriers. While the sources of capital for Alfred may not be identified, the fact that he is working while going to school, and at the same time achieving academically, signifies that there are influential resources present somewhere in his life. Additionally, it should not be presumed that his positive sense of direction did not stem from his family's influence simply because of the lack of financial resources. Interestingly, while Alfred seemed proudly aware of the value of his accomplishments, his ability to celebrate them was impacted by his fear that they were somehow overshadowed by the negative undertones surrounding his ethnic and gender identity:

> It's hard for me to be a Black male and having to guard yourself. I'm in school now. I'm struggling hard to make this grade. It's just some teachers who don't believe that you are trying just because of what other people do. (p. 233)

In this statement, Alfred further illustrated the presence of the influence of sociocultural implications on his academic achievement (Brand et al., 2006). The notion of being guarded was an expression of his need to defend himself, or arm himself against negative impressions, which seemed to be a constant for him. He ended this statement with his perceptions of what some of his teachers believed about him. Alfred, in this statement, realized that his teachers as members of society were also privy to the same social agenda purporting unfavorable opinions of him and his peer group. He knew that as an African American male, he could be viewed according to these characterizations. Distinguishing himself was important and also difficult. The complexities of navigating these courses were expressed in his dialogues. He fought to distinguish himself from any negative characterizations that could exist about him in the thoughts of others. Thus, the students who are socioculturally disadvantaged have to be concerned about more than learning the content when they enter the classroom. The messages of the hidden curriculum compound their participation as students. The failure to consider these factors in the development of agendas aimed at supporting the achievement of individuals from marginalized groups would be a major oversight. The achievement gaps, dropout rates, low percentages of individuals from underrepresented groups in science and mathematics careers and disciplines, as well as other casualties, could all be directly linked to the hidden curriculum.

Impact on Students' Identities

Findings from the eminent doll test (Clark & Clark, 1939) used in the landmark Brown v. Board of Education case, in which the Supreme Court declared laws advocating segregation unconstitutional, further evidences the impact of sociocultural implications on individuals' perceptions of themselves and their self-worth. In this research, African American children were presented with two dolls, one being White and the other Black. They were asked to respond to a series of statements related to the dolls such as "show me the doll you like the most or would like to play with" or "show me the nice doll or bad doll." Repeatedly, the children selected the White doll for all of the positive inferences and the Black doll for all of the negative inferences. Even more revealing were their responses to a final request asking them to point to the doll that resembled them. Interestingly, the children would either hesitate to put forth the Black doll or become frustrated and leave the room because they did not want to associate the bad doll with their identity. Clark concluded that the social implications harmed the children's perceptions of their identity. Later on, this study was informally replicated by a young African American woman, Kiri Davis (2005), in her documentary "A Girl Like Me" revealing very similar results. Fifteen of the twenty-one children associated the Black doll with negative inferences, signifying negatively internalized references for their identities. It can be deduced that students who have internalized negative conceptions of themselves and their worth to society will find it difficult to have a positive attitude toward life and learning. It can also be deduced that students who are struggling with these negative characterizations could become frustrated and give up.

These examples of the influence of sociocultural factors on students' conceptions of their identity are used to shed light on the mentality that some students from historically marginalized or underrepresented groups possess when they enter the classrooms, although no group is excluded from exposure to these influences. Other individuals involved in schooling have also been exposed to these characterizations to include teachers, administrators, and student peer groups. As a matter of fact, all groups have societal characterizations associated with their identities, for example, the familiar "Asians are smarter" stereotype (Yee, 1992). However, the nature of the distinction positions certain groups in either an advantaged or disadvantaged status, which ultimately influences their attitudes and dispositions. According to Giroux (1983), many times, students' response or reactions to environments that seem to align with the negative beliefs associated with their identities is resistance. This resistance, similar to Herbert Kohl's (1992) theory of "not learning," is a students' response to what they perceive to be an environment that reinforces or agrees with the negative stereotypes. Some of his students seemed to feel that if they cooperated in any way with these infrastructures, they were either compromising or abandoning their identities. According to Kohl, the state of not learning can be mistaken by teachers as an inability to learn. He asserts that students choose to not learn when they feel as though they are facing a potentially hostile situation or that the individual who is teaching is not perceived as one who respects them or cares about them. Thus, the student rejects the teacher and the educational system, and as Giroux (1983) indicates, to their detriment. Giroux points out that the unfortunate end to this form or resistance is that the students fail to liberate themselves through their own advancement and also miss out on the opportunity to establish their sense of agency. Whether this resistance is an act of surrender on the part of the student in response to feelings of never being able to measure up, or an act of defiance by refusing to learn, it is imperative that the nature of this conflict is understood.

Revealing society as a culprit minimizes the potential for "blaming the victim." While there are many individuals who overcome circumstances that may indeed cripple others, it does not negate the consequences that all historically marginalized groups confront.

The recognition and acknowledgment of society's propensity to perpetuate biases that disenfranchise groups helps to explain probable causes for students' resistance and lack of motivation, which influences withdrawal, or even lack of confidence in their abilities and potential.

Exclusivity of Science

> In science, from my experience, it's mostly the White students with the best grades... I didn't think I could do as good as they did in science maybe because of their race also. But then, like most of the White students, I thought it was hereditary that they were smart. They were smarter than me because I have to work harder to get an A than them. So, you know sometimes that will hurt your self-esteem also. You'll think well, I have to work so hard, and they're living this good life or whatever, and I'm struggling. It's very hard. (p. 232)

In this statement, Alfred is comparing himself to his White peers and, in his mind, does not appear to measure up (Brand et al., 2006). He perceived his White peers as getting the best grades and not having to work as hard. Alfred seemed to believe that they were smarter because of their ethnicity. He even considered them as being smarter due to heredity. He also mentioned how his struggle, when compared to his White peers who seemed to be having it easier, was potentially damaging to his self-esteem. The notion of inferiority when comparing himself to his White peers was evident. Alfred also expressed his opinion on the life that he perceived his White peers to be enjoying, "this good life." It appeared that he came to this conclusion merely based upon their ethnicity without really knowing the conditions in which they lived. However, this assumption is easily derived based upon the favorable representations of White people in the social media. Also significant in his statement was the fact that this very successful African American male did not seem to be able to overcome the negative associations with his identity and consequently was unable to fully celebrate his accomplishments.

According to Aikenhead (1996), students are crossing cultural borders when they enter science classrooms. Western "school science" is a microculture, representing images that only certain groups of individuals identify within this society. Crossing the borders into the microculture of school science requires students from underrepresented groups to continuously negotiate messages of inferiority transmitted through the Eurocentric worldview. The concept of border crossings provides a frame of reference for understanding how science exists for individuals who are not of European descent. They are alienated from its texts and must cross borders in order to gain access. The ease at which marginalized students cross borders into science is dependent upon their level of capital. So it should not be perceived as a simple crossover. As expressed in the dialogues of Lezly and Alfred, the border crossing experience is marked by a need to constantly challenge notions of inferiority in themselves and in what they perceive is bound in the beliefs of others about them. Socioculturally, a critique of science as taught in schools encourages an analysis of how its contexts could disenfranchise certain groups. Atwater and Riley (1993) explain that students from underrepresented groups are estranged from science due to its monocultural representations. This lack of identity with its content positions them passively, solely as recipients or patrons rather than thinkers or problem solvers, which are key characteristics of scientists.

Similar to the theory of border crossings, Tobias (1988) explains this exclusivity in terms of "insiders" and "outsiders" in science classrooms. Tobias explains that the insiders are individuals who automatically consider themselves as having the potential to do well in science, having the capacity to embrace its ideals as reported. They have no need to question the contexts of western science, because they identify with much of what is represented. They inherit capital through the images of successful scientists based upon their identities, which validates them and their potential for making contributions. They have no need to question whether or not they belong or whether success is expected of them. Alternatively, the outsiders are those students underrepresented in the disciplines of science who do not experience that same luxury. The exclusivity of science encourages students who do not belong to the dominant group to be uncertain about their ability to be successful with the demands and expectations of school science. Unlike the insiders, there are very few images resembling "outsiders" within the curriculum. Thus, when these images are presented, they are in danger of being considered exceptional or rare. Coupled with the social media, which also represents science expertise as an exclusive right, marginalized students are again bombarded with messages signifying a lack of ability and potential. These factors challenge the participation of "outsiders" in school science. Understanding these challenges is essential to interpreting the experiences of marginalized students, which is necessary for intervention. Issues like the achievement gap must be processed in terms of the influence of social inequities to properly align strategies with students needs. Teachers who are able to deconstruct society in terms of its oppressive structures and political agendas are able to confront its transgressions. An acknowledgment of how the process of marginalization of groups of individuals was used to preserve status quo positions teachers to become advocates for their students. Advocacy is a key strategy for teaching students who are socioculturally disadvantaged. It is informed by an awareness of how marginalized groups were sacrificed for the common good of the dominant class and consequently might require support to overcome the disparities. Teachers as advocates seek ways to reverse the impact of exclusion by first believing in their students' abilities to be successful and then inspiring confidence in them to overcome feelings of inadequacy. Advocacy is fueled by sociocultural consciousness.

Sociocultural Consciousness and Science Teacher Education

The acknowledgment and understanding of the process of marginalization and its consequences translate into sociocultural consciousness. Villegas and Lucas (2002) define sociocultural consciousness as "understanding that people's ways of thinking, behaving, and being are deeply influenced by such factors as race/ethnicity, social class, and language." Sociocultural consciousness is key to understanding the needs of students who have been socioculturally disadvantaged and for developing and employing strategies that align with their needs. The struggle to declare worth fought externally and internally, images and profiles of inferiority, lack of identity with the infrastructure and presentation of schooling, and limited resources are sociocultural disadvantages that could apply to all historically marginalized groups and like all other disadvantages warrant emphasis on learner accommodations. The characteristics of this learner include behaviors such as distrusting the system, lack of engagement or apathy, displacement, lack of ownership, negative self-concept, low self-efficacy, and resistance to name a few. All historically marginalized groups are susceptible, yet their levels of vulnerability vary dependent upon other factors in their lives that may have strengthened their resolve or increased their capital. Those who are most vulnerable could exhibit the following behaviors: positioning themselves somewhere in the classroom where they can go unnoticed, failing to turn in assignments, or barely putting forth effort to achieve beyond the minimal and defensive stances which could be perceived as a chip on their shoulder. Too often, a common conclusion is that these students do not care about learning or their future. At the surface, this is probably a justifiable conclusion; however, the factors that influence these dispositions and attitudes, which are hidden from plain sight, lie at the root of the problem and should not be ignored. Kohl (1992) stated the following:

> Until we learn to distinguish not-learning from failure, and respect the truth behind this massive rejection of schooling by students from poor and oppressed communities, it will not be possible to solve the major problems of education in the United States today. (p. 17)

The Individuals with Disabilities Education Act (IDEA) is an act requiring all schools that accept federal funds to provide equal access to education for all students with disabilities. Students are evaluated, and an Individualized Education Plan (IEP) is drafted with the support of the parents that outlines the needs of the students and the instructional strategies for meeting the students' needs in the least restrictive environment. The least restrictive environment implies that a student is provided with a learning environment in which he/she can be educated peers with no disabilities as much as is humanly possible to support normalcy. Accommodations are made in the learning environment based upon students' disabilities, which aids them in functioning in a manner similar to their nondisabled peers. An IEP is tailored specifically for a student with a disability and is designed to meet his or her educational needs. The goal of the IEP is to help the student achieve educational objectives by alleviating constraints caused by the handicapping condition. The IEP should help teachers, as well as others, understand the student's disability in terms of how it affects their learning. Although sometimes controversial, the evaluation component of IDEA results in an identification of the specific problem that the students possess or, in other words, a label. While there is much discussion about whether or not labeling is politically correct or in the best interest of the child, it informs teachers of the student's disability so that they can provide learning experiences that accommodate the student's needs. The "label" prioritizes the student's disability and their needs, encouraging teachers to be concerned about the student with the disability and not consider their constraints outside of their realm of responsibility.

This reference to IDEA is solely in recognition of the attention and services afforded to students who are identified with special needs. Specialists work with regular classroom teachers to develop and implement instructional plans that accommodate the needs of the identified students related to their disability. Teachers strive to create learning experiences that will allow the student to function as normal as possible, as if there was no disability. Accomplishing this objective takes a concerted effort from all involved: the regular classroom teacher, specialists, and school administrators. Legally, these services must be provided to students with disabilities.

Sociocultural disadvantages pose obstacles that limit or can even inhibit students' abilities to achieve academically as opposed to peers who do not face the same challenges. There is no identification process for students who are socioculturally disadvantaged considering social inequities exist for all members of select groups.

Students who are socioculturally disadvantaged enter classrooms under a cloud of negativity. Consequently, they may enter the classroom unsure about themselves and their potential, having low self-esteem or low self-efficacy. Or the student who is socioculturally disadvantaged may enter the classroom considering certain trajectories as appropriate paths for them and totally dismissing others. Students who are socially disadvantaged may consider academic excellence as deviant from the normal behaviors of their peer group and follow the path to least resistance. They may enter the classroom with skepticism toward their teachers, particularly those who look differently from them, provoking resistance. They also may enter the classroom having an identity influenced by popular culture and media. Teachers need to understand these hegemonies, so that they will avoid judging the student for being vulnerable, especially in comparison to those from the same disenfranchised group who achieve despite the circumstances:

> A lot of people, especially minority people could do ten times better than what they are doing now if they were just pushed, pushed by anybody. I think that the teachers have a big responsibility. They have a classroom of twenty something students, and they sit there and see people talking, and it's like, why should I bother with them? But as a teacher I think you should push those people, push them to learn…But math and science, especially math, and especially for minority women, it's not pushed at all. (p. 232)

In this statement, Lezly acknowledges that the students' abilities were not in question (Brand et al., 2006). They were unmotivated. Lezly also references students' behaviors as causing them to be ignored by their teachers. Embedded within Lezly's statement is an awareness of how being a "minority" would encourage certain choices from the students and the teacher's responsibility to motivate them beyond those choices. Lezly appears to be socioculturally aware which encourages her interpretation of students' behaviors, even though they are her peers, as having social influences that prevent them from realizing their abilities and potential.

Sociocultural Consciousness and Culturally Responsive Teaching

Villegas and Lucas (2002) list sociocultural consciousness as one of the qualities of a culturally responsive teacher. The remaining qualities are having an affirming view of students, seeing themselves as responsible for and capable of bringing about change to make schools more equitable, understanding how learners construct knowledge and are capable of promoting knowledge construction, knowing about the lives of their students, and designing instruction that builds on what students already know while stretching them beyond the familiar. Sociocultural consciousness provides teachers with the foundation for understanding the plight of marginalized students. A critical awareness of their predicaments encourages dispositions that are empathetic to their needs. Empathy, unlike sympathy, takes into account the unfortunate circumstances the students may be facing and inspires advocacy to confront and reverse their impact before causing irremediable damage. Teachers who are socioculturally aware are more likely to develop relationships with students that break down the barriers of distrust and skepticism. The students are more likely to feel that the teachers believe in them and care about them, which is critical to the achievement of students who are uncertain about their abilities and potential:

> Because if you get along with your teacher, I mean I had teachers I didn't like, and I didn't get good grades in the class. If you like your teacher, you are going to basically, not be afraid to ask the teacher questions, and if you have a good relationship with the teacher you will like to try to at least keep your grades up so that you won't disappoint the teacher, or whatever. That's how I feel about it. The teachers I didn't like, I really didn't get good grades. (p. 233)

This statement was made by Keith, an African American male student in the research study investigating sociocultural factors influencing "minority" students' achievement in science and mathematics (Brand et al., 2006). This student, also successful, expressed the significance of a positive relationship with teachers to students' academic achievement and performance. Of significance in his statement is how the relationship with the teacher would encourage and motivate students to do well. The idea that a positive relationship with teachers is significant to students' achievement is not a surprise. Most students desire to have positive relationships with their teachers regardless of their ethnic or cultural background. Conceivably, positive relationships with teachers motivate most students to achieve their best. However, teachers who are socioculturally aware recognize that there are factors that negatively impact their ability to develop relationships with students who are socioculturally disadvantaged. They recognize the damaging effect that the negative characterizations perpetuated through social media can have on their ability to establish relationships with their students. They understand that the same source has communicated to the students who are socioculturally disadvantaged, ideas about them as the teacher. Recalling Alfred's earlier statement about his teachers, he said that some teachers do not think that he is trying because of what others (African American males) are doing. In this statement, he described some of his teachers as thinking that he or maybe some of his peers are not trying to achieve. Alfred perceived some of his teachers as buying into the stereotypes about African American males. McGee and Martin (2011) conducted a research study on how successful African American mathematics and engineering college students managed racist and stereotypical comments to achieve in their disciplines. The students in their research found it burdensome and draining to constantly have to combat negative perceptions. In response they developed protective stances and dispositions to achieve in spite of the stereotypes to prove their academic worth. One of the female students in the study reported that after answering a question, her engineering professor told her that he did not expect her to be able to answer a question of that caliber and was even more surprised to find out that she answered it without help. The existence of the influence of stereotypes is evidenced in this professor's statements. Thus, classrooms could be considered arenas of cultural conflict in which mental battles occur continuously between students and teachers, students and their peers, and students with themselves.

Sociocultural Consciousness and Equitable Classrooms

Sociocultural consciousness provides depth of understanding to the concept of equitable classroom. It goes beyond simply making sure that every student sees that someone who looks like them did something great. It helps them identify the greatness within them. Culturally responsive teaching practices are asserted as strategies that align with the needs of diverse learners. Thus, there needs to be an understanding of students' needs in order to create a meaningful and productive learning environment. Sociocultural consciousness is the chief informant for culturally responsive pedagogy—instruction that takes into account the diverse needs of students as influenced by culture. It can be considered the brain that informs the instructional decisions for accommodating the needs of marginalized students. Culturally responsive practices are motivated and driven by empathy. In culturally responsive teaching, empathy is fueled by an acknowledgment and ownership of society's habits of disenfranchising groups, and a commitment to supporting students in overcoming resulting circumstances, which is especially relevant when presenting the ideals of Western science. Therefore, science teacher education programs should provide learning experiences that foster preservice and inservice teachers' critique of science undergirded by their knowledge of society's history of marginalization. Without this knowledge, preservice and inservice teachers will not recognize the inequities that exist within their classrooms. Consequently, they will lack the tools necessary for creating learning environments that will motivate these students to overcome their disparities and achieve. Science teacher educators have a major responsibility to develop programs that support preservice and inservice teachers to develop these tools. Accomplishing this goal requires the application of research-based learning practices that advance their understanding of inequities and their potential impact in the classroom.

Culturally Responsive Teaching and Social Justice

Culturally Responsive Teaching

Geneva Gay (2002) defines culturally responsive teaching as "using the cultural characteristics, experiences, and perspectives of ethnically diverse students as conduits for teaching them more effectively." Sleeter (2011) asserts that in addition to instruction, culturally responsive teaching is a political agenda addressing equity and social justice. Culturally responsive teachers understand society in terms of its history of exclusion and the structures that emerged from that exclusion. The recognition and admission of society's history of victimization acknowledges that certain groups of individuals have to play catch up due to the stifling conditions that hampered their participation and achievement. As well, their struggle is exacerbated by ongoing victimization perpetuated in the social media in addition to limited resources and capital.

Teachers who acknowledge these realities are able to develop empathy and realize that certain behaviors or dispositions from students are symptoms of oppression and that even the students may not recognize the source of their frustration. This is the awareness of the influence of social inequities that Villegas and Lucas (2002) noted as a necessary element for culturally responsive teaching. The culturally responsive teacher who has this awareness understands the rationale for instructional strategies more deeply. They understand that incorporating representative images and customs from diverse ethnicities and cultures into the curriculum is more than just adding historical facts. They recognize that integrating representative ethnicities or cultures may help to dispel messages of inferiority and empower students who are socioculturally disadvantaged to overcome feelings of inadequacy and achieve. They also realize that this strategy is beneficial to all students considering most students have had some level of exposure to negative messages about marginalized groups. Counter images and perspectives can improve all of the relationships within the classroom, and hopefully transfer outside of the classroom, which is beneficial to a democratic society. Presuppositions about groups can be accepted as factual without question if there are no counter experiences. As society becomes increasingly diverse and global over the next decades, it is becoming increasingly important for groups of individuals to work together to solve problems for the common good.

In a sense, teachers can be considered gatekeepers, holding keys that open the doors of opportunities for their students. This statement of reality, however, does not ignore the responsibilities that students have for deciding their own fate. Yet this power relationship between teachers and students positions students as subordinates, and while they are not powerless, there is heavy reliance upon their teachers. Some students are supported in their academic pursuits from other sources, yet other students, particularly from marginalized groups, are looking to their teachers to support them in believing in their potential and ability. The statements from Lezly, Keith, and Alfred in the research study investigating sociocultural factors influencing their learning in science and mathematics confirm the importance of having positive and supportive relationships with teachers. From a sociocultural perspective, the importance of these relationships, particularly for students who are socioculturally disadvantaged, makes sense. If a student feels as though their teacher buys into the myths about them being inferior, they are going to find it hard to believe that their teachers' interactions with them are more than a formality. Understandably, the potential for establishing trust is impeded. For socioculturally disadvantaged students who are taught by society to distrust most people outside of their ethnicity and culture, not being able to trust their teachers is detrimental to their educational experiences. Kincheloe (2005) stated the following:

> The more research sociocognitives produce, for example, the more it becomes apparent that a large percentage of student difficulties in school results not as much from cognitive inadequacy as from socially contextual factors. Critical teachers need a rich understanding of the social backgrounds of students, the scholarly context in which disciplinary and counter-disciplinary knowledges are produced and transformed into subject matter, and the political context that helps shape school purpose. (p. 33)

Kincheloe (2005) advocates a critical pedagogy requiring teachers to go beyond learning pedagogical strategies and content to understand schooling as a highly politicized institution that is shaped by history and special interests, and how invisible forces allow schooling to appear to be democratic, and yet behind the scenes enforce a structure that is both exclusive and oppressive. Critical pedagogists seek to develop in their students an ability to improve their lives and also society. Thus, it is conceivable to think of culturally responsive teaching as social justice teaching.

Social Justice Teaching

Social justice teachers actively pursue the resources that all students need to achieve and consider the needs of the students individually according to their life circumstances. The ultimate goal of social justice teaching is equitable education (Chubbuck, 2010). In a society where funding is allocated according to the tax base of each community, equitable education is difficult to achieve. The understanding of social justice teachers extends beyond a mere acceptance of the fact that individuals from affluent communities should have more resources accessible to them because they worked to obtain their wealth and therefore deserve it. Social justice teachers critique the distribution of wealth in terms of how marginalized groups began with unequal footing and were sequestered to impoverished conditions and inadequate resources, as well as laws and policies designed to keep them from experiencing equal opportunities. Social justice teachers understand that these conditions can be interpreted as a lack of value and internalized by the students. In other words, social justice teachers understand how attending a school in an impoverished community, laden with crime and limited supplies, could signify little worth to society. They also recognize this condition as a form of oppression, which dates back to the history of how this society originated and how these hierarchies began. Thus, they are not deluded by the occasional success of some individuals from marginalized groups, which may be considered by some as counter data for the claim that today's society is subtly oppressive. They instead divert their attention to identifying and alleviating the barriers that challenge the achievement of students who do not seem to be able to rise above it on their own. Chubbuck defines social justice teaching in three parts. Firstly, social justice teaching involves instruction in which the curricula, pedagogies, expectations, and interactions seek to improve learning opportunities for students, particularly those who are underserved. Secondly, social justice teachers understand how structural inequities can hinder student learning and as a result work to change the structures. Thirdly, social justice teachers extend their advocacy beyond the school context to promote changes in the larger society. Chubbuck points out that a major challenge to social justice teaching is the "demographic imperative." She defines the demographic imperative as racial divides between the students' ethnicities and their White middle-class teachers who possess little awareness of how the differences between their racial and cultural identities and their students' identities could impact the classroom. This awareness extends beyond teachers taking school bus trips to view the communities where the students live. This awareness critiques how the students and their families ended up in those circumstances and devotes effort to reverse the impact of their life circumstances on their achievement and futures. School bus trips to the communities in which the students reside could result in mere sympathy without these important critiques. As a matter of fact, any strategy designed to meet the needs of students from marginalized groups should be informed by an understanding of the condition that necessitated it, similar to students with disabilities as accounted for in IDEA.

Giroux (1992) advocates a critical multiculturalism, which is characterized by principles and instructional practices that offer an analysis and challenge to discriminatory ideologies that are entrenched within U.S. society and schools. The by-product of critical multiculturalism is critical awareness. Teachers who are critically aware are empowered. They are not bystanders waiting on someone to recommend strategies to them. They are initiators who develop their own tools and strategies for motivating and supporting their disenfranchised students. They also take responsibility for the society in which their students are developing, by trying to reverse the damage to their mental, physical, and emotional well-being. Taking responsibility for society does not imply accepting blame. Accepting responsibility for society is an acknowledgment of society's marginalization of groups, in its many forms, and the unfairness of these biases. According to Giroux:

> …multiculturalism is also about understanding how dominant institutions provide the context of massive unemployment, segregated schools, racist violence, and run down housing. A critical multicultural curriculum must shift attention away from an exclusive focus on subordinate groups, especially since such an approach tends to highlight their deficits, to one which examines how racism in its various forms is produced historically and institutionally in various levels of dominant culture. (pp. 9–10)

The environmental conditions that some students find themselves in are consequences from hardships that were inflicted upon them. The underlying structures contributing to the problem, along with the resulting conditions, provide a comprehensive picture of the students' circumstances. This is essential considering students living in challenging situations need to separate their identity from their life situations. As well, the students need to know that their teachers do not view them according to their life situations. Thus, social justice teaching practices take on different forms for teachers who are critically aware because the strategies that they employ are aligned with an in-depth assessment and understanding of the students' needs. This does not imply that teachers should not hold their students accountable for their actions. Instead, it implies that social justice teachers seek ways to motivate students to direct their energies toward working within the system to liberate themselves and their communities. Social justice teachers understand that the students are experiencing conflict between what society communicates about them and what is an acceptable image, considering the images represented in school could be considered an abandonment or compromise of their identity. Consequently, the student may adopt a stance that expresses their opposition in an attempt to maintain their allegiance to their cultural identity and expression. Teachers who are critically aware understand these emotions from the students' perspectives. They are able to empathize with students who feel that they are being told that they are not acceptable and seek ways to help these students realize that conforming to the expectations of the class is not surrendering their ethnic and cultural identity. Social justice teachers interpret students' dispositions in a manner that doesn't reflect negatively on the student. Kohl (1995) stated the following:

> No amount of educational research, no development of techniques or materials, no special programs or compensatory services, no restructuring or retraining of teachers will make any fundamental difference until we concede that for many students the only sane alternative to not-learning is the acknowledgment and direct confrontation of oppression-social, sexual, and economic both in school and in society. Education built on accepting that hard truth about our society can break through not-learning and can lead students and teachers together, not to the solution of problems but to direct intelligent engagement in the struggles that might lead to solutions. (p. 32)

The analysis of society that contextualizes this discussion of social justice teaching practices speaks to the exigency of this knowledge on the beliefs and philosophies of teachers in order to meet students' needs. Teaching practices stem from beliefs and philosophies. Often, when faced with the responsibility of teaching students from diverse backgrounds, teachers rely on recommendations from consultants to inform their practices. Recommendations from specialists, while they may be useful, can be mere band aids, in that the teachers can use them and not really understand what caused the problem. The needs of students who are socioculturally disadvantaged are not readily apparent. Generally, the students' issues are camouflaged behind their behaviors, which are symptoms of the problem.

Conclusion

Conclusively, the remaining culturally responsive qualities outlined by Villegas and Lucas (2002) are informed by sociocultural consciousness which adds depth to their application. Socioculturally conscious teachers understand how these qualities align with the needs of students who are socioculturally disadvantaged. For example, the quality, "affirming view of students," at the surface would be considered a desirable trait for all teachers of all students. However, socioculturally conscious teachers recognize in depth why an affirming view is necessary for students who are socioculturally disadvantaged. An affirming view is essential to countering the messages of inferiority that lead to low self-esteem and low self-efficacy. For the quality, "accepting responsibility for bringing about change to make schools more equitable," teachers who are socioculturally conscious understand the need to function as advocates to locate resources that will help students who are socioculturally disadvantaged overcome their life circumstances and not view themselves in terms of their disparities. For the quality, "understanding how learners construct knowledge and are capable of promoting knowledge construction," socioculturally conscious teachers understand why it is important to engage these students in their learning, particularly students who have been conditioned to think they have limited ability and little to contribute. The teachers also understand that these students need to be empowered and encouraged to develop and apply their skills in learning for benefit beyond their classrooms. For the quality, "knowing about the lives of their students," socioculturally conscious teachers recognize that demonstrating interest in the lives and cultures of their students who are socioculturally disadvantaged shows that they care about them, which is pertinent to establishing trusting relationships. Additionally, teachers' knowledge of their students' lives aids them in relevantly integrating aspects of the students' cultures in the curriculum, as well as provides a snapshot into their students' worlds, which would allow them to appropriately interpret their needs. Finally, socioculturally conscious teachers recognize that "designing instruction that builds on what their students already know while stretching them beyond the familiar" is validating for students who are socioculturally disadvantaged and challenges them to reach their potential. They understand that these students are motivated by teachers' expressions of confidence in their abilities to learn, especially in disciplines like science and mathematics which have been portrayed in society as beyond their reach. Without sociocultural consciousness, teachers cannot fully understand the distinctions which tailor these strategies for students who are socioculturally disadvantaged. Conceivably, the lack of awareness would impact the nature of the intervention, as well as teachers' commitments to be shaped by these qualities, as well as the amount of energy that they would devote to employing the strategies.

Thus, sociocultural consciousness could be considered a catalyst, empowering teachers to confidently and competently understand and address the needs of their students who are socioculturally disadvantaged. Science is not beyond the reach of marginalized students. It is the exclusive presentation of science in society and the reinforcement of biases in the overt and covert curriculum in schooling that discourages their participation and achievement. To increase the numbers of students from underrepresented groups achieving in science disciplines, science teacher educators should incorporate courses and learning experiences into their programs that foster sociocultural consciousness. Achieving this goal becomes more expedient with the nationwide focus on the demands for more individuals to meet the future STEM needs of this nation. More and more, efforts to meet the demand for more STEM professionals shift attention toward individuals from underrepresented groups. Successfully increasing the numbers will require science teacher educators to prepare teachers who appropriately interpret students' needs and accept the responsibility for providing the resources necessary to facilitate their achievement.

References

Aikenhead, G. S. (1996). Science education: Border crossing into the subculture of science. *Studies in Science Education, 27*, 1–52.

Anyon, J. (1981). Social class and school knowledge. *Curriculum Inquiry, 11*(1), 3–42.

Apple, M. W. (1978). Ideology, reproduction, and educational reform. *Comparative Education Review, 22*(3), 367–387.

Apple, M. W. (1986). *Ideology and Curriculum* (2nd ed.). New York, NY: Routledge.

Atwater, M., & Riley, P. (1993). Multicultural science education: Perspectives, definitions, and research agenda. *Science Education, 77*, 661–668.

Banks, J. (2010). Multicultural education: Characteristics and goals. In J. Banks & C. Banks (Eds.), *Multicultural education: Issues and perspectives* (7th ed.). Hoboken, NJ: Wiley.

Bourdieu, P. (1979). Symbolic power. *Critique of Anthropology, 4*, 77–85.

Bourdieu, P. (1986). The forms of capital. In J. G. Richardson (Ed.), *Handbook for theory and research for the sociology of education* (pp. 241–258). New York, NY: Greenwood Press.

Brand, B., Glasson, G., & Green, A. (2006). Sociocultural factors influencing students' learning in science and mathematics: An analysis of the perspectives of African American students. *School Science and Mathematics, 106*(5), 228–236.

Chubbuck, S. M. (2010). Individual and structural orientations in socially just teaching: Conceptualization, implementation, and collaborative effort. *Journal of Teacher Education, 61*(3), 197–210.

Clark, K., & Clark, M. (1939). Development of consciousness of self and the racial identification in Negro preschool children. *Journal of Social Psychology, 10*, 591–599.

Davis, K. (Director). (2005). *A girl like me* [Documentary]. United States. Retrieved from http://www.kiridavis.com/

Ferguson, R., Gever, M., Minh-ha, T. T., & West, C. (1990). *Out there: Marginalization and contemporary cultures*. New York, NY: New Museum of Contemporary Art.

Gay, G. (2002). Preparing for culturally responsive teaching. *Journal of Teacher Education, 53*(2), 106–116.

Giroux, H. A. (1983). Theories of reproduction and resistance in the new sociology of education: A critical analysis. *Harvard Educational Review, 53*(3), 257–293.

Giroux, H. A. (1992). Curriculum, multiculturalism, and the politics of identity. *NASSP Bulletin, 76*(548), 1–11.

Kincheloe, J. L. (2005). *Critical pedagogy primer*. New York, NY: Peter Lang.

Kohl, H. (1992). I won't learn from you! Thoughts on the role of assent in learning. *Rethinking schools 7*(1), 16–17, 19.

Kohl, H. (1995). *"I won't learn from you": And other thoughts on creative maladjustment*. New York, NY: New Press.

McGee, E. O., & Martin, D. B. (2011). "You would not believe what I have to go through to prove my intellectual value!" Stereotype management among academically successful black mathematics and engineering students. *American Educational Research Journal, 48*, 1347–1389.

Sleeter, C. (2011). An agenda to strengthen culturally responsive pedagogy. *English Teaching: Practice and Critique, 10*(2), 7–23.

Tobias, S. (1988). Insiders and outsiders. *Academic Connections*, 1–5.

Villegas, A. M., & Lucas, T. (2002). Preparing culturally responsive teachers: Rethinking the curriculum. *Journal of Teacher Education, 53*(1), 20–32.

Yee, A. H. (1992). Asians as stereotypes and students: Misperceptions that persist. *Educational Psychology Review, 4*(1), 95–132.

Part II
Foundations of Science Teacher Education

The Impact of Beliefs and Actions on the Infusion of Culturally Relevant Pedagogy in Science Teacher Education

Natasha Hillsman Johnson and Mary M. Atwater

Introduction

Culturally relevant pedagogy has been the subject of much debate and discussion in teacher education since Gloria Ladson-Billings' seminal book, "The Dreamkeepers," was published in 1999. Countless hours of professional learning, education courses, and conference sessions have been dedicated to increasing our understanding of this type of pedagogy. Even though K-12 classrooms and college campuses are more diverse than ever and despite all the rhetoric, most teacher educators and teachers still are not seeing culturally relevant teaching and social justice as necessary to support the needs of culturally diverse student populations (Ball & Tyson, 2011; Hollins & Guzman, 2005; Lee, 2011; Nieto & McDonough, 2011; Zeichner, 2005). Despite the first author's best intentions as a secondary science teacher for almost ten years, she has not always been successful in implementing culturally relevant pedagogy to her students. There are countless other teachers who probably share her sentiment, so this chapter will further examine this important issue in education.

We will explore how teacher beliefs and actions perpetuate the absence of culturally relevant pedagogy in science teaching in US classrooms. First, we will discuss specifically what is meant by the terms culturally *relevant* pedagogy and culturally *responsive* pedagogy. Next, we will examine the origin of teacher beliefs and how they influence teacher actions and instructional delivery. We will also discuss current

N.H. Johnson (✉)
College of Education, Department of Mathematics & Science Education,
University of Georgia, 212 Aderhold Hall, Athens, GA 30602-7126, USA
e-mail: yjohnson@uga.edu

M.M. Atwater
College of Education, Department of Mathematics & Science Education,
University of Georgia, 376 Aderhold Hall, Athens, GA 30602-7126, USA
e-mail: atwater@uga.edu

practices and changing demographics in K-12 education and the importance of equity and social justice to culturally relevant pedagogy. Finally, we will discuss the specific teaching beliefs and actions that are necessary to infuse culturally relevant pedagogy into teacher education programs for middle and high school science classrooms. The chapter will close with examples of culturally relevant teaching practices and recommendations to effectively prepare teachers for multicultural science education.

What is Culturally Relevant *Pedagogy?*

Pedagogy simply refers to the practice and the work of teachers, focusing on the many aspects of the art and science of teaching (Haberman, 1991; Smith, 1985), while culture can be viewed as a system of meanings and practices that is carried by people and produced in settings between people in moment-to-moment interactions (Nasir & Hand, 2006). Culture is not a static part of individuals, even though it is about an individual's learned beliefs, traditions, and customs for an individual's behavior that are shared among the group members with whom the individual identifies. However, individuals like science teacher educators and science teachers might be "conditioned and programmed, but they have the freedom to seek new experiences, knowledge, and skills that change their cultural framework" (Freire, 2005).

According to Ladson-Billings (2009), culturally *relevant* pedagogy is defined as a "pedagogy that empowers students intellectually, socially, emotionally, and politically by using cultural referents to impart knowledge, skills, and attitudes" (p. 20). Culturally relevant pedagogy includes not only the cultural referents to pedagogy, but these cultural referents are included in any curriculum (Grant & Ladson-Billings, 1997). This teaching practice not only addresses student performance but also helps students to accept and affirm their cultural identity while developing critical perspectives that challenge inequities that schools (and other institutions) perpetuate (Ladson-Billings, 1995, p. 469). Hence, there are three dimensions to culturally relevant pedagogy: emphasis on learning and performance, cultural competence, and engendering a sense of sociocultural-political critique. Educators, researchers, and teacher preparation programs have focused a great deal of attention on multicultural education, specifically culturally relevant pedagogy (Morrison, Robbins, & Rose, 2008; Osborne, 1996; Wortham & Contreras, 2002) in an attempt to close the achievement gap between culturally diverse student populations and their White counterparts. Common myths, which are beliefs accepted uncritically to justify actions (*Webster's Encyclopedia Unabridged Dictionary for the English Language*, 1989), can create barriers to current understanding and effective implementation of culturally relevant pedagogy. According to Irvine (2010), many educators believe that only teachers of color can practice culturally relevant pedagogy, it is not beneficial for White students, and caring teachers of diverse student populations lack the necessary classroom management skills for effective instruction. However when properly facilitated, this method of instruction works to engage, motivate, and effectively teach children being served in culturally diverse classrooms (Irvine).

Social constructivism is a theory of knowledge that argues that humans generate knowledge and meaning from an interaction between their experiences and their ideas. Another postulate of the constructivist view of learning discussed by Limon (2001) is the "importance of connecting the new knowledge to be acquired with the existing knowledge that students have, in order to promote meaningful learning" (p. 358). Similarly, culturally relevant pedagogy as a theoretical model postulates that "teachers must be able to construct pedagogical practices that have relevance and meaning to students' social and cultural realities" (Howard, 2003, p. 196). Successful teachers of students from divers groups that cultural knowledge and learning styles can be leveraged to make learning more relevant and effective (Howard).

What Is Culturally **Responsive** *Pedagogy?*

Unfortunately, culturally responsive pedagogy is viewed by many as the same thing as culturally relevant pedagogy because the former requires teachers to have more in-depth knowledge, skills, and teaching experiences than the later. Others believe that culturally responsive pedagogy is important in preparing teachers for the cultural diversity they will find in their classrooms. The success of teacher preparation programs in educating preservice teachers to meet these diversity challenges is dependent upon how their graduates are prepared with the "skills, attitudes, and knowledge necessary to enhance their ability to undertake the gigantic responsibility of creating classroom environments appropriate for achieving excellence and equality of learning for all children" (Phuntsog, 1999, p. 99).

Many use the term culturally responsive *teaching*, rather than culturally responsive *pedagogy*, to describe what teachers should do in their classrooms. Culturally responsive teaching primarily involves utilizing students' cultural experiences and backgrounds as an avenue for helping them to learn academic knowledge and develop academic skills (Phuntsog, 1999). In this framework, students' cultural identities are a dynamic blending of race, ethnicity, class, gender, region, religion, and family (Huber, 1991; Wlodkowski & Ginsberg, 1995). Many agree that teacher education programs have several essential foci if their graduates are able to be culturally responsive teachers: the development of (a) a knowledge base about cultural diversity and ethnic and cultural diversity content in the curriculum, (b) abilities to establish caring and learning environments in classrooms, (c) abilities to communicate with students from ethnically diverse backgrounds, and (d) abilities to respond to students from ethnically diverse backgrounds (Gay, 2002). Others also include transformative curriculum that bring about meaning and understanding (Atwater, 1996; Gormley, McDemontt, Rothernerg, & Hammer, 1995; Wlodkowski & Ginsberg, 1995).

Definition of Teacher Beliefs and Actions

It is our contention that teacher beliefs and actions are problematic to the effective and sustained infusion of culturally relevant pedagogy in science education. A belief can be defined as a person's reality, specifically what one accepts as truth. In his work, *How We Think*, Dewey (1933) stated that beliefs are:

> ...something beyond itself by which its value is tested; it makes an assertion about some matter of fact or some principle or law. It covers all the matters of which we have no sure knowledge and yet which we are sufficiently confident to act upon and also matters that we now accept as certainly true, as knowledge, but which nevertheless may be questioned in the future (p. 6).

Teacher beliefs are of particular importance, and educator preparation programs must encourage more preservice teachers to question their own belief systems, especially those working with culturally diverse student populations which have not experienced academic success and are at a greater risk for school failure (Bryan & Abell, 1999; Harrington & Hathaway, 1995; Hollingsworth, 1989; Olmedo, 1997; Tobin & McRobbie, 1996).

During my (first author) certification program through the Teach for America organization, I participated in a sharing session that allowed and encouraged us as preservice teachers to examine our belief systems about education, schools, and students. One of the other participants cautiously admitted that the absence of African American and other "minority" students in her higher level math and science courses had caused her to internalize the notion that these groups were indeed less capable of such work. How many educators and administrators at the K-12 and college levels who work with culturally diverse student bodies on a daily basis hold similar beliefs? This story serves as a reminder of the need to expose White students to diverse student populations, as well as the powerful message that is transmitted in their absence. For this future educator to be forced to look me, an African American woman, in the eyes and come to terms with her personal beliefs was undoubtedly a transformative experience that would serve to shape her interaction with students. If we could encourage preservice and in-service teachers, as well as teacher educators to question these current beliefs, it would take us one step closer to transforming the actions of teachers in culturally diverse classrooms.

Teacher actions refer to the way in which the process of teaching and learning is facilitated in their classroom (Cornett, Yeotis, & Terwilliger, 1990; Ross, Cornett, & McCutheon, 1992). It encompasses every minute detail that occurs from lesson planning and activity selection to communication with parents and students. Teachers, who fundamentally believe "minority" students to be less intelligent, will transmit this value system in the way of low expectations, less demanding content, and the perpetuation of cultural deficit models (Jacob & Jordan, 1993; Nieto, 2000).

Current Practices and Changing Demographics of K-12 Schools and Colleges

In 2007, African American, Asians/Pacific Islanders, Latino/as, and Native Americans composed 33 % of the US population (Aud, Fox, & KewalRamani, 2010). By the year 2050, "the nation's population of children is expected to be 62 % "minority," up from 44 % today" (Bernstein, 2008). While their enrollments in public schools are increasing, Black and Latino/a students' performances in science and mathematics are decreasing (Brand, Glasson, & Green, 2006). Based on trends in the US population, it is projected that by the year 2050, the majority of students receiving precollege education will be members of a variety of racial and ethnic groups (U.S. Census Bureau Projections Show a Slower Growing, Older, More Diverse Nation a Half Century from Now, 2012).

As demographic trends continue to change, it becomes increasingly important that not just K-12 educators but also faculty members, administrators, and teacher educators understand how to meet the needs of a culturally diverse student body (Thomas, Wilder, & Atwater, 2001). Current initiatives to increase student interest and participation in science, technology, engineering, and mathematics (STEM) majors make the establishment of effective teaching practices in the area of science education an urgent matter (National Research Council, 2012). Students who lack quality science educational experiences at the elementary, middle, and high school levels are less likely to choose such majors at the postsecondary level. Students who choose to pursue such majors will find themselves at an academic disadvantage if they have experienced less challenging coursework or have not been given access to advanced courses as a result of teacher beliefs and low expectations for Black, Latino/a, Native American, and Pacific Islander students. It is essential that current teaching practices be transformed to include the principles of culturally relevant pedagogy.

According to Irvine (2010), many "well-meaning educators often assume that culturally relevant pedagogy means simply acknowledging ethnic holidays, including popular culture in the curriculum, or adopting colloquial speech" (p. 58). Instead, we need teachers who understand diverse cultures that can use this knowledge to engage students, plan effective lessons, and accurately assess student learning in the classroom. Such teachers would also possess the ability to prepare students to navigate the mainstream culture (Ladson-Billings, 1995), in order to promote success beyond the walls of the classroom in educational measures such as the Scholastic Aptitude Test (SAT) and American College Testing (ACT) that continue to lack cultural relevance for many students.

Importance of Equity and Social Justice in Culturally Relevant Pedagogy

An understanding of science is becoming increasingly important in this technology-driven society as science has impacted every part of our lives, from our food choices at the grocery store to decisions and options for health care and to the methods and speed with which we communicate with one another. According to Barton (2002),

> science holds a uniquely powerful place in our urban society. It opens doors to high-paying professions, provides a knowledge base for more informed conversations with health care workers, educators and business and community leaders, and it demystifies key urban environmental issues like air and water quality standards, population density, and toxic dump and building regulations. (p. 1)

The Tuskegee Experiment offers a great illustration of the dire consequences that can result when there is an absence of equity and social justice in science education and scientific knowledge. Also known as the Tuskegee syphilis study, it was a long-term, no treatment observational study of 600 African American sharecroppers from Macon County, Alabama. More than half of the men had previously contracted syphilis, while the remaining men did not have the disease. The men were never told they had syphilis nor were treated for the disease. But they did receive free medical care, meals, and free burial insurance because they had "bad blood." Others were infected with this disease by those having syphilis (Remembering Tuskegee, 2002). The study continued to 1972 until a press leak terminated the study (*Special Obituary tribute: Tuskegee syphilis research study survivors*, 2004). This most infamous biomedical research study led eventually to federal laws and regulations requiring Institutional Review Boards for the protection of human subjects in studies involving human beings. Over the past 40 years since the conclusion of this study, despite federal regulations to prevent such exploitation of human subjects, the disparities in scientific knowledge have continued to increase in the United States and all over the world.

The current *Science for All* reform is working to improve the experiential science learning for nonmajors and increasing the level of comprehension and understanding for students in science fields. Science offers a unique challenge to the implementation of culturally relevant pedagogy. Science educators must not only challenge notions related to the academic potential of culturally diverse students but also eliminate discussions of "weed out" courses and notions of science as the "holy grail," not within reaches of the masses. We must learn to adopt a "science appreciation" approach to the discipline as described by Gould (1997) with this music analogy:

> Few Americans can play violin in a symphony orchestra, but nearly all of us can learn to appreciate music in a seriously intellectual way….Similarly, few can do the mathematics of particle physics, but all can understand the basic issues behind deep questions about the ultimate nature of things and even learn the difference between a charmed quark and the newly discovered top quark. (as cited in Oliver et al., 2001, p. 22)

Through the scholarship of teaching and learning, we must change the dialogue about the sciences. As the science education community changes its perspective on science attainment, changing views of the general public and students will follow. As we move this discussion to the topic of teaching at the middle and high school level, we will close by considering the specific goals of preparation programs and professional development that teaches the principles of cultural relevance.

Several months ago, a colleague of mine (second author) sent me an email message asking "What, if any, are the distinctions between culturally relevant and culturally responsive pedagogies? I would appreciate any clarification and assistance that you can provide on this matter." After much contemplation, I wrote the following response:

First, culturally relevant and culturally responsive pedagogy are both about pedagogy. According to Grant and Ladson-Billings (1997), culturally responsive pedagogy is the same as culturally relevant pedagogy. However, based on my personal understanding of relevancy and responsiveness, I view these two terms very differently. Relevancy means that cultural ideas and actions must be connected with the matter at hand in teaching science or pertinent to the science teaching, while responsiveness means making the adjustments suddenly to the science teaching. Science teachers must possess knowledge and skills related to science and their students to create a safe learning environment so that the science curriculum can be understood in such ways that their students do understand natural phenomena and develop scientific skills. Despite the best education and preparation, any experienced teacher knows that one cannot always anticipate or plan for every event or questions that arise in the classroom.

While teachers can plan for culturally relevant teaching (science activities are designed), many teachers are not culturally responsive in their science teaching. Several years ago, Rob Parks (pseudonym), a colleague of mine (first author), developed a scientific method song aligned to "Ride with Me," by rapper Nelly. It was designed to help students remember the steps of the scientific method and offers an example of culturally relevant teaching. I "admired and acquired" this song for use in my own classroom. One year after reviewing the lyrics with one of my classes, a student asked "Why they needed to know the scientific method when only White people were scientists?" and waited for my response...this is an example of the need for cultural responsiveness. This question required that I not only understand the intent and motive behind the question but also very quickly activate my knowledge of the contributions of African Americans in science and deliver my response in a way that would serve as motivation to my student population.

If I know little about the culture of certain students in my science classroom, I can plan for culturally relevant activities, but a situation can arise in which I will not be culturally responsive. Even when you know your students well, it is very difficult to be culturally responsive in teaching science. If we could prepare preservice teachers to do culturally relevant teaching, I would have accomplished a worthy goal. Culturally responsive teaching comes with many years of teaching. For teachers to be culturally responsive teachers, they must be able to immediately respond to cultural incidents in their classrooms. Beginning teachers have limited experiences with different cultural groups, so time in the classroom, diverse teaching experiences, and additional professional development can provide them the opportunity to become culturally responsive teachers (Darling-Hammond, 2006).

Teaching Science in Middle and High School

What Beliefs Are Necessary for Science Teachers?

Mackeracher (2004) contends that "what one values and believes to be true about learning is incorporated into one's philosophical orientation to learning and to learners, and determines how one is likely to facilitate learning" (p. 5). It is the latter point that becomes significant when we consider the widespread and persistent notion that students of color are intellectually inferior and not capable of meeting rigorous academic standards. According to Bryan and Atwater (2002), there exists a need to examine teacher beliefs about student characteristics, external influences on learning, and appropriate teacher responses to diversity. Little has changed since Lipman suggested in 1993 that, despite massive attempts at school reform and restructuring, teacher ideologies and beliefs often remain unchanged, particularly toward African American children and their intellectual potential (as cited in Ladson-Billings, 1995, p. 478) (Darling-Hammond, 2006).

Jones and Carter (2007) examined the multifaceted constructs of teacher beliefs and attitudes and how their beliefs and attitudes influence instructional practices. They utilized a sociocultural model of embedded belief systems where teachers' beliefs about science, beliefs about teaching science, and beliefs about learning science influence teacher actions and practices in science classrooms. Since teacher beliefs are positioned in the milieu of the existing social norms of the school community, teachers are concerned about how their enacted practices are perceived by their colleagues and administrators, parents, and community stakeholders. Some teachers feel that institutional constraints leave them little time to reflect upon, let alone change the misalignments between their belief systems and practices, especially as they relate to culturally relevant pedagogy.

According to Howard (2003), one of the central principles of culturally relevant pedagogy is an authentic belief that students from families that are low income and diverse cultures are capable learners. When I (first author) joined the staff at my current high school, I was presented with a new challenge as an educator. Most of my career I had worked with African American students in urban, Title I schools. In my first day of school, I was surprised to learn that in some of my chemistry classes the English Language Learner (ELL) population approached 25–30 %. As an inservice teacher, graduate student, and aspiring professor, I could rattle off countless facts about learning styles and effective instruction for Spanish-speaking students. How could I leverage my knowledge about language acquisition, learning styles, and parental involvement to meet the needs of my ELL students?

Despite my best efforts, my ELL students continued to score significantly lower on summative assessments, demonstrated chronic absenteeism, and several dropped out of school. On the difficult days, I was reminded of my personal belief system and my purpose as an educator and I was motivated to work a little harder. If I could only increase my understanding of the cultural differences observed in my students, I might be able to better meet their academic needs. The differences in learning styles, motivation, career aspirations, preexisting knowledge, and communication styles are all a result of their unique backgrounds and personal stories.

So, what specific beliefs are needed for culturally relevant science teaching? There exists a need for teachers who fundamentally believe that all students are capable of academic success. Teachers must understand and recognize that the K-12 educational system is not currently designed to meet the needs of culturally diverse groups of students. Despite, traditional approaches to science instruction, teachers must recognize that science instruction can be made culturally relevant. Ultimately, successful science teachers will know that the hard work, additional planning time, and out-of-pocket costs are worth the enhanced learning outcomes.

Science Teacher Actions for Equity and Social Justice

So, how do we transform these beliefs into the actions necessary for teachers for infusion? Successful science teachers working with culturally diverse groups must enter these classrooms and teaching positions by choice and not by assignment. School systems must carefully consider the background and preferences of teacher candidates during the placement process. One commonality shared between Ladson-Billings' highly effective teachers of African American children was the amount of time they had devoted to perfecting their craft, on average 12 years of classroom teaching experience. School administrators must recognize the uniqueness of science as a discipline and in turn allow additional planning time for laboratory preparation and reduce the number of different courses teachers are expected to effectively prepare for each school year. These changes could greatly improve teacher morale and reduce teacher turnover which both greatly influence student achievement (Ingersoll, 2000; Ruby, 2002). This cadre of teachers must be willing to work with introductory, general, and remedial courses where the needs are the greatest over honors and advanced courses. Additionally, these teachers should possess or actively seek knowledge about different cultural groups. Finally, these teachers must understand how vital and critically important it is to help students experience success and become excited about science. However, teachers will not seek these experiences unless they believe that these students are worthy of quality teaching and learning (Delpit, 1995).

Examples of the Importance of Classroom Culture

After an extensive literature search, I (first author) found it very difficult to identify examples of culturally relevant science instruction. Perhaps this is a reflection of how difficult middle and secondary science teachers find it to make science curriculum more culturally relevant for students. Bryan and Atwater (2002) discuss the academic culture of the sciences that has long been dominated by White males.

The primary focus remains on the transmission of knowledge, and to some knowledge is still viewed as scientific only if it is "objective" and "value free" knowledge. Many science teacher educators and teachers are still educated to view and approach science in isolation of self and social phenomena. This approach makes science less appealing to students of color and women and also makes it harder for students to connect with and retain the information because they are not actively engaged in the learning process. I (first author) recently had a conversation in my chemistry course about nature of transition metals and the rules for naming chemical compounds. We watched a short video clip from the movie *Erin Brockovich*, where Julia Roberts's character explains the different types of chromium and the dangers of high chromium levels in the drinking water for people in the community. This instructional strategy proved a lot more successful than my traditional methods to explain the importance of having a basic understanding of chemistry and more importantly the ability to educate one's self on important issues that relate to one's family and community.

With the absence of culturally relevant science examples in the literature, I (first author) decided to reflect on my own teaching experiences and observations of some truly phenomenal science teachers during my career. Many of these teachers have been recognized for their excellence with honors such as Teacher of the Year, Master Teacher, and National Board Certification. Here are a few examples of culturally relevant science instruction that have been used to teach chemistry, biology, and physics.

Chemistry

Chemistry is the study of matter and the various physical and chemical changes that occur in that matter (Whitten, Gailey, & Davis, 1988). Many students have struggled to make sense of the more abstract and mathematical concepts associated with this subject. Although it can be a challenge, there are certainly opportunities to infuse culturally relevant teaching into the chemistry curriculum. One personal example dates back to one of my first years in the classroom. A student asked me when we were going to use some chemicals. His question caught me by surprise, as we had performed laboratory experiments on a regular basis. After some follow-up questions, I came to understand that he associated chemicals with brightly colored liquids that smoked and fizzed when handled.

After reflecting on this experience, I created a project to help students connect the seemingly abstract chemistry concepts to their own experiences. The project was entitled "Chemistry in Our Daily Lives" and students were asked to use magazines, newspapers, and pictures to create a collage. The images were to reflect any examples of matter from daily life such as hair care products, food items, clothing, or consumer products. After a little thought and consideration of the definition of matter, most students realized that chemistry was a very important part of all aspects of their daily lives. This activity certainly set the stage for further study and worked to break down many barriers to the learning of science. A colleague, Karla Hill (pseudonym), developed an Atomic Theory Timeline Project, where students are asked to trace the development of the modern atomic theory and compare it to any topic that is of interest to them. Students trace the work of scientists such as John Dalton, Ernest Rutherford, and Robert Millikan and the discovery of the subatomic particles but are able to relate it to a topic of personal significance. Students have opted to do a wide variety of projects including music, fashion, and religion. This extension allows the students to better understand how knowledge evolves over time and also how innovation can transform any industry. This is an excellent way of making a very mundane topic in chemistry more personal and relevant to students.

Biology

Biology is the study of life and living organisms and their interactions with their environment from the microscopic to the global levels and a diversity of perspectives (Alters & Alters, 2005), and while some students are better able to connect to this scientific discipline, others still struggle to understand the cell and make sense of the vocabulary. Boutte, Kelly-Jackson, and Johnson (2010) share three examples of culturally relevant teaching in the science classroom in the area of biological sciences: (a) cell analogies collage, (b) extracting DNA activities, and (c) integumentary system unit. The cell analogies collage project requires students to connect the structure and function of various part of the cell to activities and experiences from their daily life. For the DNA extraction activity, teachers bridged student interest in forensic science and medical applications to introduce the concept of blood typing. The final activity explored the integumentary system in the human body and facilitated a discussion of the historical, social, and political issues behind skin color and ethnic hair. Differences in skin color and hair texture have long been discussed within and outside of the African American community. As a result many students come to classrooms with a multitude of questions about the cause of such differences. This serves as an example of how student interest can be harnessed to support academic learning goals but only if the teacher is aware and knowledgeable of the opportunity.

Physics

The *Merriam-Webster Dictionary* defines physics as the study of the interactions between matter and energy (Physics n.d.). Although many students come to class with a great deal of background knowledge and exposure to topics such as motion, forces, and electricity, students find it very difficult to navigate the mathematical and problem solving skills that are necessary in this subject. One technique that is commonly used to promote learning and understanding is hands-on laboratories and projects. One project adopted from a colleague, Mark Enon (pseudonym) that was particularly popular with my students and able to reinforce key concepts and skills, was a physics poster project. Students were asked to create a poster of an original problem, illustrate the scenario, and properly solve the problem. Students gained an understanding of the importance of such topics as measurement and relationships among variables. The experience offered students a better rationale for the importance of units and scale and connected the content to a topic of personal interest. Of all the courses that I have taught, physics was always the most difficult for me to teach. Despite the personal relevance of the subject matter, it always seemed much more difficult to infuse cultural relevance into the lessons. This could be directly related to my own personal (and negative) experiences as a physics student at the high school and college level.

As I reflect on the examples shared, sadly not one stands out as a truly excellent example of culturally relevant teaching in the science classroom. It is my hope that this chapter will serve as a starting point for further discussion and development of culturally relevant science instructional materials and practices. One observation is that many teachers are successful at incorporating culturally relevant activities into the curriculum to engage the students; of course, this is a basic practice of excellent teaching (Stepanak, 2000). Many educators fail to go a step further to incorporate the critical consciousness and sociopolitical components. This could be attributed to lack of experience, lack of time, or even lack of knowledge of the historical, social, and political cross sections of the subject. According to Boutte et al. (2010), many students are aware of scientific racism which often leads to a resistance to learning science content. "Scientific racism can be defined as the use of scientific methods to support and validate racist beliefs about African Americans and other groups based on the existence and significance of racial categories that form a hierarchy of races that support political and ideological positions of white supremacy" (Davis & Martin, 2008, p. 14). As a result, it is critical that science teacher educators become comfortable leading discussions on scientific racism and infusing this dimension of culturally relevant teaching in the science education classroom (Boutte et al., 2010).

Secondary Science Teacher Education Programs

What Beliefs Are Needed to Infuse Culturally Relevant Pedagogy in Science Teacher Education Programs?

Institutions of higher learning serve as models to K-12 education and the community at large. Although universities have become more diverse in recent years, the presence of faculty members of African, Latino/a, Native American, and Asian ancestry on many campuses is still a rare occurrence (Allen, Epps, Guillory, Suh, & Bonous-Hammarth, 2000). I (first author) began this chapter with a reminder about the message that is transformed by the absence of culturally diverse students from

STEM programs, what message is being sent by education programs that continue to lack diversity at the administrative, faculty, and student levels. When I began my doctoral program, I expected that I would find little diversity in my courses in the chemistry department, but I was shocked by how little diversity existed in my science education courses. The undergraduate and graduate courses were overwhelmingly filled with White, female students. There was also not a great deal of diversity among the university faculty. I began to wonder if this was representative of most colleges and universities. Many of the students in my courses did not have significant teaching experience or had not worked in urban settings or with diverse populations. How can we teach what we do not know or have not experienced? One belief that is needed to infuse culturally relevant pedagogy in science teacher education programs is that there is a need for all students to learn this type of pedagogy regardless of their desired teaching locations. Another belief is that this type of curricular focus will be well received by all education students, not just those destined to work in urban schools. Current science teacher preparation programs are not designed to meet the needs of culturally diverse student populations.

Actions Necessary for Science Teacher Educators

Gloria Ladson-Billings (2000) writes that African Americans are fighting for their lives when they fight for education. Few science teacher educators understand that we must prepare science teachers to fight for the science lives of their students, especially students who are underserved and unrepresented in the sciences. The social and cultural experiences of different groups of students are unique. For example, Cornel Pewewardy (1993) asserts that one reason Native Americans experience difficulties in public schools is that educators attempt to insert culture into education, not inserting education into culture. In other words, students' cultures are a vehicle for learning science. For a variety of reasons, science teacher educators do not grapple with the discontinuity between what K-12 students experience at home and what they experience in schools in their interactions with teachers and classmates. The kinds of student speech and language interactions with teachers and other school personnel influence their academic success. I (second author) have directed a science venture program in a high school. Most of the students are Latino/a and speak Spanish. As a monolingual speaker, I appreciate that these high school students have opportunities to learn science in Spanish from the teacher coordinator and guest speakers in the program. All of the Spanish-speaking students do speak some English and unfortunately must communicate with me in English. I know that I miss so much about knowing the student and knowing how the students think about natural phenomena and science because they must formulate their thoughts in English. In other words, I am not one of them even though they value my actions and thoughts. Some of them have encouraged me to learn Spanish. However, I fail to make the time to learn Spanish due to my many faculty responsibilities. So as a teacher educator, how do I get preservice teachers to value diversity among their students and understand they have a responsibility to teach all of their students, especially preservice science teachers who have not experienced diverse ways of valuing, thinking, and doing in their lives. How do we get preservice science teachers to grapple with discontinuity and these battles for K-12 student science learning?

Reassessing Admissions Policies

In most science teacher education programs, admission requirements include high school graduation, SAT or ACT scores, and declaring science education a major. A few programs require prospective teachers to write an essay, but based upon the second author's experiences, these essays are usually not used in a serious way. Due to the changing landscape in teacher employment, many teacher education programs are struggling to recruit high school graduates and graduate teachers into teacher education programs, especially students of color. Several general approaches have been utilized to have a more diverse student population in teacher education programs: (a) altering admission requirements for traditionally undergraduate and graduate teacher education programs based on academic criteria and relying more on personal factors and life experiences, (b) creating alternative teacher education programs that focus on high needs urban and rural areas, (c) creating articulation agreements with two-year and technical colleges that enroll large numbers of students in color, and (d) recruiting students into social justice teacher education programs who teacher educators think want to make a difference in the lives of underserved and unrepresented groups of students (McDonald & Zeichner, 2009). Admissions counselors and program coordinators must continue to work to attract and retain the most talented individuals to programs in education, especially in science teacher education programs, individuals who are not only qualified but committed to teaching diverse student populations.

Reexamining Coursework

A variety of instructional strategies and course assignments have been advocated and used in social justice and multicultural education courses such as autobiography, case studies, dialogue journals, literature, films, portfolios, and storytelling (Grossman, 2005; Hollins & Guzman, 2005). However, these instructional practices in teacher education programs do not guarantee that prospective science teachers have the commitment and the knowledge and skills to be culturally relevant teachers (McDonald & Zeichner, 2009). The goal of teacher educators who use these instructional practices is to expand their students' social consciousness and help their students to comprehend that there is no one worldview and their students do not have to embrace the dominant worldview to be successful in understanding natural phenomena and using their science knowledge and skills to change the world around them. Many teacher education programs require preservice science teachers to successfully pass social foundation courses that focus on (a) critical educational issues influencing the social and political contexts of educational settings and examine the nature and function of culture, (b) the development of individual and group culture, (c) the meaning of education and schooling in a diverse culture and the moral and ethical responsibilities of teaching, and (d) the influences of culture on learning, development, and pedagogy. Despite these requirements, few science teacher educators build on this foundational knowledge in their science methods, curriculum, and practicum courses. Few science teacher educators have the knowledge, skill, and commitment to do so.

Restructuring Field Experiences

Many teachers have commented that their practicum and internship teaching experiences did not prepare them for the realities of their day-to-day job as a teacher (Cochrane-Smith & Zeichner, 2005). These experiences are often not indicative of the average teaching experience and can lead to job dissatisfaction and retention issues (Ingersoll, 2000). Preservice teachers cannot be placed in ideal classrooms and school environments that lack academic, behavioral, or diversity issues and fail to reflect the true dynamic of the teaching and learning process. Many teacher education programs have restructured their programs such that courses and teaching internships occur in the school setting with the idea that the prospective teacher must understand they are cultural beings also. The goal is if they are aware they are cultural beings, then they will be more willing to teach in a culturally relevant manner. Research findings are inconclusive about the characteristics of these school-based experiences; however, it is clear from these studies that it is the quality of these experiences that matter rather than just placing prospective teachers in schools with underserved and underrepresented groups of students (Atwater & Suriel, 2010; McDonald & Zeichner, 2009). For instance, if mentor teachers are not committed to teaching all of their students and do not possess the knowledge and skills to do so, then preservice science teachers are very limited in their experiences. During one teacher internship program, preservice teachers requested classroom instruction from doctoral students who were not serving as university supervisors and had been successful in teaching underserved and underrepresented students. One of the doctoral student university supervisors sought out these doctoral students and got their commitment to work with these preservice teachers. However, the science teacher educator responsible for the teaching internship chose for this not to happen. Why was this the case? Was he uncomfortable and thought it was a reflection upon his inadequacy. Or did he think this was not essential for these preservice teachers to be successful? This is an example that science teacher educators must be committed to providing opportunities for their students to have opportunities to discuss successful teaching scenarios so they can come to understand that one attempt at culturally relevant teaching is not enough even if one is successful.

Where Do We Go from Here?

Recruiting and Retaining Faculty of Color

Research shows that the most persistent, statistically significant predictor of enrollment and graduation of African American graduate and professional students is the presence of African American faculty members (Blackwell, 1981; Milan, Chang, & Antonio, 2005; Moreno, Smith, Clayton-Pedersen, Parker, & Teraguchi, 2006). Institutions that are successful in recruiting and retaining African American faculty members do a far better job of recruiting, enrolling, and graduating African American students than do those with few or no African American faculty members (Blackwell, 1981). It is very difficult to recruit faculty of color into science teacher education programs if very few are obtaining doctorates. At this time, there are less than 40 Black science education faculty members in the United States in traditionally White and Historically Black Universities and Colleges in tenure-track positions. Many Black faculty members who are awarded doctorates in science education return to the precollege setting to teach. Hence, the academy loses many potentially capable science teacher educators.

The few faculty members of color in science teacher education programs in the United States struggle with a host of challenges to become promoted and tenured. Many felt unsupported by their colleagues and department heads (Johnson, Atwater, Freeman, Butler, & Parsons, 2012). The participants in this research study mentioned that reduced committee work, access to research assistants, and release time from instructional loads are essential to early career success; however, many possess a persistent belief that access and distribution of such resources was inequitable to them as African American female faculty members.

Balancing teaching, research, and service is difficult for most teacher educators at research intensive institutions. Black faculty members may find it much more difficult than their White colleagues. They have to deal with negative teaching evaluations, especially those faculty members that infuse multicultural education and social justice in their courses (Atwater, Freeman, Butler, & Parsons, 2012). Tenure looms in the minds of tenure-track science educators at the Assistant Professor rank. Without tenure, one cannot remain in one's science teacher education faculty position. Some academicians find the challenges too much to infuse multicultural education and social justice in their teacher education programs and leave the academy to pursue work outside of the university setting (Atwater, Freeman, Butler, & Parsons, 2011).

Investment in Research Initiatives

Research has shown that knowing teachers' beliefs and designing instruction to explicitly confront those beliefs facilitate refinement of and/or transformation of beliefs and practices (Cochrane-Smith & Zeichner, 2005) However, changes in beliefs of prospective teachers do not always translate into changes in practices. This concern demands research that examines both belief and actions. The following research questions should be explored by science teacher educators:

1. What are the beliefs of the applicants of the science teacher education programs that are likely to predict the successful teaching of marginalized, underserved, and underrepresented students? Is it possible to accept preservice science teachers who resist efforts to become critically conscious of their own beliefs and aid them to become culturally knowledgeable and skilled prior to their teaching internships?
2. What activities are successful in engaging preservice science teachers in thought-intensive activities to facilitate their making explicit their beliefs about learning and teaching of marginalized, underserved, and underrepresented students?
3. What are the connections among teacher beliefs about marginalized, underserved, and underrepresented students, their parents/guardians, and communities and their decisions and actions in science classrooms?
4. What prerequisites and requirements are necessary in the development of preservice science teachers into equitable and socially just teachers?
5. What is needed to help preservice science teachers gain beliefs that will assist them in changing their instructional practices and curriculum materials so that their students empower themselves and are motivated to take action to change their lives and take charge of their circumstances as students and adults? In other words, how can critical multicultural education and social justice be infused in science teacher education programs?

Teaching is a personal process of inquiry, and tensions arise as a result of teaching. These tensions can serve as a catalyst for developing knowledge and skills and can highlight the inconsistencies between teachers' beliefs and their practices. Learning to teach requires a shifting frame of reference. Reframing is crucial in the process of developing professional knowledge and skills and requires recurring perturbations to teachers' beliefs and recurring opportunities to practice. This reframing of beliefs and practices holds true for science teacher educators.

Teacher educators need to be concerned about the characteristics of students entering science teacher education programs. Their cultural, ethnic, gendered, linguistic, racial, and socioeconomic backgrounds are important to educating equitable and socially just science teachers. Many students enter these programs with little or no intercultural experiences and usually graduate with an encapsulated self in their sociocultural-historical backgrounds. It is necessary for science teacher education programs to provide students with experiences so that they:

1. Are knowledgeable about the lives of students from different social classes
2. Confront their own values, beliefs, stereotypes, and prejudices
3. Possess the knowledge and skills to instruct effectively as equitable and socially just teachers

Science teacher educators from underrepresented groups in the sciences bring more than physical role models; they bring diverse family histories, value orientations, and experiences to students in science classrooms. These are experiences one does not find in a science methods book but instead perspectives that these science teacher educators bring to teacher education. These science teacher educators are valuable for both prospective White students and students of color. Exposure to one teacher educator of color who has embraced multicultural education and social justice in his or her teaching is not enough for preservice science teachers.

A Call for Social Justice Education

Social justice education is not a new educational notion (Ackerman, 1980; Runciman, 1966). However, twenty-first-century social justice education does rest on the following three principles: equity, activism, and social literacy. It embraces relevance, rigor, and revolution (Ayers, Quinn, & Stovall, 2009). According to Nieto and Bode (2012), social justice is "a philosophy, an approach, and actions that embody treating all people with fairness, respect, dignity, and generosity" (p. 12). It (a) contests, defies, and disturbs misunderstandings, untruths, and beliefs that leads to structural inequality based on culture, gender, language, race, social class, and other social and human differences, (b) provides all students the resources necessary to reach their full potential, (c) draws on the abilities and assets that all students bring to the classroom, and (d) creates a learning environment that advances critical thinking and agency for social change. Social justice in science education is about teaching students to question and act to change their world using their science knowledge and skills.

James Baldwin (1979) describes Black children's language and its impact on learning and teaching in this manner:

> It is not some Black students' language that is in question; it is not their language that is abhorrent: It is their experiences. A child cannot be taught by anyone who despises him and a child cannot afford to be fooled. A child cannot be taught by anyone who demands, essentially, that the child repudiates his experiences, and all that gives him sustenance, and enter a limbo in which he will no longer be [B]lack, and in which he knows that he can never become [W]hite. Black people have lost too many children that way. (p. E 19)

How many children have been lost to our current educational system and practices? How many marginalized students have been disconnected from or denied access to a quality science education, students who have the potential to bring a fresh perspective to the field of science, create innovations in technology, or cure the world's most deadly diseases? As we close this chapter, whether you are a science teacher educator, K-12 teacher, or preservice teacher, we encourage you to examine your personal beliefs and how they might influence your actions to infuse culturally relevant teaching in science classrooms.

References

Ackerman, B. A. (1980). *Social justice in the liberal state*. New Haven, CT: Yale University Press.

Allen, W. R., Epps, E. G., Guillory, E. A., Suh, S. A., & Bonous-Hammarth, M. (2000). The Black academic: Faculty status among African Americans in U.S. higher education. *The Journal of Negro Education, 69*(1/2), 112–127.

Alters, S., & Alters, B. (2005). *Biology: Understanding life*. Hoboken, NJ: Wiley.

Atwater, M. M. (1996). Social constructivism: Infusion into multicultural science education research. *Journal of Research in Science Teaching, 33*, 821–838.

Atwater, M. M., Freeman, T. B., Butler, M. B., & Parsons, E. R. C. (2011, October). *Science teacher educators for multicultural education, equity, and social justice*. Paper presented at the Southeastern Association Science Teacher Education meeting in Athens, Georgia.

Atwater, M. M., Freeman, T. B., Butler, M. B., & Parsons, E. R. C. (2012, March). *Journeys of Black scholars in the academy: Helping to re-image research and teaching*. Paper presented at the international meeting of the National Association for Research in Science Teaching, Indianapolis, IN.

Atwater, M. M., & Suriel, R. L. (2010). Science curricular materials through the lens of social justice: Research findings. In T. K. Chapman & N. Hobbel (Eds.), *Social justice pedagogy across the curriculum: The practice of freedom* (pp. 273–282). New York: Routledge.

Aud, S., Fox, M. A., & KewalRamani, A. (2010). *Status and trends in the education of racial and ethnic groups* (NCES 2010-015). U.S. Department of Education. Retrieved from http://nces.ed.gov/pubs2010/2010015.pdf

Ayers, W., Quinn, T., & Stovall, D. (2009). Preface. In W. Ayers, T. Quinn, & D. Stovall (Eds.), *Handbook of social justice in education* (pp. xiii–xiv). New York: Routledge.

Baldwin, J. (1979, July 29). If Black English isn't a language, then tell me what is? *New York Times (1923-Current file)*. Retrieved from http://www.nytimes.com/books/98/03/29/specials/baldwin-english.html

Ball, A. F., & Tyson, C. A. (2011). Preparing teachers for diversity in the twenty-first century. In A. F. Ball & C. A. Tyson (Eds.), *Studying diversity in teacher education* (pp. 399–416). Lanham, MD: The Rowman & Littlefield Publishing Group.

Barton, A. C. (2002). Urban science education studies: A commitment to equity, social justice and a sense of place. *Studies in Science Education, 38*(1), 1–37.

Bernstein, R. (2008). *An older and more diverse nation by midcentury*. Retrieved from http://www.census.gov/newsroom/releases/archives/population/cb08-123.html

Blackwell, J. E. (1981). *Mainstreaming outsiders: The production of Black professionals*. Bayside, NY: General Hall.

Boutte, G., Kelly-Jackson, C., & Johnson, G. L. (2010). Culturally relevant teaching in science classrooms: Addressing academic achievement, cultural competence, and critical consciousness. *International Journal of Multicultural Education, 12*(2), 1–20.

Brand, B., Glasson, G., & Green, A. (2006). Sociocultural factors influencing students' learning in science and mathematics: An analysis of the perspectives of African American students. *School Science and Mathematics, 106*, 228–240.

Bryan, L. A., & Abell, S. K. (1999). The development of professional knowledge in learning to teach elementary science. *Journal of Research in Science Teaching, 36*, 121–140.

Bryan, L., & Atwater, M. M. (2002). Teacher beliefs and cultural models: A challenge for science teacher preparation programs. *Science Education, 86*(6), 821–839.

Cochrane-Smith, M., & Zeichner, K. M. (2005). Executive Summary. In M. Cochrane-Smith & K. M. Zeichner (Eds.), *Studying teacher education: The report of the AERA Panel on Research and Teacher Education* (pp. 1–36). Mahwah, NJ: Lawrence Erlbaum Associates.

Cornett, J. W., Yeotis, C., & Terwilliger, L. (1990). Teacher personal practical theories and their influence upon teacher curricular and instructional actions: A case of a secondary science teacher. *Science Education, 74*(5), 517–529.

Darling-Hammond, L. (2006). *Powerful teacher education: Lessons from exemplary programs*. San Francisco: Jossey-Bass.

Davis, J., & Martin, D. (2008). Racism, assessment, and instructional practices: Implications for mathematics teachers of African American students. *Journal of Urban Mathematics Education, 1*(1), 10–34.

Delpit, L. D. (1995). *Other people's children: Cultural conflict in the classroom*. New York: New Press.

Dewey, J. (1933). *How we think: A restatement of the relation of reflective thinking to the educative process*. New York: Heath.

Freire, P. (2005). *Teachers as cultural workers: Letters to those who dare teach*. Cambridge, MA: Westview Press.

Gay, G. (2002). Preparing for culturally responsive teaching. *Journal of Teacher Education, 53*(2), 106–116.

Gormley, K., McDemontt, P., Rothernerg, J., & Hammer, J. (1995, April). *Expert and novice teachers: Beliefs about culturally responsive pedagogy* (ERIC Document Reproduction Service No. ED 384 599). Paper presented at the American Educational Research Association Meeting, San Francisco, CA.

Gould, S. J. (1997). Drink deep, or taste not the Pierian spring. *Natural History, 106*(8), 24–25.

Grant, C. A., & Ladson-Billings, G. (1997). *Dictionary of multicultural education*. Phoenix, AZ: Oryx Press. ED 412 315.

Grossman, P. (2005). Research on pedagogical approaches in teacher education. In M. Cochran-Smith & K. M. Zeichner (Eds.), *Studying teacher education: The report of the AERA panel on research and teacher education* (pp. 425–476). Mahwah, NJ: Lawrence Erlbaum Associates.

Haberman, M. (1991). The pedagogy of poverty versus good teaching. *Phi Delta Kappan, 73*(4), 290–294.

Harrington, H. L., & Hathaway, R. S. (1995). Illuminating beliefs about diversity. *Journal of Teacher Education, 46*, 275–284.

Hollingsworth, S. (1989). Prior beliefs and cognitive change in learning to teach. *American Educational Research Journal, 26*(2), 160–189.

Hollins, E. R., & Guzman, M. T. (2005). Research on preparing teachers for diverse populations. In M. Cochran-Smith & K. M. Zeichner (Eds.), *Studying teacher education: The report of the AERA panel on research and teacher education* (pp. 477–548). Mahwah, NJ: Lawrence Erlbaum Associates.

Howard, T. C. (2003). Culturally relevant pedagogy: Ingredients for critical teacher reflection. *Theory into Practice, 42*(3), 195–202.

Huber, T. (1991, October). *Restructuring to reclaim youth at risk: Culturally responsive pedagogy* (ERIC Document Reproduction Service No. ED 341655). Paper presented at the 13th annual meeting of the Mid-Western Educational Research Association, Chicago, IL.

Ingersoll, R. (2000). *Turnover among mathematics and science teachers in the U.S. GSE Publication 1-1-2000*. National Commission on Mathematics and Science Teaching for the 21st Century. Retrieved from: http://repository.upenn.edu/cgi/viewcontent.cgi?article=1095&context=gse_pubs

Irvine, J. J. (2010). Culturally relevant pedagogy. *Education Digest, 75*, 57–61.

Jacob, E., & Jordan, C. (1993). Understanding minority education: Framing the issues. In E. Jacob & C. Jordan (Eds.), *Minority education: Anthropological perspectives* (pp. 3–14). Norwood, NJ: Able.

Johnson, N., Atwater, M. M., Freeman, T. B., Butler, M. B., & Parsons, E. R. C. (2012, March). *African American female faculty members: Factors influencing their recruitment, retention, and promotion at traditionally White institutions*. Paper presented at the international meeting of the National Association for Research in Science Teaching, Indianapolis, IN.

Jones, M. G., & Carter, G. (2007). Science teacher attitudes and beliefs. In S. K. Abell & N. G. Lederman (Eds.), *Handbook of research on science education* (pp. 1067–1104). Mahwah, N.J.: Lawrence Erlbaum Associates.

Ladson-Billings, G. (1995). Toward a theory of culturally relevant pedagogy. *American Educational Research Journal, 32*(3), 465–491.

Ladson-Billings, G. (2000). Fighting for our lives: Preparing teachers to teach African American students. *Journal of Teacher Education, 51*(3), 206–214.

Ladson-Billings, G. (2009). *The dreamkeepers: Successful teachers of African American children.* San Francisco, CA: Jossey-Bass.

Lee, T. S. (2011). Teaching native youth, teaching about native peoples: Shifting the paradigm to socioculturally responsive education. In A. F. Ball & C. A. Tyson (Eds.), *Studying diversity in teacher education* (pp. 275–294). Lanham, MD: The Rowman & Littlefield Publishing Group.

Limon, M. (2001). On the cognitive conflict as an instructional strategy for conceptual change: A critical appraisal. *Learning and Instruction, 11*, 357–380.

Lipman, P. (1993). *The influence of restructuring on teachers' beliefs about and practices with African American students.* Unpublished doctoral dissertation, University of Wisconsin, Madison.

Mackeracher, D. (2004). *Making sense of adult learning.* Buffalo: University of Toronto Press.

McDonald, M., & Zeichner, K. M. (2009). Social justice teacher education. In W. Ayers, T. Quinn, & D. Stovall (Eds.), *Handbook of social justice in education* (pp. 595–610). New York: Routledge.

Milan, J. F., Chang, M. J., & Antonio, A. L. (2005). *Making diversity work on campus: A research based perspective.* Washington, DC: Association for American College and Universities. Retrieved from http://www.wesleyan.edu/partnerships/mei/files/makingdiiversityworkoncampus.pdf

Moreno, J. F., Smith, D. F., Clayton-Pedersen, A. R., Parker, S., & Teraguchi, D. H. (2006). *The revolving door for underrepresented minority faculty in higher education: An analysis from the campus diversity.* Washington, DC: Association for American Colleges and Universities. Retrieved from http://www.aacu.org/irvinediveval/documents/RevolvingDoorCDIInsight.pdf

Morrison, K. A., Robbins, H. H., & Rose, D. G. (2008). Operationalizing culturally relevant pedagogy: A synthesis of classroom-based research. *Equity and Excellence in Education, 41*(4), 433–452.

Nasir, N. S., & Hand, V. M. (2006). Exploring sociocultural perspectives on race, culture, and learning. *Review of Educational Research, 76*(4), 449–475.

National Research Council. (2012). *A framework for K-12 science education: Practices, crosscutting concepts, and core ideas.* Washington, DC: National Academies Press.

Nieto, S. (2000). *Affirming diversity: The sociopolitical context of multicultural education.* New York: Longman.

Nieto, S., & Bode, P. (2012). *Affirming diversity: The sociopolitical context of multicultural education.* Boston: Pearson.

Nieto, S., & McDonough, K. (2011). "Placing equity front and center" revisited. In A. F. Ball & C. A. Tyson (Eds.), *Studying diversity in teacher education* (pp. 363–384). Lanham, MD: The Rowman & Littlefield Publishing Group.

Oliver, J. S., Jackson, D. F., Chun, S., Kemp, A., Tippins, D. J., Leonard, R., et al. (2001). The concept of scientific literacy: A view of the current debate as an outgrowth of the past two centuries. *Electronic Journal of Literacy Through Science, 1*(1), 1–33.

Olmedo, I. M. (1997). Challenging old assumptions: Preparing teachers for inner city schools. *Journal of Teaching and Teacher Education, 13*, 245–258.

Osborne, A. B. (1996). Practice into theory into practice: Culturally relevant pedagogy for students we have marginalized and normalized. *Anthropology and Education Quarterly, 27*(3), 285–314.

Pewewardy, C. (1993). Culturally responsive pedagogy in action: An American Indian magnet school. In E. Hollins, J. King, & W. Hayman (Eds.), *Teaching diverse populations: Formulating a knowledge base* (pp. 77–92). Albany, NY: State University of New York Press.

Phuntsog, N. (1999). The magic of culturally responsive pedagogy: In search of the Genie lamp in multicultural education. *Teacher Quarterly, 26*(3), 97–111.

Physics. (n.d.). In *Merriam Webster Dictionary online*. Retrieved from: http://www.merriam-webster.com/dictionary/physics.

Remembering Tuskegee syphilis study still provokes disbelief, sadness. (2002). Retrieved on May 23, 2012 from http://www.npr.org/programs/morning/features/2002/jul/tuskegee

Ross, E. W., Cornett, E. W., & McCutcheon, G. (Eds.). (1992). *Teacher personal theorizing: Connecting curriculum practice, theory, and research*. Albany, NY: State University of New York Press.

Ruby, A. M. (2002). Internal teacher turnover in urban middle school reform. *Journal of Education for Students Placed at Risk, 7*(4), 379–406.

Runciman, W. G. (1966). *Relative deprivation and social justice: A study of attitudes to social inequity in the twentieth-century*. London: Routledge.

Smith, B. O. (1985). Research bases for teacher education. *Phi Delta Kappan, 66*(10), 685–690.

Special Obituary tribute: Tuskegee syphilis research study survivors. (2004). Retrieved on May 23, 2012 from http://www.dollsgen.com/special_obituary_tribute.htm

Stepanak, J. (2000). *Mathematics and Science Classrooms: Building a community of learners*. Portland, OR: Northwest Regional Education Laboratory. Retrieved from http://education-northwest.org/webfm_send/749

Thomson, N., Wilder, M., & Atwater, M. M. (2001). Critical multiculturalism and secondary teacher education programs. In D. Lavoie (Ed.), *Models for science teacher preparation: Bridging the gap between research and practice* (pp. 195–211). New York: Kluwer.

Tobin, K., & McRobbie, C. J. (1996). Cultural myths as constraints to the enacted science curriculum. *Science Education, 80*, 223–241.

U.S. Census Bureau Projections show a slower growing, older, more diverse nation a half century from now (2012). Retrieved from http://www.census.gov/newsroom/releases/archives/population/cb12-243.html

Webster's encyclopedic unabridged dictionary of the English language. (1989). New York: Gramercy Books.

Whitten, K. W., Gailey, K. D., & Davis, R. E. (1988). *General chemistry with qualitative analysis*. Philadelphia: Saunders College Publishing.

Wlodkowski, R. J., & Ginsberg, M. B. (1995). A framework for culturally responsive teaching. *Educational Leadership, 53*(1), 17–21.

Wortham, S., & Contreras, M. (2002). Struggling toward culturally relevant pedagogy in the Latino/a Diaspora. *Journal of Latinos/as and Education, 1*(2), 133–144.

Zeichner, K. M. (2005). A research agenda for teacher education. In M. Cochrane-Smith & K. M. Zeichner (Eds.), *Studying teacher education: The report of the AERA Panel on Research and Teacher Education* (pp. 737–760). Mahwah, NJ: Lawrence Erlbaum Associates.

Motivation in the Science Classroom: Through a Lens of Equity and Social Justice

Melody L. Russell

Motivation and Science Learning

The degree of student motivation in a particular content area is often driven by the instructional strategies that teachers implement in the classroom. Oftentimes science teachers discuss lack of motivation in the science classroom as a concern particularly relative to participation from traditionally underrepresented and marginalized groups. Students that are not motivated are disengaged and often disenfranchised with their science learning experiences which often results in their underrepresentation in the STEM (science, technology, engineering, and mathematics) pipeline on the college level and beyond. Historically, there has been a "leak in the science" pipeline particularly relative to the participation of females and other traditionally underrepresented groups (e.g., African Americans, Latinos/as, and American Indians).

Moreover, because many science topics are often taught from a traditionally Westernized perspective and teachers rarely have the skills from teacher preparation programs to teach science from a culturally relevant perspective, students from traditionally underrepresented groups are discouraged from persisting in the science pipeline (or viewing themselves as scientists). This underrepresentation results in inequities and perpetuates the culture of hegemony and status quo that exist in the STEM fields. Furthermore, the lack of role models for students of color and women in higher education STEM areas is an issue of equity and social justice. When science is taught in a manner that does not emphasize how it connects to the student's daily life, this further marginalizes students making it easy to lose interest and

---

M.L. Russell, Ph.D. (✉)
Department of Curriculum and Teaching, College of Education,
Auburn University, 5004 Haley Center, Auburn University, AL 36849, USA
e-mail: russeml@auburn.edu

develop negative attitudes towards science. One key factor for encouraging students to persist in the sciences is to design science lessons that are culturally relevant and challenge inequities in the way science is presented, while motivating students interest towards participation in the STEM pipeline.

This chapter based on research in motivational theory and achievement motivation aims to address how teachers can promote student interest in science. Moreover this chapter provides insights into how teachers can motivate their students to achieve in the science classroom through a lens of equity and social justice. Specifically, since many teachers express dissatisfaction with their students' lack of motivation, this chapter attempts to provide insights and strategies to help them focus on transforming their classroom environments to better motivate students and engage more students in the science pipeline.

What Is Motivation?

Although there are a number of motivational theories and definitions of motivation in the literature, the consensus is that motivation is an internal condition or state that serves to drive or direct an individual towards completing a task or goal (Cavas, 2011). According to Palmer (2005), motivation can in essence be applied to any process that triggers learning and maintains the intended learning behavior. Motivation in educational research is a broad and complex topic relative to teaching and learning, and to those outside the field of motivational research, this topic can seem fragmented (Murphy & Alexander, 2000) and overwhelming. However, motivation is one of the best predictors of an individual's persistence for the long term in a particular area of interest (e.g., educational interest) (Harackiewicz, Barron, & Elliot, 1998; Tauer & Harackiewicz, 2004). According to Ormrod (2008), as cited in Lei (2010), motivation can be characterized as an internal state that enhances or arouses the learner, guiding her/him in a particular direction and keeping them engaged, towards completion of a task.

Over the past few decades, research on motivation has flourished and more is known about what motivates students (Guvercin, Tekkaya & Sungur, 2010; Wigfield, 1997) on both the precollege level and the college level. Moreover, the extent to which and whether or not a student engages in a challenging task is often determined by her/his degree of motivation (Lei, 2010). Theories that drive research on motivation are typically centered on motivation being defined as the "energization" and "direction of behavior" (Pintrich, 2003). Essentially theories in motivation try to explain what actually drives an individual towards a specific activity or task (Pintrich, 2003). In order for knowledge construction to occur, the learner must first be motivated to put forth effort in completing a task (Palmer, 2005). The Expectancy-Value theory of motivation is used as a theoretical framework for much of the research in motivation (Wigfield & Eccles, 2000; Weinberg, Basile, & Albright, 2011). The essence of Expectancy-Value theory outlines student expectations for success and the value they place on completing a set or assigned task (Wigfield &

Eccles, 2000; Weinberg et al., 2011). Moreover, a student is more likely to engage and show interest in a particular topic or activity if they perceive value in completing the task (Wigfield & Eccles, 2002 as cited in Weinberg et al., 2011).

Motivation to learn, motivation to learn according to Brophy (1998) is the tendency to see an academic task or activity as meaningful towards a specific academic goal or benefit. In science, the degree of student motivation is often determined by the level of engagement the student has in the science-related activity in an effort to better understand the content (Lee & Brophy, 1996).

Motivation is an important factor in the science classroom because it essentially promotes the construction of knowledge and conceptual understanding of science concepts (Cavas, 2011). Moreover, being motivated to learn science is beneficial to students in the early school years as it inspires them to become future scientists (Bryan, Glynn, & Kittleson, 2011). It is also important for students to be motivated in science as it promotes scientific literacy for all students (Bryan et al., 2011). Consequently, if science teacher educators are to prepare preservice teachers for an increasingly diverse classroom, it is paramount that strategies for motivating students are clearly delineated to help promote scientific literacy for all students while increasing the scientific pool of applicants in a global society.

There are a number of major reforms in science education geared towards increasing the motivation of students intrinsically and extrinsically (NRC, 1996, 2012). Whether or not students elect to learn a challenging task or engage in science can often be determined by motivation (Ormrod, 2008 as cited in Lei, 2010). Students may be interested in science because of external factors (i.e., parents, teacher praise, grades, rewards) or internal factors (i.e., desire to attend college, or self-efficacy). Motivation is typically referred to in two major categories which include either intrinsic or extrinsic (Lei, 2010). There has often been much debate as to which is the preferred strategy for motivating students.

Intrinsic motivation is defined as the internal satisfaction a student feels about completing a particular task (Lei, 2010) and students typically complete an activity for enjoyment (Brewer & Burgess, 2005). Extrinsic motivation is characterized by the use of external rewards or incentives (often in the form of grades) to stimulate a student to complete a task (Brewer & Burgess, 2005). There is also a third category called "motivation to learn" which addresses the overall benefit or degree of meaningfulness of the academic task (Marshall, 1987) There are numerous pros and cons for intrinsic and extrinsic motivation. However, according to research on intrinsic motivation, it is important that students that are intrinsically motivated are encouraged not to lose track and become "too consumed" in completing a task (Lei, 2010).

Another one of the drawbacks of intrinsic motivation is when students experience tunnel vision and do not complete other essential tasks (Ormrod, 2008 as cited in Lei, 2010) which can impact their achievement and persistence in the other tasks. However, intrinsically motivated students learn better and tend to be more creative than students that are extrinsically motivated (Niemiec & Ryan, 2009). On the other hand, the extrinsically motivated students typically complete a task based on grades, or other tangible rewards that can represent they succeeded in learning a task (Lei, 2010). The benefit of this type of motivation is that students will take initiative to complete the task, however, the challenge for teachers is that they must ensure that the incentive is something that is valuable to the student otherwise they will lose interest in completing the task (Covington, 2000 as cited in Lei, 2010). Overall, research by Areepattamannil, Freeman, and Klinger (2011) demonstrates that learning for the enjoyment of science (intrinsic motivation) plays a very important role in students learning in the science classroom. Subsequently, it is important to promote students' intrinsic motivation to enhance their scientific literacy and thinking processes in science (Areepattamannil et al., 2011).

Motivation and Participation in Science

There are a number of factors that impact the motivation and participation of students from underrepresented groups in the sciences. Some of these variables include ability level, attitude, self-perception, socioeconomics, peer and parental influence, school factors, and home factors (Singh, Granville, & Dika, 2002). Research by Markowitz (2004) has even demonstrated the positive impact of outreach programs in biomedical research on precollege students interest in participating in science as well as enhancing the desire to pursue careers in science. Oftentimes students from marginalized and underrepresented groups lack motivation because of the curriculum and low expectations from teachers encountered in the science classroom. Recent studies demonstrate how interventions that integrate more academic rigor (Ruby, 2006) and inquiry-based instruction (Pickens & Eick, 2009) can motivate students in science. Oakes (1985, 1990a, b) describes the following factors as specifically impacting the participation of females and people of color from underrepresented groups in science: (a) access to resources, (b) cultural barriers, (c) socioeconomic status, (d) interest, and (e) lack of encouragement.

Decades after research by Butler-Kahle (1992) research by Norman, Ault, Bentz, and Meskimen (2001) examined the historical and sociocultural implications that impact science participation relative to the Black/White achievement gap. For the most part, a large number of students from traditionally underrepresented groups typically attend schools in urban settings where there is often a lack of resources in comparison to Whites that attend schools in more suburban areas (Norman et al., 2001). Norman et al. posit that throughout history there were various racial/ethnic groups (e.g., Polish, Jewish, Italian) that were immigrants (though voluntary) and relegated to the bottom of the social caste system in the United States. However, as time passed, these immigrant groups were able to assimilate into the mainstream culture in the United States. Subsequently, they moved out of the more impoverished urban areas and achievement differences between the new immigrants and European Americans diminished to the point where the differences in achievement were almost nonexistent (Norman et al.). This demonstrates that there are also sociocultural in addition to ethnic/racial factors that impact student achievement. Consequently, since today many African Americans and Latino/as are situated in more urban and high-poverty areas, they still maintain a relatively low status in society. This is not to say that students from high-poverty areas cannot and do not achieve at high levels, but that living in such areas has an impact on access to resources, quality of education, exposure to a challenging curriculum, and additional educational opportunities that many students in suburban areas are afforded. As a result, the stigma of inferiority, and a type of bigotry that translates into low expectations, limits access to resources that promote interest, positive attitudes, and achievement in science.

Relative to females, and students from traditionally underrepresented and underserved populations sociocultural factors (Butler-Kahle, 1982; Butler-Kahle & Lakes, 2003) also impact their participation in science, as they are often subjected to stereotypes that subsequently have the same lasting impact on self-concept in science. Although there are a number of research studies on motivation and achievement in science, there is still little research on methods for motivating and encouraging students particularly from traditionally underrepresented groups to achieve and persist in science (Brown, 2000; Pickens & Eick, 2009).

One reason that motivation is of interest to science teachers educators is because attitudes and motivation towards school as a whole are predictors of high school adaptation and performance (Murdock, Anderman, & Hodge, 2000). Research by Hill, Atwater, and Wiggins (1995) on seventh graders in life science in urban classrooms suggested that students who possess positive attitudes towards science were more likely to take more science classes. In addition, when these students were asked about their career goals, the students with positive or undecided attitudes in science were more likely to say they would choose science careers (Hill et al., 1995). Since it is clear that motivation is critical to participation, attitude, and interest in science, it is also important to examine the role that teachers and pedagogical strategies play on motivating students in the science classroom. Simply put, science teachers play a significant role in their students' achievement and how they teach can have a profound impact on student motivation in the science classroom. Essentially, because teachers often teach the way they were taught (typically in a traditional and didactic manner), they fail to integrate more culturally relevant pedagogy into their lessons especially for those teaching in high-poverty school districts where there are larger populations of traditionally underrepresented and marginalized students. Interest and affinity towards science can have a significant impact on student motivation and without interest in science (especially during the early grades) students from underrepresented groups will not have opportunities to persist in science to pursue STEM careers.

Teachers can have profound impact on a student's motivation to learn (Blumenfeld & Meece, 1987). This being said, it is essential to encourage and promote positive attitudes towards science for students from underrepresented groups, especially if we are to increase and diversify the scientific pool and enhance the participation of underrepresented groups (e.g., African American, Latinos/as, and females) in STEM fields. Additional research in science education has described not only attitudes as impacting participation or interest in science but motivation, achievement styles, and other affective variables (Atwater & Simpson, 1984).

Motivation and Equity in the Science Classroom

Motivation plays a critical role in achievement and significantly influences learning, as well (Ames, 1992). Though reform efforts in science education address equity in STEM education, the reality is that there is little equity relative to access and opportunities for students from traditionally underrepresented groups in STEM areas. The National Assessment of Education Progress delineated average science scores of students for age levels 9, 13, and 17 years old have increased minimally over the past three decades (Campbell, Hombo, & Mazzeo, 2000).

It is well known that teacher effectiveness is directly correlated with increased science achievement for students. However, unfortunately many of the lower-achieving students (particularly students from high-poverty, rural, and urban areas) typically encounter the least effective teachers (Lynch, Kuipers, Pyke, & Szesze, 2005). Oftentimes the most ill-prepared teachers, who may lack certification are more likely to have teaching assignments outside of their content area teaching outside of their content area, and are assigned to teach in the most challenging school districts and lowest-performing school systems. This situation only contributes to an unfortunate term often referred to by researchers a "pedagogy of poverty," and even teachers that have learned about the benefits of hands-on inquiry-based learning in their teacher preparation programs revert back to the more traditional teaching strategies. Consequently, many teachers in the low-performing schools kick into "survival mode," and expectations are lowered for students as many teachers who are not prepared for the challenges of a beginning teacher revert back to what is familiar (i.e., worksheets, lectures). Attitudes are also extremely important during the middle school years as they tend to become more negative, and self-concept and perceptions of competence tend to decline around this time (Anderman & Maehr, 1994). Moreover, research by Weinburgh (1995) determined that positive attitudes in science can lead to high achievement. Research also indicated that this was especially so for the low performing girls in science (Weinburgh, 1995).

Science career choice and goals for attainment are often attributed to early choices students make since there are specific "gate-keeping" science and mathematics courses that students must take in the junior high school and early high school years to continue in a career trajectory in science (Lavigne, Vallerand, & Miquelon, 2007). Research by Bryan et al. (2011) recommend that to promote motivation for students to learn science, researchers must examine ways to increase students' participation in AP courses in high school, as well as investigate and assess students' motivation to learn science in high school science courses. There also needs to be more emphasis on role models to interest students in science careers (Bryan et al.). In addition, for females, science participation is often attributed to achievement and subsequent enrollment in math courses, as well (Butler-Kahle & Lakes, 2003). If we go back even further than the impact of middle and high school science experiences on science career choices, we can closely examine the elementary years. Maximizing the number of quality science experiences during the early grades can have a positive impact on attitude, interest, and motivation in the sciences.

Motivation has a direct impact on academic achievement and promotes interest and engagement in completing academic tasks to further learning (Singh, Granville, & Dika, 2002). As a result, it is critical that factors that motivate students from underrepresented groups to participate in science be clearly delineated so that teachers and teacher educators can address these factors in their science and science education classrooms to promote equity. Subsequently, the problem of the pervasive achievement gap in science has been the center of research that addresses equity and diversity in over the past few decades (Atwater, 2000; National Science Foundation, 1994, 1998, 2000, 2012). Unfortunately, the state of low achievement in science among students from underrepresented groups who are often primarily African American or Latino/a pose significant barriers to their persistence in science. In an increasingly technological society, it is important to address the long-term, adverse impact that limiting access to STEM careers for any student will have long-term, adverse impacts on the national and global economy. Studies on the achievement gap in science has specifically highlighted the gap relative to Black students and White students (Braun, Chapman, & Vezzu, 2010; Norman et al., 2001; Simms, 2012). Additional studies investigate the science achievement gap relative to socioeconomic status in urban or rural areas (Lee & Madyun, 2009; Pickens & Eick, 2009; Ruby, 2006) or track placement (Oakes, 1985, 1990a, b; Pickens & Eick, 2009).

Self-Efficacy, Self-Concept, and Attitude: Motivation in Science

A vast amount of research in counseling psychology has examined the role of self-efficacy and self-concept in predicting student career goals and aspirations (Gainor & Lent, 1998). Social cognitive theory asserts that an individual's career aspirations are attributed to their self-efficacy and ability (Nauta, Epperson, & Kahn, 1998). Bandura (1977) describes self-efficacy as an individual's personal judgment regarding one's ability to perform a specific behavior or task or their self-perceived confidence to be successful in a science-related task, activity, or course (Britner & Pajares, 2001, 2006). Self-concept is defined as an individuals' perception of their academic ability (Bong & Skaalvik, 2003 as cited in Areepattamannil et al., 2011) or how they view themselves (i.e., I see myself as a good student in science). Furthermore, confidence in the content affects student motivation and achievement (Mamlok-Naaman, 2011). Moreover, according to Nelson & Debacker (2008) positive peer relationships and the extent to which students feel valued and accepted also positively impacts achievement motivation. This kind of information is critical relative to science teaching and learning (Arzi, Ben-Zvi, & Ganie as cited in Mamlok-Naaman, 2011). Furthermore, when students are both interested in a science concept and understand it, they tend to have better attitudes towards science as opposed to students that have difficulty with the concepts (Mamlok-Naaman, 2011). Several researchers assert that student motivation in science can increase or decrease how a student learns, or if they want to learn a concept (Barila and Beet, 1999; Fairbrother, 2000; Pintrich, Marx, & Boyle, 1993 as cited in Mamlok-Naaman, 2011).

There is a significant amount of research that focuses on the obvious relationship between achievement and academic motivation both in the United States and abroad (Ames, 1992; Bryan et al., 2011; Gottfried, 1985; Nolen, 2003; Oliver & Simpson, 1988). Studies have also found and discovered a correlation between achievement in science and student attitudes (Butler-Kahle, 1982; Sorge, 2007 as cited in Milner, Templin, & Czerniak, 2010). Over the past years, there has been a steady decline in student academic motivation and motivation can be attributed to both school-related and home-related factors (Gottfried, Marcoulides, Gottfried, & Oliver, 2009). Student attitude and interest in science play a significant role in motivation, and interest in science often results in increased attention during formal instruction as well as participation in science-related activities or courses (Farenga & Joyce, 1999; Farmer, Waldrop, & Rotella, 1999; Germann, 1988 as cited in Farmer, Waldrop, & Rotella, 1999; Marcowitz, 2004 as cited in Farmer, Waldrop, & Rotella, 1999).

This being said, it is imperative that science is taught so that students from traditionally underrepresented groups see that science is a topic in which they can be successful. Teachers must make it clear that they hold high expectations for all of their students and encourage them to engage in complex learning of abstract science topics that challenge students. Furthermore motivation has a direct impact on student achievement, engagement, and the process of conceptual change (Wentzel & Wigfield, 2007; Wigfield & Wentzel, 2007). Motivational factors play a significant role in future career goals and plans of individuals relative to self-efficacy and self-concept (Singh, Granville, & Dika, 2002) and these factors especially play a critical role in motivating students from traditionally underrepresented groups in STEM areas (Wentzel & Wigfield, 2007; Wigfield & Wentzel, 2007).

Strategies that Enhance Motivation in the Science Classroom

Factors that have been examined relative to motivation look at the impact of attitude, achievement, teaching strategies, and professional development. In order to continue investigating strategies for making the science pipeline more inclusive, researchers need to continue to focus on how high achievement and interest in science and mathematics are known predictors or indicators of students' persistence in the science and mathematics pipeline (Powell, 1990; Thomas, 1986), as well as gateways to careers and degrees in science.

Research by Brewer and Burgess (2005) on the college classroom showed the following results that could be transferable to the precollege setting which include personal qualities and good classroom management. On the secondary level, research demonstrates inquiry-based and interactive instruction (Pickens & Eick, 2009) as a primary motivator in the science classroom. Much of the research on motivational strategies is embedded in what we already know about teaching and learning. More specifically, Williams and Williams (2012) posit that there are five keys that teachers can implement to improve student motivation. These include the teacher, student, pedagogy/methods, environment, and content. Listed below is a synopsis of five primary motivation factors according to Williams and Williams (2012). They refer to these factors as the five keys ingredients that have an impact on motivating students in the classroom (Williams & Williams, 2012).

1. Teacher should have a good mastery of content, qualifications in content area/pedagogical content knowledge, and motivational level (Williams & Williams, 2012)
2. Content should be accurate and relevant (Williams & Williams, 2012)
3. Pedagogical approach/methods should be both experiential and engaging (Williams & Williams, 2012)
4. Environment should be one of quality and conducive for motivating and learning, available, and accessible (Williams & Williams, 2012)
5. Students should not be in a traditional mode of receiver of knowledge but they should come to class motivated whether intrinsically or extrinsically (Williams & Williams, 2012)

Certainly, this list is not mutually exclusive and there are a number of other factors; however, these are considered critical in fostering an environment that promotes learning and achievement towards motivating students. In addition, other aspects of pedagogy or methods relative to motivating were the teachers' enthusiasm, addressing learning styles, and setting goals and objectives.

According to Pickens and Eick (2009), students' benefited from a class that fostered a positive learning environment and high teacher expectations a result of the use of hands-on instruction and interactive teaching strategies. Milner et al. (2010) describe the structure of the learning environment as a key motivating factor for students and use a constructivist classroom context. Essentially their research addressed the impact of incorporating a life science laboratory into the classroom to increase motivation through constructivist teaching and learning practices. As a result of the implementation of the life science laboratories, students were able to investigate science in a more "authentic" environment (Milner et al.). Students interviewed revealed that they experienced science in more relevant ways, which allowed them to apply what they learned to the traditional classroom and the life science laboratories enhanced the students constructivist learning and engaging them in science via an inquiry-based continuum (Milner et al.).

Additional research by Nolen (2003) also reported that the learning environment was a significant predictor of satisfaction and achievement in science particularly when the teacher promoted independent thinking and student understanding. The use of technology and media in the science classroom has also been shown to be effective in motivating students (Liu, Horton, Olmanson, & Toprac, 2011). Researchers in this study implemented a media approach through problem-based learning (PBL) for the middle school science classroom and results indicated that students' knowledge on the content covered and motivation increased (Liu et al., 2011). Moreover, students expressed that they enjoyed the activities and results indicated positive relationships between motivation scores and science content knowledge (Liu et al. 2011).

One key factor in motivating students in science looks at how teachers design their curriculum and structure their lessons to be more relevant to students' daily lives (Bryan et al., 2011). In addition, students emphasized that they are motivated by good grades, teacher competence in content area, teacher enthusiasm, teachers caring ethic and hands-on activities (Bryan et al.). Moreover, students prefer less PowerPoint-oriented lectures and more inquiry, autonomy, field trips, labs, collaboration in class projects, and social interactions in class (Bryan et al.). These are important factors that have been highlighted and provide a platform for teachers and teacher educators to work from in efforts to transform their own classrooms and promote achievement for their students. More specifically, implementation of these strategies is a beginning towards involving more students from underrepresented groups in the STEM areas on both the secondary and post secondary level.

It is paramount for science teacher educators to investigate key motivating factors that encourage and enhance the participation of traditionally underrepresented groups in STEM to promote more equitable representation in the STEM pipeline. Though some believe that there is already equal access to science participation, the reality is that this is not true for many students from culturally diverse backgrounds, females, and individuals with disabilities. It is essential that science teachers and science teacher educators examine their pedagogical strategies and provide enriched science learning experiences that enhance interest, attitudes, and motivation in science. These types of inputs will better level the playing field and promote equity in outcomes relative to who persists in science throughout the middle and high school years and who pursues STEM degrees and STEM careers.

Policy Changes and Motivation in the Science Classroom

The primary impetus for this chapter was to highlight the critical role that teachers can play in motivating their students to learn science. High expectations, inquiry-based lessons, and teacher competence in content area are just a few factors central to motivating students in the sciences, particularly students from traditionally underrepresented and marginalized groups. We live in an increasingly diverse and technologically advancing society and it is paramount that the students in the United States compete relative to technological innovations in this global market. There is a significant amount of "untapped talent and unlimited potential" (Russell, 2005) and teachers need to raise their expectations for students from traditionally underrepresented groups, females, and students with disabilities in STEM so more students are given access to opportunities that promote achievement and interest in science. Educators need to focus on equity relative to the outcomes for students from traditionally marginalized and underrepresented groups in the STEM areas relative to jobs and careers in the STEM areas. Increasing participation of traditionally underrepresented groups also has implications from an economic standpoint since these students will go on to degree programs and careers in the STEM areas which can positively impact their financial stability and economic mobility.

Lastly, I have included several recommendations that I have as a science teacher educator for facilitating teachers in motivating students in science and promoting equity and social justice through their teaching: (a) professional development for teachers to prepare them to teach more culturally relevant curricula, (b) more systemic mentoring programs for beginning teachers in high-poverty school districts in rural and urban areas, (c) collaborative grants between Colleges of Education and school districts that provide opportunities for science teachers and science teacher educators to develop programs to facilitate new teachers with the transition from college into the first years of teaching, and (d) required core courses in equity in teaching. Since new teachers are often overwhelmed with teaching schedules, and more likely to teach out of their content area, there is a need for systemic reform in how teachers are prepared to teach in culturally diverse settings relative to motivating students. Unfortunately, teachers in urban and rural areas (where you would typically find more students from traditionally underrepresented groups) are less likely to implement and design lessons that are relevant to the students' daily lives and provide an enriched science curriculum.

Many schools in high-poverty, rural, or urban areas are less likely to offer advanced science and mathematics courses (essential gatekeepers to careers in the STEM areas); students at schools in these areas are less likely to encounter a rigorous curriculum and enough content background to persist at higher levels in the STEM pipeline. Moreover, a challenging curriculum with support and high expectations from the teachers is also an important factor for motivating students. These aforementioned recommendations for changes are one step in the right direction relative to increasing participation and motivation in students from traditionally underrepresented groups. Until science teacher educators and teachers are better prepared to promote equity and social justice in their science classrooms through instructional strategies that foster motivation in science, many students from groups traditionally marginalized and underrepresented in the STEM areas will never realize their full potential in order to participate in the science pipeline and we will never plug the leak in the STEM pipeline and promote more opportunities and pathways to STEM degrees and careers.

References

Ames, C. (1992). Classrooms: Goals, structures, and student motivation. *Journal of Educational Psychology, 84*(3), 261.

Anderman, E. M., & Maehr, M. (1994). Motivation and schooling in the middle grades. *Review of Educational Research, 64*, 287–309.

Areepattamannil, S., Freeman, J., & Klinger, D. (2011). Influence of motivation, self-beliefs, and instructional practices on science achievement of adolescents in Canada. *Social Psychology Education, 14*, 233–259.

Arzi, H. J., Ben-Zvi, R., & Ganiel, U. (1986). Forgetting versus savings: The many facets of long-term retention. *Science Education, 70*(2), 171–188.

Atwater, M., & Simpson, R. (1984). Cognitive and affective variables affecting Black freshman in science and engineering at a predominantly White university. *School Science and Mathematics, 84*(2), 100–112.

Atwater, M. (2000). Equity for Black Americans in precollege science. *Science Education, 84*, 154–179.

Bandura, A. (1977). Self-efficacy: Toward a unifying theory of behavioral change. *Psychological Review, 84*, 191–215.

Blumenfeld, P., & Meece, J. (1987). Task factors, teacher behavior, and students involvement and use of learning strategies in science. *The Elementary School Journal, 88*, 235–250.

Braun, H., Chapman, L., & Vezzu, S. (2010). The black-white achievement gap revisited. *Education Policy Analysis Archives, 18*(21), 1–98.

Brewer, E., & Burgess, D. (2005). Professor's role in motivating students to attend class. *Journal of Industrial Teacher Education, 42*(3), 23–47.

Britner, S., & Pajares, E. (2001). Self-efficacy beliefs, motivation, race, and gender in middle school. *Journal of Women and Minorities in Science and Engineering, 7*, 271–285.

Britner, S., & Pajares, E. (2006). Sources of science self-efficacy beliefs of middles school students. *Journal of Research in Science Teaching, 43*, 485–489.

Brophy, J. (1998). *Motivating students to learn.* Madison, WI: McGraw Hill.

Brown, M. (2000). *Factors that impact African American college students' educational pursuits and career aspirations in science.* Unpublished dissertation, The University of Georgia.

Bryan, R., Glynn, S., & Kittleson, J. (2011). Motivation, achievement, and advanced placement intent of high school students learning science. *Science Education, 95*, 1049–1065.

Butler-Kahle, J. (1982). Can positive minority attitudes lead to achievement gains in science? Analysis of the 1977 National Assessment of Educational Progress toward science. *Science Education, 66*(4), 539–546.

Butler-Kahle, J., & Lakes, M. (2003). The myth of equality in science classrooms. *Journal of Research in Science Teaching, 40*(supplement), 58–67.

Campbell, J., Hombo, C., & Mazzeo, J. (2000). NAEP 1999 Trends in academic progress: Three decades of student performance. *Education Statistics Quarterly, 2*(4), 31–36.

Cavas, P. (2011). Factors affecting the motivation of Turkish primary students for science learning. *Science Education International, 22*(1), 31–42.

Cummins, J. (2001). Empowering minority students: A framework for intervention. *Harvard Educational Review, 71*(4), 656–675.

Eccles, J. S., Wigfield, A., Midgley, C., Reuman, D., Iver, D. M., & Feldlaufer, H. (1993). Negative effects of traditional middle schools on students' motivation. *The Elementary School Journal, 93*(5), 553–574.

Farenga, S. J., & Joyce, B. A. (1999). Intentions of young students to enroll in science courses in the future: An examination of gender differences. *Science Education, 83*(1), 55–75.

Farmer, H. S., Wardrop, J. L., & Rotella, S. C. (1999). Antecedent factors differentiating women and men in science/nonscience careers. *Psychology of Women Quarterly, 23*(4), 763–780.

Gainor, K., & Lent, R. (1998). Social cognitive expectations and racial identity attitudes in predicting the math choice intentions of Black college students. *Journal of Counseling Psychology, 45*, 403–413.

Gottfried, A. E. (1985). Academic intrinsic motivation in elementary and junior high school students. *Journal of Educational Psychology, 77*(6), 631.

Gottfried, A. E. (1990). Academic intrinsic motivation in young elementary school children. *Journal of Educational Psychology, 82*(3), 525.

Gottfried, A. E., Marcoulides, G. A., Gottfried, A. W., & Oliver, P. H. (2009). A latent curve model of parental motivational practices and developmental decline in math and science academic intrinsic motivation. *Journal of Educational Psychology, 101*(3), 729–739.

Grolnick, W., Farkas, M., Sohmer, R., Michaels, S., & Valsiner, J. (2007). Facilitating motivation in young adolescents: Effects of an after school program. *Journal of Applied Developmental Psychology, 28*, 332–344.

Guvercin, O., Tekkaya, C., & Sungur, S. (2010). A cross age study of elementary students' motivation towards science learning. *Hacettepe University Journal of Education, 39*, 233–243.

Harackiewicz, J., Barron, K., & Elliott, A. (1998). Rethinking achievement goals: When are the adaptive for college students and why? *Educational Psychologist, 33*(1), 1–21.

Hill, G., Atwater, M., & Wiggins, J. (1995). Attitudes towards science of urban seventh grade life science students over time and the relationship to future plans, family, teachers curriculum, and school. *Urban Education, 30*(1), 71–92.

Jones, G., Howes, A., & Rua, M. (2000). Gender differences in students' experiences, interests, and attitudes towards science and scientists. *Science Education, 84*, 180–192.

Lavigne, G., Vallerand, R., & Miquelon, P. (2007). A motivational model of persistence in science education: A self-determination theory approach. *European Journal of psychology of Education, 22*(3), 351–369.

Lee, M., & Madyun, N. I. (2009). The impact of neighborhood disadvantage on the Black–White achievement gap. *Journal of Education for Students Placed at Risk, 14*(2), 148–169.

Lee, O., & Brophy, J. (1996). Motivational patterns observed in sixth-grade science classrooms. *Journal of Research in Science Teaching, 33*(3), 303–318.

Lei, S. (2010). Intrinsic and extrinsic motivation: Evaluating benefits and drawbacks from college instructors' perspectives. *Journal of Instructional Psychology, 37*(2), 153–160.

Liu, M., Horton, L., Olmanson, J., & Toprac, P. (2011). A study of learning and motivation in a new media enriched environment for middle school science. *Educational Technology Research Development, 59*, 249–265.

Lynch, S., Kuipers, J., Pyke, C., & Szesze, M. (2005). Examining the effects of a highly rated science curriculum unit on diverse students: Results from a planning grant. *Journal of Research in Science Teaching, 42*(8), 912–946.

Markowitz, D. G. (2004). Evaluation of the long-term impact of a university high school summer science program on students' interest and perceived abilities in science. *Journal of Science Education and Technology, 13*(3), 395–407.

Mamlok-Naaman, R. (2011). How can we motivate high school students to study science. *Science Education International, 22*(1), 5–17.

Marshall, H. (1987). Motivational strategies of three fifth grade teachers. *The Elementary School Journal, 88*, 133–150.

Milner, A., Templin, M., & Czerniak, C. (2010). Elementary science students' motivation and learning strategy use: Constructivist classroom contextual factors in a Life Science laboratory and a traditional classroom. *Journal of Science Teacher Education, 22*, 151–170.

Murphy, P., & Alexander, P. (2000). A motivated exploration of motivation terminology. *Contemporary Educational Psychology, 25*, 3–53.

Murdock, T., Anderman, L., & Hodge, S. (2000). Middle-grade predictors of students' motivation and behavior in high school. *Journal of Adolescent Research, 15*, 327–351.

National Research Council (Ed.). (1996). *National science education standards*. Washington, DC: National Academy Press.

National Research Council. (2012). *A framework for K-12 science education: Practices, crosscutting concepts, and core ideas*. Washington, DC: Academy Press.

National Science Foundation. (1994). *Women, minorities, and persons with disabilities in science and engineering* (pp. 94–333).

National Science Foundation. (2000). *Women, minorities, and persons with disabilities in science and engineering: 2000*. Arlington, VA: ERIC Clearinghouse.

Nauta, M., Epperson, D., & Kahn, J. (1998). A multiple-groups analysis of predictors of higher level career aspirations among women in mathematics, science, and engineering majors. *Journal of Counseling Psychology, 45*, 483–496.

Nelson, R. M., & DeBacker, T. K. (2008). Achievement motivation in adolescents: The role of peer climate and best friends. *The Journal of Experimental Education, 76*(2), 170–189.

Niemiec, C. P., & Ryan, R. M. (2009). Autonomy, competence, and relatedness in the classroom applying self-determination theory to educational practice. *Theory and Research in Education, 7*(2), 133–144.

Nolen, S. (2003). Learning environment, motivation, and achievement in high school science. *Journal of Research in Science Teaching, 40*(4), 347–368.

Norman, O., Ault, C. R., Bentz, B., & Meskimen, L. (2001). The black-white "achievement gap" as a perennial challenge of urban science education: A sociocultural and historical overview with implications for research and practice. *Journal of Research in Science Teaching, 38*(10), 1101–1114.

Oakes, J. (1985). *Keeping track: How schools structure inequality*. New Haven, CT: Yale University Press.

Oakes, J. (1990a). *Lost talent: The underparticipation of women, minorities, and disabled students in science*. Santa Monica, CA: The Rand Corporation.

Oakes, J. (1990b). *Multiplying inequalities: The effects of race, social class, and tracking on opportunities to learn mathematics and science*. Santa Monica, CA: The Rand Corporation.

Oliver, J. S., & Simpson, R. D. (1988). Influences of attitude toward science, achievement motivation, and science self concept on achievement in science: A longitudinal study. *Science Education, 72*(2), 143–155.

Palmer, D. (2005). A motivational view of constructivist- informed teaching. *International Journal of Science Education, 27*(15), 1853–1881.

Pickens, M., & Eick, C. J. (2009). Studying motivational strategies used by two teachers in differently tracked science courses. *The Journal of Educational Research, 102*(5), 349–362.

Pintrich, P. R. (2003). A motivational science perspective on the role of student motivation in learning and teaching contexts. *Journal of Educational Psychology, 95*(4), 667.

Powell, L. (1990). Factors associated with underrepresentation of African Americans in mathematics and science. *Journal of Negro Education, 59*(3), 292–298.

Rodriguez, A. (1997). *Counting the runners who don't have shoes: Trends in student achievement in science by socioeconomic status and gender within ethnic groups*. National Institute for Science Education University of Wisconsin-Madison, Research monograph 3, National Science Foundation.

Ruby, A. (2006). Improving science achievement at high-poverty urban middle schools. *Science Education, 90*(6), 1005–1027.

Russell, M. L. (2005). Untapped talent and unlimited potential: African American students and the science pipeline. *The Negro Educational Review, 56*(2), 167–182.

Simms, K. (2012). Is the Black-White achievement gap a public sector effect? An examination of student achievement in the third grade. *Journal of At-Risk Issues, 17*(1), 23–29.

Singh, K., Granville, M., & Dika, S. (2002). Mathematics and science achievement: Effects of motivation, interest, and academic engagement. The *Journal of Educational Research, 95*(6), 323–332.

Sorge, C. (2007). What happens? Relationship of age and gender with science attitudes from elementary to middle school. *Science Educator, 16*(2), 33–37.

Tauer, J., & Harackiewicz, J. (2004). The effects of cooperation and competition on intrinsic motivation and performance. *Journal of Personality and Social Psychology, 86*(6), 849–861.

Thomas, G. (1986). Cultivating the interest of women and minorities in high school mathematics and science. *Science Education, 70*(1), 31–43.

Weinberg, A. E., Basile, C. G., & Albright, L. (2011). The effect of an experiential learning program on middle school students' motivation toward mathematics and science. *Research in Middle School Education Online, 35*(3), 1–12.

Weinburgh, M. (1995). Gender differences in student attitudes toward science: A meta-analysis of the literature from 1970–1991. *Journal of Research in Science Teaching, 32*, 387–398.

Wentzel, K. R., & Wigfield, A. (2007). Motivational interventions that work: Themes and remaining issues. *Educational Psychologist, 42*(4), 261–271.

Wigfield, A. (1997). Reading motivation: A domain-specific approach to motivation. *Educational Psychologist, 32*(2), 59–68.

Wigfield, A., & Eccles, J. S. (2000). Expectancy–value theory of achievement motivation. *Contemporary Educational Psychology, 25*(1), 68–81.

Wigfield, A., & Wentzel, K. R. (2007). Introduction to motivation at school: Interventions that work. *Educational Psychologist, 42*(4), 191–196.

Williams, K. C., & Williams, C. C. (2012). Five key ingredients for improving student motivation. *Research in Higher Education Journal, 12*, 1–23.

Part III
Pedagogical and Curricular Issues in Science Teacher Education

Negotiating Science Content: A Structural Barrier in Science Academic Performance

Barbara Rascoe

I begin with the premise that teaching science requires a different kind of knowing than teaching college science; hence, science educators are to do more to help science teachers negotiate their science content knowledge if these science teachers are to, in turn, enhance students' science performance (Haak, HilleRisLambers, Pitre, & Freeman, 2011; Orstein, 2010; Valadez, 2010). Science teachers who are most successful at helping science students negotiate science content, pardon the pun, *know* their science. Knowing, truly knowing (Deng, 2007; Graeber, 1999), means that science teachers are able to take the science content and reframe it, reposition it, contextualize it, and transform it (Chisholm, 2008; Hurd, 1991; Webster-Wright, 2009; Wood, 2007) so that their students can feel the essence of science core ideas. Students learn science best via experiences that stimulate positive emotional responses to science content. Other considerations are that telling and perfunctory practice do not necessarily translate into their knowing and understanding science.

To illustrate strategies for negotiating science content, the newly developed National Research Council's [NRC] (2012) "A Framework for K-12 Science Education: Practices, Crosscutting Concepts, and Core Ideas" is used to introduce science core ideas. However, the major emphasis of this chapter involves a depiction of how to negotiate science content using crosscutting concepts (NRC). Finally, I position negotiating science content relative to science and engineering practices (NRC) and revisit NRC (1996) science standards, which include science and technology and history and nature of science.

B. Rascoe, Ph.D. (✉)
Tift College of Education, Mercer University,
1400 Coleman Avenue, Macon, GA 31201, USA
e-mail: rascoe_bj@mercer.edu

Framework for K-12 Science Education

The latest framework for K-12 Science Education (NRC, 2012) advocates for K-12 science instruction that is structured around three major dimensions: (a) core ideas, (b) science and engineering crosscutting concepts, and (c) science and engineering practices.

Core Ideas

The core ideas have broad importance across multiple science and engineering disciplines, are key organizing principles of a specific discipline, and provide a key tool for understanding or investigating ideas that are more complex and solving problems. They also relate to the interests and life experiences of science students and connect to societal or personal concerns that require scientific or technological knowledge. Table 1 contains the disciplinary core ideas for physical science, life science, Earth and space science, and engineering, technology, and applications of science (NRC, 2012).

Crosscutting Concepts

The crosscutting concepts embody the understandings and abilities that are essential for students' understanding the essence of the core ideas of science and engineering (NRC, 2012) and may be used to reframe, reposition, contextualize, and transform science content. The components of the crosscutting concepts are (a) patterns; (b) cause and effect: mechanism and explanation; (c) scale, proportion,

Table 1 Summary of science and engineering disciplinary core ideas

Science disciplines	Summary of core ideas		
Physical science	Matter and its interactions	Motion and stability of forces and interactions	Energy, waves, and their applications in technologies
Life science	Structures and processes of molecules and organisms	Interactions, energy, and dynamics of ecosystems	Heredity and biological evolution
Earth and space science	Earth's place in the universe	Earth's systems	Earth and human activity
Engineering, technology, and applications of science	Defining and delimiting engineering problem	Developing and optimizing solutions	Interdependence and influence of engineering, science, technology, and the natural world

Fig. 1 The relationships among science crosscutting concepts

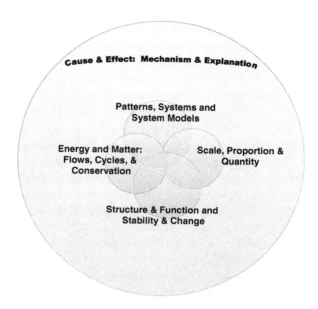

and quantity; (d) systems and system models; (e) energy and matter: flows, cycles, and conservation; (f) structure and function; and (g) stability and change (NRC). These crosscutting concepts describe integrative schemes that bring together participants' (teachers and students) many experiences in science classes across all grade and content levels. *Cause and effect: mechanism and explanation*, as a crosscutting concept, not only supports core ideas in science but also serves as an anchor for the remaining crosscutting concepts, given that all of the crosscutting concepts intercept. Figure 1 represents this aspect of the relationships among the unifying, crosscutting concepts.

Per Fig. 1, one inference is that all of the crosscutting concepts intersect and are contingent upon and use cause and effect: mechanism and explanation as foundational support for tenets that allow students to see linkages among the various fields of and core ideas of science. In this regard, may we infer that all knowledge in science is dependent upon evidence and explanations of the cause and effect mechanism crosscutting concept, which (a) is essential to understanding scientific core ideas and (b) serves as evidentiary support for each of the other six crosscutting concepts?

"Evidence consists of observations and data on which to base scientific explanations. Using evidence to understand interactions allows individuals to predict changes in natural and designed systems" (NRC, 1996, p. 117). I think I can safely say that most precollege science classes begin with how science is done as teachers readily emphasize the importance of the scientific method—notwithstanding safety and welfare. I contend that many science educators misappropriate its utilitarian value as a general and universal science method to garner and support evidence. The way the scientific method is frequently presented is not only superficial and naïve;

it is not how science is done. This is the political element. The use of the scientific method appears to be a social contrivance. The scientific method can be viewed as a social method that helps students construct scientific arguments and communicate the logistics of evidence (Driver, Newton, & Osborne, 2000).

Medawar (1963/1990) argues that although scientific papers are written in a manner to best communicate and persuade readers of the logic behind the reported work, the implication to nonscientists is that researchers actually follow a systematic method. The caveat is that frequently many students come to believe that experiments are the only route to understanding the natural world rather than accepting the idea that imagination and creativity play a huge role in research (Clough, 2000). The success of science is not due to the one method, which, in turn, demarcates science from other human endeavors. Students should understand that scientists tend to use whatever methods and approaches that will produce insight into a research problem.

Conversely, some educators aspire to hold tightly to the "scientific method," given that the NRC (2012) states:

> Although there is no universal agreement about teaching the nature of science, there is a strong consensus about characteristics of the scientific enterprise that should be understood by an educated citizen. For example, the notion that there is a single scientific method of observation, hypothesis, deduction, and conclusion—a myth perpetuated to this day by many textbooks—is fundamentally wrong. Scientists do use deductive reasoning, but they also search for patterns, classify different objects, generalize from repeated observations, and engage in a process of making inferences as to what might be the best explanation. Thus, the picture of scientific reasoning is richer, more complex, and more diverse than the image of a linear and unitary scientific method would suggest. (p. 78)

Cowens (2006) proposes that when scientists try to solve problems, they usually search for an answer in an orderly and systematic manner. Cowens also asserts that anyone who is inquisitive can be a scientist and that scientists answer questions by following a simple, logical, and straightforward prescription called the scientific method. Cowens appears to encourage teachers to allow their students to utilize the scientific method in all investigations. Nevertheless, what is the rest of the story that students need to understand? In science classrooms, the scientific method is utilized for two things: (a) to follow the logic in predesigned investigations of others and (b) to write up one's investigation so that others can follow one's logic. As we think about scientific methods, Feyerabend (1993) proposes that most science is done "against the method." This means that most scientists' ways of thinking are not always linear. As a scientist and educator, I begin with the conclusions in my head based on my theories regarding how I think the world works. Then, I play with experiments (minus some of the steps proposed in the scientific method) that I think will provide the evidence to support my conclusions. When I think I have a design that will convince you that I am "right," I will write it up for you—to convince you. This is clearly against the method. I challenge science teachers to go into scientists' labs and ask to see their notes. These notes may have little if any resemblance to the scientific method. They probably do not include all the tenets of the scientific method (inclusive, specifically, of observing, defining the problem or question, forming a hypothesis, testing the hypothesis, observing and recording results, forming conclusions, and reporting results) unless, of course, they are preparing to write, to publish, or to seek funding for their research.

Sometimes, blunders, per my experiences in science learning environments, can provide unexpected learning experiences. So do we, as science teachers, inquire as to how these occurrences are analogous to our students' vicarious experiences that inform their knowledge and inform students' paradigm shifts? These conversations can further direct other conversations relative to how science is done as we ask: *What is the role of accidental discovery in science? Are there accidents in science? Can we ascribe Fleming's discovery of penicillin as an accident? Can scientific phenomenon be accidental and a discovery simultaneously* (Garcia, 2009)? *Can there be accidents if we, like scientists, use our knowledge along with our observational skills to discern intuitively that something is different about an anomaly to render it a significant discovery?* In the article "A death in Antarctica," Mervis (2009) alleges that discovery by serendipity is a new context for an old riddle, which infers the many methods of science. Additionally, Kuhn (1996) would infer that science methodologies are attributable to new contexts that are (a) different science disciplines and (b) paradigms shifts in particular science disciplines.

The next crucial step for science students is analyzing science evidence or our scientific observations. Frequently, science instruction suggests that scientists make unproblematic observations, that interpretations of data are straightforward, and that science methodology results in objective knowledge about the natural world (Deutsch, 2011; Weinberg, 2003). "Teaching that scientists possess these characteristics is bad enough but it is abhorrent that science educators should actually attempt to mold children in the same false image" (Gauld, 1982, p. 118). Moving students toward a more realistic understanding of how science and scientists work is a most desirable goal (Clough, 2000). So, after the experiment is done and evidence via observations is determined, how do students structure explanations?

Scientific explanations incorporate existing scientific knowledge and new evidence from observations, experiments, or models into internally consistent, logical statements. Different terms are used to describe various types of scientific explanations. Knowing the difference between terms such as *theory* and *law* can be problematic relative to students' understanding the different products of science and addressing two science myths as proposed by McComas (1996). The following activity allows science apprentices (teachers and students) an opportunity to reexamine the definitions in their heads regarding scientific theories and scientific laws. Again, as students examine these differences, they are to posit how the terms *theory* and *law* are used in their daily lives. A theory is an explanation for an observation or series of observations (evidence). Scientific laws describe the behavior of matter and energy (McComas). Do these definitions always work as one examines various scientific theories and laws? Table 2 below provides specific examples of theories and laws that may be used to augment students' analysis of the differences between a science *theory* and a science *law*.

Table 2 Examples of science theories and science laws

Science theories	Science law
The Big Bang Theory refers to the idea that the universe has expanded from a primordial hot and dense initial condition at some finite time in the past and continues to expand to this day	Law of conservation of energy states that energy cannot be created or destroyed. It may be transformed from one form to another, but the total amount of energy does not change
Darwin's theory of evolution states that species change over time	Newton's first law of motion states that a physical body will remain at rest, or continue to move at a constant velocity, unless an outside net force acts upon it
The atom theory of matter states that all matter is made up of tiny indivisible particles (atoms)	Ohm's law states that the current in a circuit varies in direct proportion to the voltage across the circuit and inversely with the circuit's resistance
The theory of plate tectonics states that the Earth's surface layer, or lithosphere, consists of seven large and 18 smaller plates that move and interact in various ways	The law of segregation (Mendel) states that the members of a pair of alleles separate when the sex cells (sperms and eggs) are formed. A sex cell will receive one allele or the other

Dialogue or inquiry that may ensue from the information in Table 2 may, perhaps, revolve around examining relationships and differences between laws and theories and include the following:

1. A law may be a theory. Yes or No? If yes, how? If no, explain.
2. A theory may not be a law. Yes or No? If yes, how? If no, explain.
3. Most laws are universal and are accepted by all members of the science academy. Is this an accurate assessment? Explain.
4. There are some theories that are not accepted by all members of the science academy. Explain.
5. What are examples of laws described in science textbooks that may be viewed as outliers?

As students develop and as they understand more science concepts and processes, their explanations should become more sophisticated (Chiou & Anderson, 2010; Kincaid, 2009)—that is, their scientific explanations should increasingly include a rich scientific knowledge base. The crosscutting concepts as they relate to explanations not only require evidence of logic and high levels of analysis but also require that science teachers have a greater tolerance for criticism and uncertainty (NRC, 1996).

One tale of caution, in this instance, ought to be the use of cookbook or recipe laboratory activities (Mohrig, 2004). These kinds of laboratory activities are dangerous because teachers frequently will garner all the materials for the laboratory activity—having students use the scientific method. However, problems may occur when students do not get the annotated teachers' edition rendition of the *correct* results. Teachers may react in two ways in this case scenario. The first reaction usually is reproaching the students as they retort, "What did you do wrong?" The second reaction could be a teachable moment (Giberson, 2005) as teachers ask, "What procedures did you use?" "What were your results?" "Can you explain the variables relative to your results?" "Did you get your expected results? Explain." "What were your observations?" "What, if any, inferences can you make?" When each student is instructed to explain their observations in light of other laboratory groups having different observations and different explanations and when teachers have a greater tolerance for criticism and uncertainty (NRC, 1996), lively discussions occur and students have opportunities to negotiate science in black, white, and gray (Sweeney, 2008; Venkatramani, Keinan, Balaeff, & Beratonk, 2011).

Evidence infers relationships between and among natural phenomena. This is not only social but also political (Ivan, 2010). It is social in that we tend to reanalyze evidence by others' reactions to it. It is political where classroom dynamics serve to devalue background speculation that may emasculate science academic achievement (Beauboeuf-Lafontant, 1999). Science educators devalue background speculation when considerations are not given to science learners' different ways of thinking and knowing, which includes life experiences, perceptions, learning styles, gender, ethnicity, culture, as well as social, cognitive, and social sensibilities (National Science Teachers Association [NSTA], 2003). An example would be a student's life experience might have taught him or her that a tomato is a vegetable. This life experience is the background speculation that makes it difficult to visualize that a tomato is, scientifically, a fruit. This entails influencing students' insights and courses of actions relative to using their expertise and skillfulness as they learn to use science as a tool to provide explanations. This works best when explanations are situated in discourse and in passionate inquiries that are conversational, yet considers students' silent, salient questions, such as "but, what if," "suppose we," or "have you ever wondered or thought about"? Science is similar to life and similar to rules—sometimes black (opaque), sometimes white (transparent), but mostly gray (vague or translucent) (Ivan, 2010; Wolfram, 2002).

Other procedural dynamics relative to the attitudes and values of processes of science include rules of evidence in which science instruction should position what is *in vogue*. Specifically, *proof* has no place in scientific descriptions or scientific assertions and may be viewed as a scientific myth (McComas, 1996). *Proofs* are not done in science. They are done in mathematics, which is a human creation in abstraction. Secondly, there are no absolutes in the universe. All that we know relative to natural phenomena are subject to contingency, such as paradigm shifts and time and space differences. For example, the virus may be viewed as a contingency to the cell theory and quantum mechanics may be viewed as a contingency to Newton's laws of motion. The more appropriate phrases for use instead of *proof* are *evidence supports* or *the evidence does not support*. Scientists no longer seek to *prove* theories; they endeavor to falsify them with verification (Feyerabend, 1993; Zeide, 2010); hence, the rationale for repeated testing is apposite—the use of the scientific method notwithstanding.

Other science myths or attitudes that we neglect to correct include (a) whether theories become laws, (b) whether a hypothesis is an educated guess, and (c) whether

Table 3 Science inquiry with tolerance of criticism and uncertainty examples using opaque, transparent, and translucent to symbolize what students may understand

Science explanation with inquiry		
All living things are composed of cells. The structure of a virus is not analogous to that of a cell. Is a virus a living thing?		
Tolerance of criticism and uncertainty		
Opaque	Transparent	Translucent
Do we have to modify our definition of what a cell is?	Are all living things composed of cells?	What scientists consider viruses to be living things? What scientists do not consider viruses to be living things?
Plants need insects to aid in reproduction. What would happen if all of the insects were killed? Would plants not be able to reproduce?		
Tolerance of criticism and uncertainty		
Opaque	Transparent	Translucent
Do scientists know enough about how some insects aid in reproduction to answer that question about all plants?	Let us position that most plants reproduce sexually. If all the insects were killed, what alternative means could be utilized to aid in reproduction (sexual and nonsexual)?	Are insects required for reproduction in all plants?

evidence accumulated carefully will result in sure scientific knowledge (McComas, 1996). Evidence is accumulated using your six senses—observations. Given that we can extend our senses using microscopes, telescopes, and other tools, all observations are based on the theories in our heads. Our theories in our head determine what we observe. Do our students know this?

Science classrooms are not to be places where students experience the raining of cold, hard scientific facts. Let me elaborate. Students should not have to accept any science information without further ado (Feyerabend, 1993). Subsequently, students need opportunities to analyze vague areas of core ideas presented in light of what they know and/or what they learned previously regarding the science content. What would this look like in the classroom? To illustrate such an activity, I am positioning *opaque* to symbolize the unknown. *Transparent* symbolizes the known, while translucent represents what one is unsure of or what is vague. When something is opaque-transparent, the issue at hand may be viewed as dichotomized—having two, seemingly, opposing sides with no middle ground. To illustrate this process, students are given an example of how analyzing the opaque, transparent, and translucent relative to science content may work. The idea is that science is not only a way of knowing. It is a way of thinking in view of the fact that information and evidence are analyzed repeatedly, linearization notwithstanding. We learn the science from experiences that are inquiry in nature. Students are given opportunities to analyze the science they think they know while learning using inquiry as a tool. Table 3 presents examples that illustrate how science explanations may be examined with a tolerance for criticism and uncertainty (opaque, transparent, and translucent).

The rationales for science inquiry in opaque, transparent, and translucent are twofold. First, our students need states of consciousness for and toward science to help them understand that science is like life and what we do in science in relation to inquiry is comparable to what we are asked to do, daily, pertaining to issues in our lives as citizens. Secondly, this exercise would provide more opportunity/time for students to examine, cognitively, their ideas about specific science concepts. The more time information is analyzed in working memory, the higher the probability the information is transferred to long-term memory (Santrock, 2010; Woolfolk, 2011). Moreover, given that students' understandings and abilities are grounded in the experience of inquiry, which is the foundation for the development of understandings and abilities in science content, what other strategies may we use to contextualize science (Coll, Dahsah, & Faikhamta, 2010; Gilbert, Bulte, & Pilot, 2011)?

To introduce the contextual elements of science, let us consider the "social contract," which is a phrase coined circa 1837 (Merriam-Webster, 2011) and is viewed as an agreement between society and science. The social contract for science was in reaction to the concern with the way economics, politics, and cultures interact in a society and the impact of science on the society. Lubchenco (1998) urges that present environmental and social changes demand that scientists define a new social contract. She says:

> I see the need for a different perspective on how the sciences can and should advance and also return benefit to society. This different perspective is firmly embedded in the knowledge of specific, identifiable changes occurring in the natural and social worlds around us...to provide the basis for new technologies. (p. 479)

Science and Engineering Practices

The NRC (2012) proposes eight (8) science and engineering practices that include (a) asking questions and defining problems, (b) developing and using models, (c) planning and carrying out investigations, (d) analyzing and interpreting data, (e) using mathematics and computational thinking, (f) constructing explanations and designing solutions, (g) engaging in argument from evidence, and (h) obtaining, evaluating, and communicating information. Each of these processes informs science inquiry for scientists and engineers. Table 4 shows how each practice may inform inquiry for (a) scientists and for (b) engineers.

Other Science Standards

The nature of science embraces other science standards that we tend to gloss over in science instruction—namely, science and technology and the history and nature of science (NRC, 1996). The science and technology standard (NRC) has been

Table 4 How science and engineering practices inform science and engineering inquiry

Science and engineering practices	Scientific inquiry	Engineering inquiry
Asking questions and defining problems	What exists and what happens? Why (cause and effect)? How does one know?	What can be done to address a particular human need? How can the need be better specified? What tools/technologies are available or can be developed to address this need?
Developing and using models	What mental and conceptual models can help us better understand the science? How do these models stand up to real world predictions? How does the model need to be adjusted?	Under what conditions do flaws develop in the design? How do the designs need to change due to space and time difference? What are limitations of the model?
Planning and carrying out investigations	What are the variables? How do I collect data under different conditions?	What can be done to address a particular need or want?
Analyzing and interpreting data	How do I organize and interpret the data? What counts as evidence? What patterns and relations can I infer?	What prototypes or models can I design given the patterns and relationships among the variables considering economic feasibility, alternatives, and failures?
Using mathematics and computational thinking	How do I numerically represent the relationship(s) among the variables?	What simulations and mathematical models can be designed?
Constructing explanations and designing solutions	How can the theory be revised or refined based on new evidence to enhance its predictive value?	How can I improve the design? What design problems need to be solved?
Engaging in argument from evidence	How do I make my case?	Considerations: weaknesses and strengths of the design? Cost/benefit analyses? Risks: appeal to aesthetics? Market receptions?
Obtaining, evaluating, and communicating information	How do I communicate how the natural world works?	How does one communicate about phenomena, evidence, explanations, and design solutions?

restructured as science and engineering practices (NRC, 2012). I have included both to provide different examples of inquiry that may be used to help students negotiate science content.

Science and Technology. Science and technology place emphases on helping students nurture scientific abilities and science understandings. Students should be able to understand connections between the natural world and the human-designed world. This standard encompasses the process of identifying scientific problems, determining risks and benefits of natural phenomena and not-so-natural phenomena, designing solutions, and being able to evaluate those solutions. First, however, students are to understand the difference between science and technology. Krone (2005) says:

> Science is the process of identifying and converting unknowns to knowns. It does so by creating knowledge through systematic observation, experiment, and reasoning. Technology is the part of applied science that transforms the understandings and discoveries of science into applications for society. (p. 556)

An important goal for science education is to educate scientifically literate students who are able to engage in discourse and debate about matters of scientific and technological concern (NRC, 1996). As students examine issues relative to science, technology, and society, they are privy to more advanced scientific perspectives of technology that transcend electronics, such as TV, cell phones, and computers (DiGironimo, 2011).

Hodson (2010) proposes a four-level approach to science and technology instruction, which includes students' examining (a) how society, culture, and technological changes mutually influence each other; (b) how technology may not be beneficial to all members of society; (c) their views and value positions; and (d) possible actions regarding environmental issues. Per core ideas or the science content, the following may be used as inquiry to help students evaluate science, technology, and society:

1. How does the not-so-natural technology compare with the technology designed by nature?
2. What science knowledge is associated with the technology?
3. How has the technology advanced/hindered science?
4. What new knowledge was generated with the technology?
5. What risks are members of society taking with the development of this technology?
6. What environmental issues do we need to address given the development of this technology?

Supporting the need for students' examining issues relative to science, technology, and society, the NRC (2009) states:

> New discoveries and new technologies do not guarantee that discovery will accelerate. The world must be ready for change, and the tools and resources must be available to capitalize on new capabilities or knowledge (p. 40).

Inquiry regarding science, technology, and society should also be science specific. Table 5 includes examples of how science educators may embrace methodologies that allow core ideas and science learning to be expanded (Martin, Sexton, & Gerlovich, 2000) with science and technology inquiry.

The central idea is to provide students with opportunities to engage in discourse and debate about science and technological concerns (NRC, 1996, 2012).

History and Nature of Science. In learning science, students need to understand that science reflects its history and is an ongoing, changing enterprise. The history and nature of science standard is embraced to help students appreciate and understand the human aspects of science and the role that science has played in the

Table 5 Science expansion with inquiry

Science content	Science and technology expansion with inquiry
Properties of water	How has the knowledge of capillary action been used to design more efficient car engines?
Adaptation and diversity	Hunters wear bright orange vests and bright orange hats. What reasons can you give for this design?
Phases of the moon	What technology did scientists use to determine the distance between the moon and the Earth?
Insects	What kind of impact will continued use of chemicals have on future insect populations?
Biomolecules	What kinds of products have modern industries invented to make use of starches?
Earthquakes	How are bridges built in earthquake zones?
Water cycle	What technology reduces the effect of droughts?

development of various cultures (NRC, 1996). The history and nature of science content standard is divided into five major categories, which include (a) science as a human endeavor, (b) nature of science, (c) history of science, (d) nature of scientific knowledge, and (e) historical perspectives. Table 6 describes the tenets of the five major categories of the history and nature of science content standard (NRC).

It is important to emphasize science as a human endeavor so that all students—irrespective of their culture, ethnic group, gender, or self-perceptions of academic ability—have an opportunity to discuss their stereotypes regarding who does science. Examples of inquiry might include the following:

1. Were all scientists White males?
2. Were all scientists, initially, perceived as smart when they were young?
3. Did all scientists all grow up in privileged environments?
4. How were scientists perceived by their peers when they added to our scientific knowledge base new information and/or posited new perspectives relative to how the natural world works?

Additionally, students are provided with opportunities to see scientists as they see the people in their daily lives—human. Examples of inquiry might include the following:

1. Were they honest?
2. Were they pleasant?
3. Were they unkempt?
4. Were they loved? To what extent were they loved?
5. Were they callous?
6. Were they humble?
7. Were they all diligent or were some of them in the right place at the right time?
8. Did some of them fail miserably before making a scientific breakthrough?
9. What turning points did they have in life that caused them to choose science as a career?

Table 6 Descriptions for each sublevel of the history and nature of science standard

History and nature of science substandards	Characteristics of substandards
Science as a human endeavor	
Who does science?	Why it is important to know?
Nature of science	
Way science works—why (cause and effect) questions	Methods of science
Theory-laden observations	*Observations as evidence*
Careful measurements	Role of accidental discovery
Theory acceptance contingent upon explanatory power	Use of language and symbolic Representation
History of science	
Concrete examples of science concept without the role of culture	Specific scientist without how culture may have influenced his ideas.
Nature of scientific knowledge	
Science in an ongoing, changing enterprise	Crosscutting scientific and engineering concepts:
	1. Patterns
	2. Cause and effect: mechanism and explanation
	3. Scale, proportion, and quantity
	4. Systems and system models
	5. Energy and matter: flows, cycles, and conservation
	6. Structure and function
	7. Stability and change (NRC, 2012)
Historical perspectives	
Role of cultures in expression of scientific ideas	Episodes in history—milestones in the development of Western civilization
Ability of women to do outstanding work in science	

Here, too, is the opportunity to have discussions as to how the culture and the times played a huge role in the acceptance of scientific ideas. Examples of this inquiry include the following:

1. How did this new information play a role in the development of various cultures relative to the development of all thought?
2. Was the contribution a product of science or a process of science?
3. How did the cultural values and/or assumptions of the culture affect the scientific contribution?
4. How was the contribution initially received juxtaposed to how the contribution is perceived today?

The history and nature of science standard also includes information relative to careers in science and specific knowledge and skills scientists need to know to be effective in their fields. According to McComas, Clough, and Almazroa (1998), a better understanding of scientists and the scientific community will enhance (a) understanding of science's strengths and limitations, (b) interest in science,

(c) social decision-making, (d) instructional methodologies, and (e) the learning of science concepts. The human element of science is perhaps the most elemental—yet most often overlooked aspect of science. The development of scientific knowledge involves human creativity and human subjectivity versus the idea that scientists are completely objective. Creativity and subjectivity are unavoidable, given the human aspect of science, and provide the nutriment for presenting science as an ongoing, changing enterprise. Increased creativity in science, which drives the scientific enterprise, is possible because scientists are different individuals from different backgrounds who bring different lenses to the interpretation of the same data. The same is true of our students who need to see the possibility that they, too, can be a part the scientific enterprise (Clough, 2006; Crowther, Lederman, & Lederman, 2005).

Conclusions

So what? At the end of any lesson, whether I am in science methodology classes or whether I am in K-12 classrooms, I ask "So what?" What else do we know given the lesson? Moreover, what does this knowledge enable us to do? What is the big picture? This inquiry embraces Feyerabend's (1993) *further ado* and is necessary for students getting it (the science or core ideas [NRC, 2012]). Figure 2 is symbolic of science classrooms providing students opportunities to examine the universe using the cause and effect: mechanism and explanation crosscutting concept and science and engineering practices in conjunction with science and technology and the history and nature of science to augment science performance. The charge of science educators must embrace helping science teachers negotiate not only science pedagogy but also science.

In summary, using evidence and explanations of the cause and effect mechanism and expanding the core ideas with (a) science and technology and (b) history and nature of science allow science educators to address "What does science allow us to do?" It allows the apprentices (teachers and students) an opportunity to examine science up close and personal, which plays important roles in students' getting science. The "how" references lifting students up and providing them with telescopic experiences or views (Fig. 2) designed to have each (a) reexamine, understand, and act on personal and social issues (personal theories regarding how the world works, growing pains, hormones, and hating science) and (b) learn science per life, physical, or Earth and space science core ideas.

Now what? We, as science educators, operate in political arenas where our students must be able to be competitive on state, national, and/or global assessments. This is also an example of what knowing the content should allow our science teachers and their science students to be able to do. Students should have opportunities to apply their new knowledge prior to the assessment(s) using the crosscutting concepts described in this chapter as lens to understand the essence of science, which requires many different ways of knowing, which, in turn, require

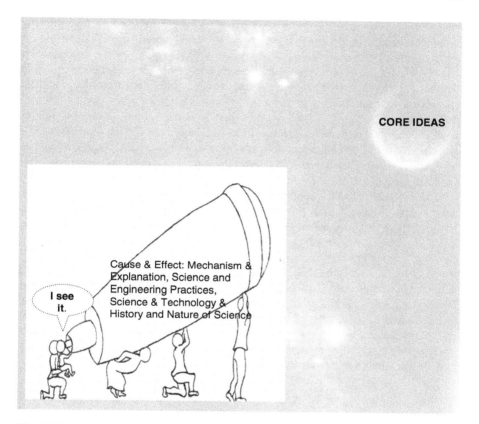

Fig. 2 How classrooms in the universe position students for telescopic experiences that augment their getting the science

different ways of thinking relative to natural phenomena (Rascoe, 2005). Assessments have three roles. One function is to discern what students know in science. The second function is to enhance their thinking, while the third function is to reinforce what they know while transforming it. How do students transform science? Grossman (2005) presents seven knowledge transformations that also help students reject the notion that science is difficult. These include procedural, conceptual, contextual, analogical, symbolic, metaphorical, and arbitrary knowledge transformations. Now is the time to help students transition from the regurgitation of science facts to (a) negotiating procedures and operations; (b) problem solving via abstract conceptualizations; (c) applying science in new and different contexts; (d) extending science knowledge to situations that are more cognitively complex; (e) translating science algebraically and in pictorial, mathematical, and graphical forms; and (f) understanding the language of science (Grossman).

The academic achievement (performance) gap in science is problematic. Subsequently, I am proposing a paradigm shift that transcends and yet embraces academic achievement. Science is personal and must be used as a milieu that empowers science students, derails barriers to enhanced performance, and promotes opportunities for all students to participate. As a science educator and as an agent of change, I am proposing that science teachers embrace excellent, equitable, democratic practices that include critical thinking, intellectual curiosity, and the participation of all students in science learning as a community effort (Bianchini, 2011; Connors & Perkins, 2009; Merrett, 2004; Moore, 2008). I want my science teachers and their science students to enjoy the experience of learning science. In reference to social justice, I want them to understand how science is an integral part of their lives as they are given the opportunity to make connections between science in the classroom and the science that is portrayed in their everyday lives (Chamany, 2006). I want them to confront assumptions, biases, and consider multiple perspectives (Mensah, 2009). I want them to know and understand that science is not something that only smart, White men do. I want my science students to develop a deeper, richer understanding of science as they develop their investigative and problem-solving capabilities. I want to position new values that are driven by their needs-to-know. I want the sheer challenge of the sport to propel them. Strong study habits and the motivation to access people resources are secondary effects (Rascoe, 2005). I do not want science students to view themselves as outsiders in science (Aikenhead & Jegede, 1999). I want them to become members of a seemingly, elite group in science who, by virtue of their shared training and experience in science, possess the rules of the game of science (Kuhn, 1996). It is illusive. It is work. It is play. It is equitable. Now, what do you want?

References

Aikenhead, G., & Jegede, O. J. (1999). Cross-cultural science education: A cognitive explanation of a cultural phenomenon. *Journal of Research in Science Teaching, 36*(3), 269–287.

Beauboeuf-Lafontant, T. (1999). A movement against and beyond boundaries: "Politically relevant teaching" among African American teachers. *Teachers College Record, 100*(4), 702–723.

Bianchini, J. A. (2011). How to foster student to student learning in science? The student, the teacher, and the subject matter. *Cultural Studies in Science Education, 6*(4), 871–882.

Chamany, K. (2006). Science and social justice: Making the case for case studies. *Journal of College Science Teaching, 36*(2), 54–59.

Chiou, G.-L., & Anderson, O. R. (2010). A multi-dimensional cognitive analysis of undergraduate physics students' understanding of heat conduction. *International Journal of Science Education, 32*(6), 2113–2142.

Chisholm, L. (2008). Re-contextualising learning in second modernity. *Research in Post-Compulsory Education, 13*(2), 139–147.

Clough, M. P. (2000). The nature of science: Understanding how the game of science is played. *Clearing House, 74*, 13–17.

Clough, M. P. (2006). Learners' responses to the demands of conceptual change: Considerations for effective nature of science instruction. *Science Education, 15*, 463–494.

Coll, R. K., Dahsah, C., & Faikhamta, C. (2010). The influence of educational context on science learning: A cross-national analysis of PISA. *Research in Science and Technological Education, 28*(1), 3–24.

Connors, M. M., & Perkins, B. (2009). The nature of science education. *Democracy and Education, 18*(3), 56–60.

Council of Chief State School Officers. (2011). *Interstate Teacher Assessment and Support Consortium (InTASC) model core teaching standards: A resource for state dialogue.* Washington, DC: Author.

Cowens, J. (2006). The scientific method. *Teaching PreK-8, 37,* 44–46.

Crowther, D. T., Lederman, N. G., & Lederman, J. S. (2005). Understanding the true meaning of nature of science: Teaching suggestions to help you highlight the meaning of science. *Science and Children, 4,* 50–52.

Deng, Z. (2007). Knowing the subject matter of a secondary-school science subject. *Journal of Curriculum Studies, 39*(5), 503–535.

Deutsch, D. (2011). The source of all progress. *New Scientist, 210*(2809), 30–31.

DiGironimo, N. (2011). What is technology? Investigating student conceptions about the nature of technology. *International Journal of Science Education, 33*(10), 1337–1352.

Driver, R., Newton, P., & Osborne, J. (2000). Establishing the norms of scientific argumentation in classrooms. *Science Education, 84*(3), 287–312.

Feyerabend, P. (1993). *Against method* (3rd ed.). London: Verso.

Garcia, P. (2009). Discovery by serendipity: A new context for an old riddle. *Foundations of Chemistry, 11*(1), 33–42.

Gauld, C. (1982). The scientific attitude and science education: A critical reappraisal. *Science and Education, 66,* 109–121.

Giberson, K. (2005). Teachable moments. *Science and Spirit, 16*(5), 6–7.

Gilbert, J. K., Bulte, A. M. W., & Pilot, A. (2011). Concept development and transfer in context-based science education. *International Journal of Science Education, 33*(6), 817–837.

Graeber, A. (1999). Forms of knowing mathematics: What preservice teachers should learn. *Educational Studies in Mathematics, 38*(1–3), 189–208.

Grossman, R. W. (2005). Discovering hidden transformations: Making science and other courses more learnable. *College Teaching, 53*(1), 33–40.

Haak, D. C., HilleRisLambers, J., Pitre, E., & Freeman, S. (2011). Increased structure and active learning reduce the achievement gap in introductory biology. *Science, 332*(6034), 1213–1216.

Hodson, D. (2010). Science education as a call to action. *Canadian Journal of Science, Mathematics, and Technology Education, 10*(3), 197–206.

Hurd, P. D. (1991). Why we must transform science education. *Educational Leadership, 49*(2), 33–35.

Ivan, H. (2010). Single-crystal-to-single-crystal reactivity: Gray, rather than Black or White. *Crystal Growth & Design, 10*(7), 2817–2823.

Kincaid, H. (2009). A more sophisticated Merton. *Philosophy of the Social Sciences, 39*(2), 266–271.

Krone, R. M. (2005). Science and technology for what? *Review of Policy Research, 22*(4), 555–569.

Kuhn, T. S. (1996). *The structure of scientific revolutions* (3rd ed.). Chicago: The University of Chicago Press.

Lubchenco, J. (1998). Entering the century of the environment: A new social contract for science. *Science, 279*(5350), 491–497.

Martin, R., Sexton, C., & Gerlovich, J. (2000). *Teaching science for all children.* Boston: Allyn and Bacon.

McComas, W. F. (1996). Ten myths of science: Reexamining what we think we know about the nature of science. *School Science and Mathematics, 96,* 10–16.

McComas, W. F., Clough, M. P., & Almazroa, H. (1998). The role and character of the nature of science in science education. *Science and Education, 7,* 511–532.

Medawar, B. (1963/1990). *Is the scientific paper a fraud? In the threat and the glory: Reflections on science and scientists.* New York: Harper Collins.

Mensah, F. M. (2009). Confronting assumptions, biases, and stereotypes in preservice teachers' conceptualizations of science teaching through the use of book clubs. *Journal of Research in Science Teaching, 46*(9), 1041–1066.

Merrett, C. D. (2004). Social justice: What is it? Why teach it? *The Journal of Geography, 103*(3), 93–101.

Merriam-Webster. (2011). Retrieved from: http://www.merriam-webster.com/dictionary/social%20contract

Mervis, J. (2009). A death in Antarctica. *Science, 323*(5910), 32–35.

Mohrig, J. R. (2004). The problem with organic chemistry labs. *Journal of Chemistry Education, 81*(8), 1083–1085.

Moore, F. M. (2008). Agency, identity, and social justice education: Preservice teachers' thoughts on becoming agents of change in urban elementary science classrooms. *Research in Science Education, 38*(5), 589–610.

National Research Council. (1996). *National science education standards*. Washington, DC: National Academy Press. Retrieved from: http://www.nap.edu/openbook.php?record_id=4962

National Research Council. (2009). *A new biology for the 21st century*. Washington, DC: The National Academies Press.

National Research Council. (2012). *A framework for K-12 science education: Practices, crosscutting concepts and core ideas*. Committee on a Conceptual Framework for New K-12 Science Education Standards. Board on Science Education, Division of Behavioral and Social Sciences and Education. Washington, DC: The National Academies Press.

National Science Teachers Association. (2003). *Standards for science teacher preparation*. Arlington, VA: Author.

Orstein, A. (2010). Achievement gaps in education. *Society, 47*(5), 424–429.

Rascoe, B. (2005). Black male students' academic achievement in science. *Journal of Women and Minorities in Science and Engineering, 11*(5), 311–325.

Santrock, J. W. (2010). *Adolescence* (13th ed.). New York: McGraw-Hill.

Sweeney, M. O. (2008). The world is not Black and White. More like Black and Gray. *Journal of Cardiovascular Electrophysiology, 19*(1), 28–31. Retrieved from DOI: 10.1111/j.1540-8167.2007.01004.x

Valadez, J. R. (2010). Explaining the science achievement gap. *Leadership, 40*(1), 30–38.

Venkatramani, R., Keinan, S., Balaeff, A., & Beratan, D. N. (2011). Nucleic acid charge transfer: Black White and Gray. *Coordination Chemistry Reviews, 255*(7/8), 635–648.

Webster-Wright, A. (2009). Reframing professional development through understanding authentic professional learning. *Review of Educational Research, 79*(2), 702–739.

Weinberg, M. (2003). A leg (or three) to stand on. *Science and Children, 40*(6), 28–30.

Wolfram, S. (2002). *A new kind of science*. Champaign, IL: Wolfram Media.

Wood, D. R. (2007). Professional learning communities: Teachers, knowledge, and knowing. *Theory into Practice, 46*(4), 281–290.

Woolfolk, A. (2011). *Educational psychology: Active learning edition* (11th ed.). Boston: Pearson Education.

Zeide, B. (2010). Falsification and certainty. *Mathematical and Computational Forestry and Natural Resource Sciences, 2*(2), 161–162.

Internationally Inclusive Science Education: Addressing the Needs of Migrants and International Students in the Era of Globalization

Charles B. Hutchison

Given the globalized nature of education, educators must not only prepare themselves, but also their diverse students to become globally-competent workers—a responsibility that becomes clear when teachers and students find themselves in classrooms and communities with people who are culturally and linguistically different from themselves. What kind of education should students receive in order to be proficient in an international, multicultural society? Equally importantly, what kind of knowledge should educators have in order to teach in an increasingly globalized world? (Wiggan & Hutchison, 2009, pp. 1–2)

The main purpose of this chapter is to illuminate the issues facing migrant populations who are involved in education in the United States (USA) and discuss ways in which science teacher education can better incorporate their needs in order to help them reach their fuller scientific potentials. In this chapter, I propose what one may call *internationally inclusive teaching*, in consonance with the concept of culturally responsive teaching. In order to invest authenticity in this chapter, I will draw on my experiences as an international student and professional who has lived and worked in Africa, Europe, and the United States and has done research and committee work on internationalization of education.

This chapter will address the following activities and subtopics, in the context of science education:

- Introductory activity: Assessing a Ghanaian student for scientific proficiency
- Teacher education and global change: A call in need of a response
- Globalized education as a social justice issue in U.S. education
- The flat world of education: Globalized education in practice
- Refugees as an international education issue
- International education: Process and product cycles
- Globalization of education: The process and the tools

C.B. Hutchison (✉)
Department of Middle, Secondary, and K12 Education, University of North Carolina at Charlotte, 9201 University City Blvd., Charlotte, NC 28223, USA
e-mail: Chutchis@uncc.edu

- The interface between globalization and science teacher education
- The nature of immigrants in U.S. classrooms
- Internationally inclusive teaching and its implications for teacher training
- Conclusion and summary
- Culminating activity: Creating an internationally inclusive lesson plan—the whys and whats

Introductory Activity: Assessing a Ghanaian Student for Scientific Proficiency

Case #1

Kofi Mensah was a good student in Ghana, West Africa, a country known for its intellectual prominence. He already had his bachelor's degree (with honors) before arriving in the United States. After working as a scientific researcher in molecular biology, he decided to become a certified teacher. He therefore enrolled in a Lateral Entry program in order to be certified to teach science.

Kofi arrived in the methods class full of confidence in his science content knowledge; therefore, he was rather surprised that his methods instructor assessed his first assignment as poor on many levels. First was the issue of spelling: Kofi spoke three languages, one of which was British English. Neither he nor his instructor knew that British English spelling was different from "American English" spelling and that several expressions and science vocabulary terms were spelled differently. For this reason, both Kofi and his instructor totally misunderstood each other: Kofi, himself, thought that his instructor did not speak properly and was lazy in speech and writing (because Ghanaians believe that educated people should not make petty mistakes—a sign of intellectual inferiority); on the other hand, his instructor thought that Kofi did not speak or write properly. This was a bad start for both of them, because they began the class with a misperception of each other, leading to a negative, self-fulfilling prophecy. In the end, Kofi, a promising science teacher, left the program frustrated.

Case #2

Chu-li was a teaching assistant in a Chinese university when she got the opportunity to move to the United States. Because of her science background, she easily got a job in a large private school, where she taught physics. Her school, however, had the policy that all their teachers needed to be certified, and so she enrolled in a certification program at the local university.

It must be noted that although Chu-li was a good teacher of content knowledge, she had serious classroom management problems. It was partly for this reason that her school thought her enrolment in a certification program would help her to become more proficient in working with her students. Meanwhile, the school used the formal system of classroom observation and supervision to assess Chu-li's work. They found her to be very traditional, yet unassertive. These characteristics, they believed, may have contributed to her loss of control in her class.

While in the licensure program, Chu-li's instructors realized that although she could speak English very well, her writing skills were not strong. For this reason, she could not articulate her content knowledge well enough for student understanding. However, she was very capable of engaging in a solid conversation about the content matter—so well so that they were surprised that there was such a gap in her English writing skills.

Chu-li had another issue: she would not openly engage with the class or the professor, although outside the classroom, she was relatively more gregarious.

Questions

1. *From your perspective, what are some of the factors at play in the teaching lives of Kofi and Chu-li?*
2. *What are some of the factors that may distinguish Kofi's issues from Chu-li's?*
3. *In terms of their backgrounds, what do you know about their cultures that could (a) enhance and (b) impede their lives* as teachers *in the United States?*
4. *Similarly, what do you know about their cultures that could (a) enhance and (b) impede their lives* as students *in the United States?*
5. *Based on the responses in questions 3 and 4 above, what are some of the solutions you would suggest for those working with both Kofi and Chu-li?*

Teacher Education and Global Change: A Call in Need of Response

In October 2010, Arthur Levine wrote an article titled "Teacher education must respond to changes in America." The subtitle of this article, however, provided the major thrust of the article; it was captioned "Teacher education must adapt to the same changes in the economy, demographics, globalization, and more that are prompting change in K-12 education." In this article, Levine noted that the current world transformation prompted by deep demographic, economic, technological, and global changes is rather rare and that such magnitude of change was last seen as far back as the Industrial Revolution. He asserted that teacher education, in its current form, was created for a different era in time, but that time has passed, and that "even if the nation's teacher education programs had been perfect, the best in the world, they would still need to change today" (p. 20). His rationale for this assertion is that when change occurs, social institutions are the last to respond and make appropriate changes; therefore, they get left behind. In establishing the global connection to the need for change in "America's" teacher education programs, Levine implied that globalization has necessitated a change not only in the school curriculum but also in universities in general. He reasoned that global transformation will force universities to work across national boundaries. By extrapolation, therefore, Levine makes the argument that globalization demands a response from teacher education, and soon.

Globalized Education as a Social Justice Issue in U.S. Education

Globalization or internationalization of education is a process that is rather nebulous in U.S. education because it lacks a clear definition and objectives. For this reason, it is mostly conflated with student or faculty travel abroad, with no systemic implications or benefits across programs. It is often added—not incorporated—into programs as remote afterthoughts and is left to students to decide as to whether they would like to invest any efforts in it or not. For the most part, faculty lack the incentive to become globally proficient, and since one cannot teach what one does not know, such faculty cannot incorporate international proficiency in their instruction. This is no different from the issue of multicultural education. Even now (decades after the Civil Rights Movement and the birth of multiculturalism), it is still a contentious course for many instructors—at a time when many school districts are populated by mostly students of color. Because some educators do not see the need for multicultural education, are not conversant with the contents, they cannot adequately prepare their students for diverse schools. The sad result, however, is that such teacher candidates are cheated by not being capable of working effectively in diverse schools.

It can be argued that just as Free Appropriate Public Education (FAPE), a part of Individuals with Disability Education Act (IDEA) (passed in Congress in 1975 as Public Law 94-142 and was later reiterated in 1990 as IDEA), had the objective of ensuring that all U.S. students were properly educated, a U.S. teacher who is not prepared to work with diverse learners has not been given appropriate education. Furthering this argument to globalization, one can argue that in an era when U.S. students are expected to compete with students from all over the world for their livelihoods, "appropriate" education should include adequate exposure to globalization and related issues. Furthermore, their teachers should be knowledgeable enough in globalized educational concepts so as to be proficient in working with (a) U.S.-born students and (b) people from different parts of the world.

The Flat World of Education: Globalized Education in Practice

Since time immemorial, trans-regional or international education has been a feature of education (compare, for example, the intercultural education of Daniel and his three colleagues in Daniel 1:1–5 in the Old Testament of the Bible. In this narrative,

Nebuchadnezzar, king of Babylon, besieged Jerusalem and sent four Israeli boys to be educated in the language and literature of the Chaldeans). In recent times, however, the phenomenon of globalization has sparked an interest in international education, as the shrinking world has accelerated the mobility of teachers and students across the globe. Because of the immensity of this phenomenon, there is a large body of literature that addresses issues of migrations across the world. In terms of Africa alone (whence the author hails), there is a general agreement that at least tens of thousands of skilled professionals, such as medical doctors, nurses, engineers, and teachers emigrate each year, and many of these emigrants move to Western countries. Take, for example, Ghana, a small West African country:

> The Ghanaian population in the United States has grown rapidly over the last decade and a half, particularly between 1990 and 2000, when the population jumped from 20,889 to 65,570, or 210 percent. Family reunification, refugee resettlement, and the strong economy of the 1990s are the factors driving this increase. Many believe these figures to be undercounts, and nonofficial estimates reach as high as 300,000. (Bump, 2006, paragraph 48)

Taking South Africa, as another example, about one-third of all the emigrating professionals were somewhat involved in education (Bailey, 2003). A National Education Association (NEA) November 2003 report estimated that 10,000 international teachers were working in U.S. public school systems on nonimmigrant or cultural exchange visas. Although the teachers mostly come from English-speaking countries such as India, Nigeria, Ghana, the Philippines, Canada, and the like, there are also international teachers from many non-English-speaking countries including France, Germany, Russia, and Mexico (NEA, 2003).

In higher education, professorial exchanges are much more common and widely documented, since the concept of "university," a term that hints at "universality" of knowledge and program transferability is often taken for granted. In K-12 education, however, the literature on teacher migration is scanty, especially in the context of U.S. education. Notwithstanding, this area of research is becoming more interesting especially due to the shortage of mathematics and science teachers in the United States (e.g., Hutchison, 2005).

Student migration has also been a perennial part of the landscape of international education, since higher education (primarily the universities and colleges)—the chief instruments of enlightenment and modernization—have sought to spread human wisdom partly through the agencies of colonization, benevolence, and opportunities for social advancement. In this connection, it is noteworthy that:

> In 2002, countries like the United States (U.S.), United Kingdom (U.K.), Germany, France, Australia, Japan, and Spain were, respectively, the leading host countries for international students seeking higher education. Conversely, students from these countries chose China, India, Greece, Turkey, Morocco, Algeria, Malaysia and South Korea as some of their top destinations for study abroad (Davis, 2003). These trends were consistent in 2008, where the U.S. and U.K. outpaced all other nations as the leading host countries for international students, while India and China led the non-Western nations as the choice destinations for international study (Institute of International Education, 2008). (As cited in Wiggan and Hutchison, 2009, p. 1)

Refugees as an International Education Issue

In addressing the issue of globalization and internationally inclusive teaching, the growth of refugees across the world—and their consequent need for education—has become an issue of global interest. Refugees are people who have been forced to leave their home country because of one or several reasons. Often, they leave because of political tyranny, wars, or natural disasters. Because refugees often leave their countries under traumatic conditions, they often leave their native countries with little personal effects, and so experience poverty, and are also vulnerable to mental health issues. In recent decades, the many conflicts around the world have increased the number of refugees in the United States, including Bosnians from Eastern Europe, Sudanese, and Southeastern Asia's Vietnamese, Cambodians, Laotians, and the Hmong.

Besides financial poverty, one of the landmark characteristics of refugees across the world is the lack of educational preparation. Many enter their new countries without appropriate educational foundations, especially when they arrive from countries engaged in long-term wars. For this reason, many of them have serious educational deficiencies, even if they are excited about learning. Many are illiterate even in their own languages, and like the child soldiers in many parts of the world (e.g., Afghanistan, Colombia, Liberia, Sudan, Sri Lanka, and Vietnam), cannot read or write. In cases where refugee children have experienced an interruption in their education, they are referred to as *students with interrupted formal education* (SIFE). Besides, these children are also candidates for English as Second Language (ESL) programs in schools.

Whereas the refugee issue is a secondary aspect of what may be called voluntary migration in search of better economic and life outcomes, the pedagogical needs of this population have some overlaps with conventional migrants, of which there are many, since the United States has a generous immigration policy (The Center for Immigration Studies [2011] noted that in the decade of 2000–2010, the United States absorbed over 13 million legal and illegal immigrants). First, both populations have some level of anxiety as they navigate their new social environments, and therefore can benefit from the beneficence of their host country. Second, they often have different pedagogical orientations with some concomitant, related matters (issues which are addressed later in this chapter). Notwithstanding such apparent concerns, they are generally disposed to hard work and are willing to learn hard in order to succeed in their new environments. Teachers often note that these are generally hard-working students who are grateful to be in school, granted their past experiences.

In summary, global migrations of teachers and students are not only a part of our past, but will continue to be a significant part of the future of U.S. education. Since these migrations are already a significant part of America's higher educational landscape, science teacher education should address the opportunities and the challenges they present.

International Education: Process and Product Cycles

It can be argued that international education has been one of the primary fuels of accelerated globalization. This acceleration has occurred partly because globalization is a process that feeds on itself, and, once begun, is not only self-sustaining, but apparently accelerative. In the next sections, the world economy and its relationship to globalized education will be explored. These two processes will be used to illustrate why it is important to address issues related to international education in today's world.

Internationalized Education as an Appendage of the Globalized Economy

The global economic recession that was observable in the U.S. economy from 2007 started more tamely and regionalized; however, because large international companies like AIG, Bear Stearns, and Lehman Brothers had globalized interests and obligations, it did not take long before the economic problem that initially appeared to ordinary U.S. citizens as a local (national) recession rippled across the globe, along the axis of international trade. In 2011, as the recession took its toll on inordinate numbers of jobless U.S. citizens, it became more obvious to ordinary U.S. citizens that companies can easily move their headquarters to any part of the world that offered them tax advantages and incentives. The lesson was also learned that the world economy is one giant, interconnected matrix. It is ever-growing, ever-connecting, and very complicated. More importantly, the sheer inertia of globalization compels whatever is in its way to succumb to its forces, and that includes education.

To international education observers, the events noted earlier come as no surprise, because the world economy has a necessary appendage: internationalized education, without which it cannot function. It is through globalized education that globalization of the economy is possible. In its fundamental form, globalization is a means of homogenization, a process by which world ideas and technologies can be articulated across national, cultural, and other conceivable barriers without undue impediments. The mechanism for achieving this homogenization is education: it is the mediating, interconvertible currency for global transactions. In other words, education is a kind of lingua franca for getting peoples of the world to be able to talk with each other, notwithstanding cultural, religious, philosophical, and other differences. Given the dominance of the world economy and human will to migrate in pursuit of better economic and life opportunities, globalized education is a process that deserves its fair intellectual space.

Globalization of Education: The Process and the Tools

If internationalized education is indeed an appendage of the globalized economy, then the question arises: "By what process did globalization of education take place?" The worldwide revolutionary events of the last few years, especially 2011, have taught us that significant amounts of teaching and learning are taking place over the Internet, at amazing speeds. Within a matter of weeks, globalization tools such as television, cell phones, and personal computers had spread the revolution started by the self-immolation of a lone Tunisian man into a Northern African and Middle Eastern revolution. Thus far, this revolution has toppled or shaken the leaderships of Tunisia, Egypt, Libya, and several others such as Syria, Yemen, Saudi Arabia, Qatar, and others appear to be in progress. More importantly, it has educated peoples of the world in fundamentally different ways. In this revolution, English was the dominant language (the lingua franca), and specific kinds of content knowledge and pedagogical tools were used. These issues will be further discussed in this section so as to illustrate the notion of globalization as a process that is promoting the homogenization of education.

As noted earlier in this chapter, in the past, globalization of education took place primarily along the axis of international travel of students and professors (or K-12 teachers, in limited cases), either in exchange programs or in formal learning arrangements. With the advent of the Internet, however, different technologies have created vast opportunities for people across the world who aspire to the lifestyles of the West to interface with Western education. As an African immigrant, the author is very familiar with this phenomenon, whereby many people in the developing world try to "dress like Americans" and speak with what they call "American accent," often heard in U.S.-based movies. Whereas, in the past, the movies transmitted U.S. culture across the world, the Internet goes far beyond this role: it also educates, for better or for worse.

The English language as the global lingua franca. As an African student in Hungary in the late 1980s, it was interesting to note that the English language had already made its mark as the global lingua franca. There were Hungarian institutions of higher learning (including the one where I was a fellow—Hungarian Academy of Sciences) that were internationalized and taught their courses in English only. These programs attracted people from all over the world, and the instructors were better paid. As a relatively good speaker and writer of the English language, I was something of a super star, because most of the less-capable speakers of English wanted to get the opportunity to practice their spoken English with me. In fact, people viewed the English language as a tool for professional progress. In this connection, I had the opportunity to work with a professor of Pharmacology in the international program. In the same vein, there are English language programs in schools across the world, from Saudi Arabia and Egypt's "American Schools" to China, Korea, and Japan, and many specialized schools where Teaching English as a Second Language (TESL) programs flourish.

Globalized content knowledge. In terms of content knowledge, one can convincingly argue that, with all its limitations, Western knowledge and values, especially those of the United States, are perceived in the world as the ideal, for several reasons. First, the United States, with all its unique challenges, is still the world's most powerful nation. Although China is a large nation in terms of population and land mass, the world still looks to the United States as its main Superpower. U.S. movies and entertainment industries have long been the major exporter of United States' ideas and culture; therefore, it is one of the top attractions for migrants. Second, Western knowledge is valued because it is the gateway to opportunities in the Western world. Until recently, when China, India, Brazil, and other emerging economies began to offer migrants attractive opportunities, most educated migrants viewed Western, developed countries as not only economically promising but also inviting. This is so because many migrants were themselves from past colonies of such Western countries and had been educated under the corresponding colonial systems. For this reason, it was easier to transfer educational credentials to the receiving country. Another reason Western education and values are perceived in the world as the ideal to migrants is that the West has long embraced relatively more transparent, democratic values and egalitarianism. Such human values appear to resonate with most humans, thus explaining several historical revolutions and revolts in environments where there is a paucity of such basic human values. Not surprising, therefore, international migrants prefer to move to such environments.

The notion that a significant part of globalized content knowledge is Western is not difficult to illustrate. For example, from January through September of 2011 (a time period captioned "Arab Spring") when large masses of people in North Africa and Middle East were clamoring for democracy in their countries, it was notable that international broadcasting networks across the world featured several ordinary protesters who spoke English with U.S. accent and were disposed to U.S. values. They wanted Western-type, democratic governments and tacitly invited Westernized, democratic nations, such as the United States, Britain, France, and Westernized nations, such as Turkey, to support their revolutions. They proffered egalitarian views, and asserted that they have the right to self-rule and self-determination. Needless to mention, many of these youth are products of Western, or home-grown Westernized education.

Globalized pedagogical tools. Not surprisingly, the Internet is, by far, the singular medium by which the world is learning. Even in the developing world, it is notable that the use of the computer and computer-based products and applications (including smart phones) is commonplace. Over the last decade, as many previously captioned "developing nations" such as China, India, Vietnam, and Ghana have become "emerging economies," there has emerged a reverse migration "back home." This is a phenomenon whereby people from Western countries who migrated from these emerging economies are going back to their native lands with strong technological knowledge. These people set up Internet cafes and accelerate the technological know-how in their countries. For this reason, it is common for even farmers in eastern Africa to manage their farming accounts on their cell phones—a technology that is used at a relatively higher functionality in such parts of the world, due to the high cost of actual computers. Besides, CNN and BBC news outlets are common in the homes of remote villages of Ghana, for example, and, for that reason, there are many Ghanaians villagers who are very conversant with world affairs. Lastly, the Arab Spring revolutions mentioned earlier were captioned by some as the "twitter revolution," because of the high usage of social media such as tweets and Facebook to spread information regarding revolutionary activities.

In summary, accelerated internationalized education and globalization have gone hand-in-hand, and the former may have even accelerated the latter. In his book, *The World is Flat: A Brief History of the Twenty-First Century*, Thomas Friedman emphasizes that globalization has created a world in which individuals of the world are in competition with each other, and time and space are no longer limiting factors to this competition. For all practical purposes, Friedman is right, and it is internationalized education that helps to make him so.

The Interface Between Globalization and Science Teacher Education

As noted earlier, human migration is a natural, unstoppable process. For this reason, teacher education programs are likely to have immigrant students whose needs should be understood in order to better help them to become effective teachers in the United States. This section will explore the nature of immigrants, the specific issues facing this population, and discuss different ways to incorporate their needs into teacher education.

The Dichotomy of Who Does Science: Capable Students Who Are Disabled by Systemic Challenges

In this era of globalization, if science education is to effect change in teachers' and students' worldviews of "who can do science" and "who does science," an understanding of the issues that face a significant teacher and student constituencies—that is, international teachers and students in the United States—may need to be addressed in order to increase their teaching and learning capacities. For example, whereas international and immigrant students constitute a large portion of science majors and recipients of science degrees in the United States, these same often students face formidable challenges in their education. Similarly, immigrant teachers in the United States face issues which mitigate their contribution to science education. In a sense, one may argue that there is a systemic disabling process that selects those who can do science and those who do science.

The systemic disabling process that selects those who can do science and those who do science occurs because there are capable students who are challenged by structural issues. Such students include unconventional learners, students of color, and students from misunderstood cultures. The latter population includes international students in general, and in recent decades, refugee students and students with interrupted formal education (SIFEs). In both theory and practice, the thousands of immigrant teachers who often seek recertification in the United States in order to participate in science education should be included in this population. In order to fully include these populations in science education, their challenges must be understood so as to mitigate them.

The Nature of Immigrants in U.S. Classrooms

A Comparison of International Teachers and Students and U.S. People of Color

Although immigrant populations are different from U.S. citizens in many ways, they share some significant similarities with marginalized populations in the United States. This section will compare these two populations in order to illuminate how an expansive view of culturally responsive instruction can serve the needs of both populations.

Similarities Between Immigrant Populations and U.S. Students of Color

- *Existential difference.* By definition, all immigrants are de facto foreigners, at least for first-generation immigrants. The notion of being a foreigner instills a sense of difference or "otherness" of immigrants from the "standard" population. Since U.S. people of color have a sense of difference from the White majority population (the population that is the tacit, cultural standard), immigrant populations share the common characteristic of "otherness" with them.
- *Linguistic difference.* One rather unsuspecting difference between immigrant populations and U.S. marginalized populations is that often they speak with some linguistic departures from "United States Standard English." Whereas immigrants often speak with their own peculiar, regional, or continental accents, certain U.S. marginalized students may speak with strong influence of the local vernacular or accent. In this connection, it may be interesting to note that low-income Whites may be included in this micropopulation.
- *Nonverbal communication.* Closely related to linguistic difference is the way in which different populations use nonverbal communication. In this regard, it is helpful to understand Edward Hall's (1976) notions of high- and low-context cultures. Whereas the "standardized" European-American's nonverbal

communication is tilted towards low context, people of color in the United States are more oriented towards high context. In functional terms, **high-context** cultures are generally found in traditional societies with long histories and long-held assumptions. Many things are left unspoken, but well understood. Therefore, silent, body language has evolved to become a significant part of communication. On the other hand, **low-context** behaviors are prevalent in the West and are important in pluralistic societies where the need for clear, unambiguous verbal explanations for behaviors and actions is necessary (ibid.). A related example is the issue of eye contact, which is often avoided by subordinates in traditional, hierarchical societies as a sign of respect for their superiors, but the reverse is taught in the United States, interestingly, as a sign of respect and attention. In effect, immigrant populations are likely to find themselves as users of one form of nonverbal communication or the other, depending on their country of origin.
- *Cultural and worldview.* Because many U.S. micro-populations are still connected to their ethnic cultural values, U.S. people of color maintain cultural traditions that are distinct from that of the White majority. Since different cultural traditions create different worldviews, it can be argued that students of color in the United States are likely to share a differentiated worldview with immigrant populations, even if in different degrees.

Differences Between Immigrant Populations and People of Color in the United States

The differences between immigrant populations and U.S. people of color are better discussed in the context of the challenges these teachers face when working within U.S. classrooms. The purpose of this section is to address such issues.

To Friedman's (2005) assertion noted earlier (that the world is flat, thus insinuating a world of equality where there is equitable competition), a caveat must be inserted, that in the context of teacher education, traditional teaching and learning mostly involve personal migration of people, and once immigrants are in the new, local context, teacher education faces peculiar challenges that must be resolved. Many international students and prospective immigrant teachers encounter several challenges in their pursuit of education in U.S. classrooms. For example, in her study of Indian students in U.S. classrooms, Kaur (2007) found that these students encountered challenges that were culturally specific. For example, the students were more reserved in the classroom and were reticent in engaging in classroom discussion. They also had different learning habits: they learned in groups, mostly among themselves. Hutchison's (2005) research on international teachers in United States has shown that there are several peculiar issues (largely sociocultural and pedagogical shocks) facing this population in their attempts at working in U.S. classrooms. His findings were corroborated by Washington-Miller (2009), who noted that Caribbean immigrant teachers in London had similar challenges, including shock, loss of confidence, impairment of self-esteem, lack of support, financial constraints, and perceptions of abuse by their own students. From both research and personal experience standpoints, these are indeed general immigrant educational issues and affect both immigrant teachers and students. Therefore, they are important considerations for the implementation of what one may call internationally inclusive teaching. Hutchison's (2005) findings include the following:

- *Culture shock.* Immigrant teachers (and students) are likely to experience social and culture shock by dint of differences of lived experiences in different countries. Culture also extends to the differences in teaching approaches across cultures—an issue that was termed *pedagogical shock*—explained in this section.
- *Systemic barriers.* Different school systems are set up differently, based on different educational philosophies. For example, in many parts of the world, there are national standards, and the administrative set-ups are different. Students may have assigned seats all day, even in high schools, and it is the teachers who move around different classes during the school day. Besides, science teachers may have the specific help of laboratory technicians or assistants who order supplies, prepare, and set up laboratory experiments. Such teachers may be rather surprised that, as science teachers in the United States, they need to assume the role of the laboratory technician, as well as teach their classes, and even help clean up.
- *Assessment issues.* Unlike the U.S. assessment system where it is easy to earn an A grade, many school districts around the world make it much more difficult to earn an A. Hence, differences in assessment philosophies across national barriers can potentially become an issue for immigrants.
- *Communication issues.* Besides the fact that international teachers may have different accents, there are different spellings, expressions, and idioms that can pose as instructional barriers when teaching across national barriers.
- *Teacher-student relations.* The U.S. society is relatively more egalitarian: it is free and open and has much less hierarchy. Partly for this reason, teacher-student relations are relatively unencumbered by social rules, and students communicate with their instructors at ease. Conversely, however, teachers from traditionalist societies have a problem with students being too close and not honoring the teacher-student hierarchical gap.
- *Pedagogical approaches.* Perhaps, contingent on their teacher-student relations, U.S. students expect their teachers to be relatively active and hands-on. On the other hand, the teaching cultures of traditionalist societies are more lecture-based and follow the sage-on-stage approach. Immigrant teachers are therefore more likely to experience a kind of teaching-based culture shock: a pedagogical shock. In the same vein, migrant students who are used to the sage-on-stage approach to teaching would also experience the corresponding shock, from the perspective of the learner in a different pedagogical culture.

The abovementioned challenges facing immigrant teachers and students signify one major point: migrants from different parts of the world may enter the United States with significant differences in their pedagogical experiences and expectations that U.S. teacher education must address. The literature on the internationalization of teacher education in the United States has been critical of the fact that

U.S. teacher education has not been adequately responsive to the need for creating what may be termed internationally inclusive pedagogy. In addressing this issue, Kissock and Richardson (2010) note that "it is time that we heed the extensive literature calling on us to internationalize our teacher education programs and bring a global perspective to decision-making, in order to prepare globally minded professionals who can effectively teacher any child from, or living in, any part of the world" (p. 89).

Internationally Inclusive Teaching and Its Implications for Teacher Training

In their 2008 article, "Developing into similarity: global teacher education in the twenty-first century," Loomis, Rodriguez, and Tillman rightly propose (albeit in lamentation of the forceful, bulldozing effects of globalization) that "the information systems of markets—economic, political, and social—are converging under the pressure of the rule-making function of institutions" (p. 233). They note that globalization has the effect of forcing both private and public institutions to create standards which are often blind to local needs. Notwithstanding this critique, they yield to the fact that the power of globalization—an unstoppable process—has forced teacher education standards to become more homogenized in order to serve what may be perceptively viewed as common globalized standards. Although Loomis, Rodriguez, and Tillman do not appear to argue in favor of globalization per se, they raise an important point: context matters, and when the teacher education is not responsive to local needs, the power of education to create local change is lost.

Embracing the fact that globalization is an inextricable part of U.S. education, more researchers are taking an interest in related issues, and that is being reflected more in the literature. Reyes Quezadas (2010), in an editorial comment in *Teacher Education* journal, noted that internationalization of teacher education is a part of the skills we need to offer our preservice teachers in order to become competent in the twenty-first century. He poses several questions that he deems critical in creating that competent teacher, including the following:

1. How do we define internationalization in teacher education, and what does it mean to have international competence in education?
2. How can colleges and schools of education ensure that all teacher education candidates are competent and have the knowledge, skills, and dispositions to be effective intercultural teachers in an era of globalization?
3. What is the role of teacher education curricula and programs in promoting teaching about world cultures and their peoples as they work with P-12 students? (pp. 1–2)

These questions will partly guide the discussion of what I propose for internationally inclusive teacher education, which are guidelines that address the needs of immigrant teachers and international students alike. I will use Geneva Gay's culturally responsive teaching as the yardstick for proposing this model.

Gay (2000) defined *culturally responsive teaching* (CRT) as the use of cultural characteristics, experiences, and perspectives of ethnically diverse students for effective teaching. Gay based CRT on the assumption that when academic knowledge and skills are situated within the lived experiences and frames of reference of students, the knowledge is more meaningful at a personal level. Besides, CRT has a higher interest appeal, and is therefore learned more easily and thoroughly. Gay later refined her concept of CRT by adding that it should include several other considerations, including the capacity to demonstrate cultural knowledge and caring, the capacity to build a learning community within the cultural context, knowledge of cross-cultural communications, and the propagation of cultural congruity in classroom instruction. These elements are in harmony with the research regarding international teachers and students (c.f. Hutchison, 2005).

In light of the research noted thus far, internationally inclusive teaching should embrace several considerations, including (a) international, cross-cultural sensitivity, (b) internationally sensitive pedagogy or andragogy, (c) communication sensitivity, and (d) orientation to social, classroom, and educational cultures present in the United States. These considerations will be discussed next, with the tacit question, "How can international students be incorporated into the mainstream United States classroom?"

International, Cross-Cultural Sensitivity

The research on cultural knowledge and skills and their relationship to teaching and learning is well established (e.g., Atwater & Riley, 1993; Banks, 1993; Gay, 2000). In fact, the very concept of multicultural education rests on the foundation that, because United States is a multicultural nation, the interests of the different comprising cultures should be represented in schooling practices. In the same vein, since international students and teachers emigrate from countries with different cultural backgrounds, it is important to be considerate of their cultural differences. The question, however, is: How can instructors learn about all the cultures of the world so as to accommodate all international students? Whereas the direct answer to this question is that it is virtually impossible to learn about all the different world cultures, instructors can begin with certain fundamentals: learning the basics of cross-cultural etiquette. In an era of globalization, it should be considered a merely modest requirement to require instructors in higher education to have some currency in world affairs. There are basic materials that offer cultural information and cross-cultural etiquette. There are also several websites on the Internet that offer free information for international travelers, including common expressions and "dos and don'ts" in different cultures. Another means to achieve international, cross-cultural proficiency is to start small, by taking short diversity courses or workshops. When well taught, many diversity, cultural anthropology, and ethnic studies courses include information addressing the composition of United States' different ethnic groups, which are mostly foreign in origin. As a long-time instructor of diversity courses and workshops, I would even add that there is a shortcut to such courses: respect for all humans. After all, "the similarities across all humanity have endowed all teachers with a universal human language, a kind of *pedagogical lingua franca* which resonates with all, and can be successfully used to reach all learners" (Hutchison, 2011, p. 244). This language includes kindness, smile, and a helping disposition.

Internationally Sensitive Andragogy or Pedagogy

In their discussion of foreign teachers in Chinese classrooms, Yao and Lu (2011) observed that foreign teachers taught Chinese students using the pedagogical styles of their home countries. They noted that the foreign teachers did not follow the Chinese curriculum as strictly as the Chinese teachers would: they used the prescribed textbooks as guides, instead of teaching from them, and augmented their teaching practices with their own personalities, knowledge, and experiences. Such a pedagogical style, although is embraced in the West, is rather frustrating to students who are "trained" to look for what may be perceived as "solid learning" of factual material so as to perform well on their standardized examinations, a high-stakes matter for themselves and their families.

International students and immigrants in general, especially those from emerging economies, reside in educational contexts which are often ruled by high-stakes assessments or examinations that are used for selection of the best students for better educational and life opportunities. In such educational environments, students are "trained" to listen for long periods of time to fact-based lectures from sage-teachers. Often, these students also come from what may be called *listening cultures*, where children are seen, but not heard. They can listen for long periods of time and are reticent at asking questions in class. If they have any questions, they instead consult either their own classmates or their textbooks. These students can benefit from a "buddy system," whereby a willing U.S.-born student (or preferably a veteran foreign student) may be paired with them to provide mentoring. This mentoring partnership would also help the new student in the areas of language and cultural exposure, so as to mitigate the effects of culture shock.

As a former international student from a listening culture in both Europe and the United States, the author is very familiar with the frustration that international students face in U.S. classrooms. For a start, it takes several months at least, before international students begin to get used to the pedagogical styles found in U.S. classrooms—a time span that covers one semester (and thus a whole course). For many students, the practical consequence is the loss of the first set of courses, as they struggle to navigate them.

Such students would benefit from specific orientations, a topic that is addressed in this section. Many students are also self-conscious, especially if they speak with an accent, and are the only foreigners in the course or program. Such students may exhibit what is termed the *minority effect*, a process whereby they try to become invisible in class, a part of which is to keep quiet (Hutchison, 2009). Ultimately, students experiencing the minority effect are disadvantaged in their learning. Internationally sensitive pedagogy therefore involves finding ways to incorporate the emotional and psychological needs of international students in the course. This may take the form of lowering the *affective filter* (Krashen, 1982) or the shame factor of the students, so as to make them comfortable in speaking out or participating in course discussions. Another way to show sensitivity to international students is not to call them out in class, unless one is familiar with their levels of confidence in speaking out in class. In a research on international teachers, one of the participants from Britain noted that, as compared to U.S. students, British students are relatively shy and are not eager to speak out in class. In a part of research interview addressing this issue, this British teacher, pseudo-named Mary, noted the following:

> Mary: Definitely. It is different: They [U.S. students] will answer back. I find that United States students or students who've been to an American school are much happier to stand up and say things, like they would stand up in front of an assembly and speak and they have the confidence to do it. I have some British students [and] I can't get them to say anything in the class. It's like me. I don't like standing up and speaking in front of people. There's a natural reticence, certainly with Brits, to get to stand up and speak, but the American would either answer back or they would contribute, or they're happy to have a dialogue with you.
>
> And it's a confidence [issue], I think. I don't know where it comes from, but we have noticed it—the other British teachers and I always agree that getting one of the American students to stand up to do a presentation [is easy], but you try to get a Brit to do it and … they're much more reserved. And I think other nationalities are the same. I don't know about other Europeans, [but] some of our Asian students are quite reserved; some of our African students are quite reserved: They won't do it. They find it difficult to stand up and deliver. Whether it's a language thing or not, I don't know … There's no reason for Brits not to do it, but they won't say anything. On the whole, they're quiet. (Hutchison, 2005, p. 169)

Communication Sensitivity

Effective teaching and learning are processes that are mediated by communication. It is partly for this reason that in many truly international programs, foreign students who speak different languages are often provided with significant language immersion experiences and tutelage before they begin taking content courses. Given that different parts of the world have different communication styles, it should come as no surprise that communication barriers are indeed a major issue in international education. Fortuijn (2002) noted that "the problem of language is a problem of understanding" and that language involves "finding the right words, the right idioms, and the right nuances; it is a problem of pronunciation and audibility, tempo, tone and tune" (p. 266). He added that even people who speak good English may have problems with idioms and nuances. Communication issues for international students and teachers therefore involve challenges ranging from differences in the meaning of individual words, accents, differences in the meanings of expressions, and even styles of communication.

Another level of communication barriers in international education has to do with communication etiquette. Phillip Gin (2004) observed that it is important to pay attention to the unspoken standards of communication rules across cultures, since they may differ significantly. For example, in many Asian countries, it is impolite to be disagreeable. Therefore, asking "yes" or "no" questions may often lead to misunderstandings. Gin illustrates his point by noting that during an international conference in the United States, when explaining expectations and regulations to Asian guests, they usually said "yes" when asked if they understood what was explained—even if language barriers prevented them from truly understanding. For them, to reply "no" would show disrespect for the instructor, implying that the explanation was incompetently provided (ibid.). This same observation is somewhat applicable to people from listening cultures, who are often concerned with face-saving, or shame avoidance. These cultures are more apt to use indirect forms of communication, especially when addressing challenging, personal issues, and these styles are a part of the process of being polite and for avoiding conflict.

Given the communication issues raised above, therefore, in the classroom, educators need to be on the lookout for differences that may pose as impediments to instruction. For example, in working with students of German origin—who are known to be more direct in the expression of their feelings—they may sound rude to the unsuspecting "American" (Kuhn, 1996), and even more so in potential interactions with Asian or African students who tend to be more indirect in speech; therefore, there is likely to be unspoken conflicts and misunderstanding. Educators should therefore assume the role of communication managers so as to mitigate any potential conflicts that may arise in their pedagogical efforts. The point here is that during instruction, the focus should be on the content, not the means of its delivery.

One of the rather slippery areas of communication challenges in international education is that of differences in spelling. International students, especially those from emerging economies and where examinations are used for their selective function, are generally more particular about correct spelling. From personal experience, I had a significant problem with one of my first U.S. instructors who graded my work as poor because he thought that I had several spelling errors. On the other hand, I was surprised that the professor spelled poorly. In time, I came to realize that there are significant differences between U.S. and British spelling. The British teacher, Mary, in the research noted earlier, corroborates this point, on being asked a question regarding spelling differences and related issues she had noted in her U.S. teaching:

> Researcher: "... How about when you have differences in spelling?"
>
> Mary: Oh, how do I do like colo[u]r? Yeah, I spell it my way. I spell h[a]emoglobin my way. And I say you don't get the [letter] "a" in it. I don't care how you spell it. Just spell it the same way every time, and I'd say I'm not changing because I've been doing this for too long, and they laugh about that. C-o-l-o-u-r [spelling it]; Colour is one. Humo[u]r.

> Yeah, hemoglobin. Things like [o]estrogen…which doesn't have the 'o' in front of it. All sorts of things like that. But it's OK. I don't think it's a problem as long as they appreciate—you know; it's not a spelling mistake (Hutchison, 2005, p. 144).

Although the communication issues can be confounding if teachers and students are taken captive of them unaware, there are several solutions to them once they surface. For a start, internationally inclusive teaching calls for the recognition that there are regional and national differences in language and that when students are self-conscious about their language issues, be it accent, spelling, or otherwise, proficient instructors can find effective means to diffuse the issues. To some of the communication barriers raised in this section, internationally proficient instructors may consider using similar strategies such as those noted in the section captioned, *internationally sensitive andragogy or pedagogy*, to address any concerns. More notably, the lowering of affective filter can be a good start for students, so as to eliminate excessive concerns about self-presentation or shame. In fact, Mary noted that she was able to resolve her communication differences by using humor, self-deprecation, and consistency to manage the resulting issues. For example, she spelled her words consistently in British English, and in time, her students came to take her spelling for granted. On the other hand, she did not impose her spelling on others, provided her students followed the British or U.S. convention.

Orientation to the Social, Classroom, and Educational Cultures Found in the United States

Just as culturally responsive teacher education often involves clinical visits to culturally diverse schools so as to familiarize oneself with the issues, internationally inclusive teaching should involve some level of familiarity with several aspects of the U.S. education system. Programs that admit significant numbers of international students or that prepare international people as prospective teachers should consider equivalents of immersion into the U.S. education system early in their programs. The rationale for this is that such an exposure would help to contextualize the contents of the program. In considering the aspects such an orientation should involve, different types of orientations may be considered. They include systemic, philosophical, and pedagogical orientations. Besides, personal or group mentoring should be considered.

- *Systemic orientation*. Ideally, this would involve a visit to a local school in order to see how the school space is physically set up and spending a day with a teacher. In the least, it would involve watching a video of the same.
- *Philosophical orientation*. Students in the United States have the right to free education, paid for by the people, through local taxation. For this reason, U.S. students think of education as a right, not a privilege. This philosophical view of education is absent in many emerging economies, where education is a privilege,

not a right. Consequently, teachers wield significant power over their students, and the citizens view the teaching profession significantly more positively than in the United States. Immigrant students are often surprised at the disregard of the teaching profession and the treatment of teachers in the United States. Immigrant students need to understand the history and philosophy of education in the United States (even if briefly addressed) as a part of their cultural exposure. This would mitigate their later disappointments when they begin working as practicing teachers.
- *Pedagogical orientations.* This would involve watching how teaching is done in a variety of U.S. classroom settings. International, prospective teachers would benefit from watching U.S. teachers who are adept at using effective hands-on and cooperative teaching approaches because these are pedagogical approaches that are generally less familiar to immigrants.

Conclusion and Summary

In the context of a globalized world, education may be defined as the process whereby the peoples and cultures of the world interface in such a way as to create mutual understanding and progress. It is a job that largely falls on the shoulders of educational systems. Because different nations have different cultures, rules of etiquette, religions, philosophies, and worldviews, the very idea of globalization of education is challenging. However, the effects of the Internet as the primary globalized learning tool—coupled with human curiosity and the will to learn—has unleashed the powers of globalization, even in traditionalist countries that are ruled by dictators. Recent history, such as the Arab Spring of 2011, teaches us that even when world leaders oppose the compelling effects of globalization, their national boundaries are no longer impervious to sweeping ideologies, such as democratic and egalitarian thoughts. For this reason, much as political and institutional leaders may be forced to exhibit their modernity by paying mere lip service to the virtues of globalization but concurrently surreptitiously try to sabotage the local effects of globalization through inaction and passive resistance, the tangible outcomes of globalization are within our gates and are staring us in the eyes. Much in the same vein, teacher education cannot turn blind eyes to the effects of globalization.

In response to the now-natural and ever-progressing effects of globalization—a process that is largely driven by economics—education has become its captive, and must have an adequate response or become obsolete. We have arrived at a time when we need to heed our history, as humans: Since time immemorial, the migrations of humans across cultural barriers have elicited some forms of necessary cross-cultural accommodations, a form of cross-cultural education. As the world continues to shrink into a small community of learners, the onus rests on educators to unveil the processes by which cross-cultural education have taken place in the past—that is, across national, religious, cultural, philosophical, and other related boundaries—refine them, and make them work well for education in an era of globalization. The ideas prescribed for internationally inclusive education is just a first step in this process.

Activity: Creating an Internationally Inclusive Lesson Plan: The Whys and Whats

The purpose of this activity is to help students to begin thinking about how the issues raised in this chapter can be translated into internationally inclusive lesson plan. One way to do this effectively is to think about how a conventional lesson may be changed into one that is internationally relevant and inclusive.

- *Step 1:* In consideration of the contents of this chapter, select a lesson plan you have already created for a conventional lesson. Make sure that your lesson has your local and national standard objectives represented.
- *Step 2:* Highlight four issues you think are important for inclusion into the lesson so as to connect with international students.
- *Step 3:* Describe ways in which you would incorporate these ideas into the lesson under "Differentiated Action" in the lesson plan format table provided below.
- *Step 4:* Explain your rationale as to why your strategy for inclusion would make a difference under "Rationale" in the lesson plan format table.

Your final product may be presented in a table form as follows:

Internationally Inclusive Lesson Plan Format

Grade			
Topic			
Rationale			
Focus questions			
Intended learning outcomes			
Standards			
Materials and equipment			
Classroom demographics	For example, 5 Asians, 4 Africans, 3 Australians, 5 Europeans, *besides* conventional U.S. students		
Activities	Time	Differentiated actions	Rationale
Bell ringer:	5 min		
Lesson element #1			
Lesson element #2, etc.			

References

Atwater, M., & Riley, J. P. (1993). Multicultural science education: Perspectives, definitions, and research agenda. *Science Education, 77*(6), 661–668.

Bailey, T. (2003). *Skills migration. HRD review.* Available at www.hrdreview.hsrc.ac.za. Retrieved August 21, 2005.

Banks, J. A. (1993). Multicultural education: Development, dimensions, and challenges. In J. W. Noll (Ed.), *Taking sides: Clashing views on controversial educational issues* (9th ed., pp. 88–103). Guilford: Dushkin/McGraw-Hill.

Bump, M. (2006, March). *Ghana: Searching for opportunities at home and abroad.* Institute for the Study of International Migration, Georgetown University. Retrieved March 14, 2012, from http://www.migrationinformation.org/Profiles/display.cfm?ID=381

Friedman, T. (2005). *The world is flat: A brief history of the twenty-first century.* New York: Farrar, Straus, & Giroux.

Fortuijn, J. D. (2002). Internationalizing learning and teaching: A European experience. *Journal of Geography, 26*(3), 263–273.

Gay, G. (2000). *Culturally responsive teaching: Theory, research, & practice.* New York: Teachers College Press.

Gin, P. (2004). *International guests: Differences in culture and custom.* Retrieved December 30, 2005, from http://accedi.colostate.edu/imis_web/StaticContent/3/Pub/jan04.htm#INTERNATIONAL%20GUESTS:%20DIFFERENCES%20IN%20CULTURE%20AND%20CUSTOM

Hall, E. (1976). *Beyond culture.* New York: Anchor/Doubleday Books.

Hutchison, C. B. (2005). *Teaching in America: A cross-cultural guide for international teachers and their employers.* Dordrecht, The Netherlands: Springer.

Hutchison, C. B. (Ed.). (2009). *What happens when students are in the minority: Experiences that impact human. performance.* New York: Rowman and Littlefield.

Hutchison, C. B. (2011). *Understanding diverse learners.* Acton, MA: Copley.

Kaur, D. (2007). *International students and American higher education: A study of the academic adjustment experiences of six Asian Indian international students at a Research Level II university.* (Published dissertation). Department of Educational Leadership, The University of North Carolina at Charlotte.

Kissock, C., & Richardson, P. (2010). Calling for action within the teaching profession: It is time to internationalize teacher education. *Teaching Education, 21*(1), 89–101.

Krashen, S. D. (1982). *Principles and practice in second language acquisition.* Oxford, UK: Pergamon.

Kuhn, E. (1996). Cross-cultural stumbling block for international teachers. *College Teaching, 44*(3), 96–100.

Levine, A. (2010). Teacher education must respond to changes in America: Teacher education must adapt to the same changes in the economy, demographics, globalization, and more that are prompting change in K-12 education. *Kappan, 92*(2), 19–24.

Loomis, S., Rodriguez, J., & Tillman, R. (2008). Developing into similarity: Global teacher education in the twenty-first century. *European Journal of Teacher Education, 31*(3), 233–245.

National Education Association. (2003). *Trends in foreign teacher recruitment.* Retrieved February 22, 2004, from http://www.nea.org/teachershortage/0306foreignteacher.html

Quezadas, R. L. (2010). Internationalization of teacher education: Creating global competent teachers and teacher educators for the twenty-first century. *Teacher Education, 21*(1), 1–5.

Washington-Miller, P. (2009). Reconstructing teacher identities: shock, turbulence, resistance and adaptation in Caribbean teacher migration to England. *Education, Knowledge and Economy, 3*(2), 97–105.

Wiggan, G., & Hutchison, C. B. (Eds.). (2009). *Global issues in education: Pedagogy, policy, school practices and the minority experience.* New York: Rowman and Littlefield.

Yao, L., & Lu, H. (2011). Cross-cultural management of foreign teachers in Chinese colleges. *International Forum of Teaching and Studies, 7*(1), 23–29.

Using Problem-Based Learning to Contextualize the Science Experiences of Urban Teachers and Students

Neporcha Cone

Recent research in science education in urban school contexts necessitates systemic reform to create a more transformative process of learning for both students and teachers. Within this process is also the importance of making science instruction meaningful for students by helping them to make connections between their lived experiences and the content to be learned (Buxton, 2003; Sterling, Matkins, Frazier, & Logerwell, 2007). Problem-based learning (PBL) provides a context in which students and teachers experience meaningful science investigations over an extended period of time to solve real-world problems in an authentic context that is consistent with students' diverse cultures and perspectives. Instead of decontextualizing the science experiences of urban students from underrepresented groups, PBL might also serve as a conduit for science teacher educators to promote their preservice and inservice teachers to examine ways to support students' scientific understanding.

Students who participate in PBL outperform students who participate in traditional programs (Sterling et al., 2007). Specifically, underrepresented groups appear to gain the most from PBL programs (Gordon, Rogers, Comfort, Gavula, & McGee, 2001). Using critical theory as a lens of analysis, this chapter examines the following questions: How can PBL foster equitable learning experiences in science? How can PBL be used to create contextually authentic learning environments? How can PBL assist teachers and students in developing the skills and capabilities needed to be successful in science? How can PBL be used to give students a critical voice, thus providing educators with the tools to create meaningful science learning environments? As a new wave of science standards begins to emerge in many states across the country, this work has implications for helping students in urban settings gain an interest in STEM-related careers.

N. Cone (✉)
Curriculum and Instruction, Kennesaw State University,
1000 Chastain Road, MD 0121, Kennesaw, GA 30144, USA
e-mail: ncone@kennesaw.edu

Equity in Science Education

The marginalization of urban students in science education, because of unequal access to quality educational opportunities, has become an important issue for policy makers and the US public. To address this concern, the National Research Council (National Research Council [NRC], 1996, 2012) and the American Association for the Advancement of Science (American Association for the Advancement of Science [AAAS], 1990), respectively, released the following statements:

> Science is for all students. This principle is one of equity and excellence…All students, regardless of age, sex, cultural or ethnic background, disabilities, aspirations, or interest and motivation in science, should have the opportunity in science to attain higher levels of scientific literacy. [This principle] has implications for program design and the education system…to ensure that the *Standards* do not exacerbate the differences in opportunities to learn that currently exist between advantaged and disadvantaged students. (NRC, 1996, p. 20)

> Science and engineering are growing in their societal importance, yet access to a high-quality education in science and engineering remains determined in large part by an individual's socioeconomic class, racial or ethnic group, gender, language background, disability designation, or national origin…Arguably, the most pressing challenge facing U.S. education is to provide all students with a fair opportunity to learn. (NRC, 2012, pp. 280–281)

> Race, language, sex, or economic circumstances must no longer be permitted to be factors in determining who does and does not receive a good education in science, mathematics, and technology. To neglect the science education of any (as has happened too often to girls and minority students) is to deprive them of a basic education, handicap them for life, and deprive the nation of talented workers and informed citizens—a loss the nation can ill afford. (AAAS, 1990/1989, p. 214)

One of the most compelling equity-related concerns is the underrepresentation of racially/ethnically diverse groups and women in science-related fields (NRC, 2012). The second key concern is that of student grouping. Research studies show that students who are grouped and enrolled in lower ability classes, the majority of whom are African Americans and Latino/as, are less likely to be given equitable opportunities to learn quality science (Gilbert & Yerrick, 2001; Oakes, 2005; Parsons, 2008). The third concern is that of academic achievement. Analysis of the persistent academic gap between the science scores of White students and students of color indicates that the United States is not a meritocracy. More specifically, disaggregated data from the 2007 Third International Mathematics and Science Study (TIMSS) showed that the scores of urban students were below the international average (Martin, Mullis, & Foy, 2008). At the national level, disaggregated data from "Nation's Report Card" (NAEP) revealed that the academic gap between students of color and White students persists at the middle school and secondary levels (National Center for Educational Statistics [NCES], 2009, 2011; NRC, 2012). In addition, those students who were eligible to receive free and reduced lunch because they live in low-income families performed well below those who were not eligible on the NAEP science assessments.

Reasons put forth to explain these disparities include disproportionate numbers of students in poor urban settings who are exposed to curricula designed for low-ability students, limited access to the best qualified science teachers, preconceived stereotypes about diverse student groups which affect teacher expectations, and an educational system that is structured to benefit the hegemony (Banks, 2006; Darling-Hammond, 2004, 2006; Darling-Hammond & Bransford, 2005; Oakes, 2005; Parsons, 2008). For example, much of the science content is presented to students using the hegemony of Western Science. This limited perspective is problematic because it fails to consider the lived experiences and different ways of knowing diverse student groups bring to the science classroom (Atwater, 1996, 2000; Lee, 1999). This invalidation may be equated to the rejection of students' cultures—simply put, when policy makers, educators, and science curricula reject students' cultures, experiences, and ways of knowing and students themselves feel invalidated, silenced, and marginalized (Atwater, 2000; Buxton, Lee, & Santau, 2008; Lee, 1999). When diverse students fail to see themselves in authentic and meaningful ways, the consequences include unjust outcomes.

Within science education, the idea of educational equity is embedded in the phrase of "science for all," an important tenet of science education reform (AAAS, 1990; NRC, 1996). In addition, these documents state that teacher educators are to build on the preexisting attitudes of preservice teachers yet fail to acknowledge that these attitudes may be underpinned by beliefs that limit "all" students from learning science in an equitable way. Lee and Lukyx (2006) underscore this point by stating that equitable learning opportunities in science will occur when school science [including the science teacher] acknowledges, values, and respects the experience diverse student groups bring from their homes and community. When provided with these opportunities, "students are capable of demonstrating science achievement, interest, and agency" (Lee & Lukyx, p. 4).

Inquiry-Based Science Teaching and Learning

> The principal goal of education is to create men [sic] who are capable of doing things, not simply of repeating what other generations have done …the second goal of education is to form minds which can be critical, can verify, and not accept everything they are offered…So we need pupils who are active, who learn early to find out by themselves, partly by their own spontaneous activity and partly through material we set up for them. (Piaget, 1973, p. 36)

The educational community must abandon the "banking" ideology and pedagogy of poverty that relegates learners, especially those in urban settings, as empty vessels to be filled. Instead, teachers must find ways to tap into the cultural knowledge and understandings these diverse groups of youths bring with them into the science classroom. Spaces must be created that invite students "in" and involve them in actively constructing meaning. By empowering urban teachers and their students to be involved in the active construction of knowledge, learners are provided with opportunities to generate and share explanations and analyses, thus creating agency.

Standards for science education promote the implementation and use of inquiry-based instruction in science classrooms. Inquiry-based instruction refers to "experiences that help students acquire concepts of science, skills and abilities of scientific inquiry, and understandings about scientific inquiry" (NRC, p. 116). This approach to learning might serve as a vehicle for social change. However, many science teachers have not experienced inquiry-based instruction prior to entering the teaching profession. Thus, teachers are being asked to teach in new ways while supporting the science learning and academic achievement of *all* students.

Although it is laudable to explicate the need to prepare students in more instructionally responsive ways, remiss in the standards are explicit strategies for helping teachers to provide equitable learning opportunities for all students. As a result, many teachers experience fear, frustration, anxiety, and a decrease in self-confidence as they attempt to make sense of what it means to be an effective teacher. Within poor urban settings, these feelings are further exacerbated when teachers are confronted by limited resources, time, or decision-making authority to do anything systematic about what or whom they are accountable for teaching (Darling-Hammond, 2004, 2006; Giroux, 2003; Lee & Bowen, 2006). Therefore, urban teachers often feel stifled in their desires and abilities to make the science curriculum meaningful and relevant to students' lived experiences.

Problem-Based Learning (PBL)

As delineated above, some of the challenges faced in science education are entrenched in the ongoing struggle for equity. Specifically, teachers are being challenged to incorporate the voices, understandings, and values of urban students into the pedagogical practices that underpin their science curriculum. Thus, in order to achieve a more equitable science education for all students, educators should consider PBL as a pedagogical model because it allows learning to emerge from the lived experiences of both teachers and students.

Problem-based learning was originally developed to aid medical school students in developing content knowledge and clinical reasoning skills. Since then, PBL has been adapted for teaching science (Bouillion & Gomez, 2001; Gallagher, Stepien, Sher, & Workman, 1995; Goodnough & Cashion, 2006; Sterling et al., 2007). Contrary to direct instruction, PBL adopts an inquiry-based approach to prepare students to be productive citizens in a global society. In PBL, students learn science content through collaborative problem solving, reflecting on their experiences (both the process and the solution), and engaging in self-directed inquiry. They are stakeholders in their own learning and participate in relevant, meaningful problem solving. Students are cognitively engaged in sensemaking, developing evidence-based explanations as they present their ideas. The teacher acts as a facilitator of the learning process, providing content knowledge as needed. Outcomes of PBL include the development of critical thinking and creative thinking skills, increased communication and problem-solving skills, the application of knowledge to new and real-life situations, and increases in motivation (Goodnough & Cashion, 2006; Strobel & van Barneveld, 2009). Characteristics of PBL include the following:

- Learning is driven by open-ended, ill-structured problems or scenarios
- Problems are context specific
- Students work as self-directed, active investigators and problem-solvers in small collaborative groups
- After a key problem is identified, a solution is agreed upon and implemented
- Teachers become metacognitive facilitators of learning, serving as guides in the learning process while promoting an environment of inquiry (Gallagher et al., 1995).

Real-world problems with no clear answer are ideal for engaging urban students in learning science (Basu & Calabrese Barton, 2007; Bouillion & Gomez, 2001; Buxton et al., 2008). The reason for this is that they allow students to make connections between their "funds of knowledge" (Moll et al., 1992) and their science experiences.

Critical Theory and PBL

Although the diverse ways of knowing and learning that urban students bring with them to the classroom can be capitalized on to support their science learning, a major weakness of public schools is that they are engaged in schooling students, rather than educating them. For example, critical theorists argue that schools should be sites of social transformation, rather than sites of social reproduction (Freire, 2004; Giroux, 2001, 2003; McLaren, 1993). Students should be educated not only to be critical thinkers but also to become change agents. In short, education should be the source of social transformation and the mechanisms for bringing the inequities embedded in the schools and community to the forefront of student consciousness. Although PBL was not specifically designed to meet the needs of critical pedagogy, one cannot help but see the alignment between the two approaches.

Like critical theory, PBL places the learner at the center of the curriculum. Instead of serving as passive receptacles of knowledge, students are active participants in the learning process. It is within this methodology that the quest for meaning takes precedence. Dialogic conversations, lived experiences, and the posing of real-world problems emerge from the lives of students or the needs of the community (Bouillion & Gomez, 2001; Gallagher et al., 1995; Giroux, 2001, 2003). One of the most important points in regard to PBL is that students are not presented with facts and concepts through expository teaching methods, afterwards being expected to recall this information via memorization. Instead, students are presented with contextualized, ill-structured problems and are asked to investigate, discover, and negsotiate meaningful solutions. During this process, a learning community, undergirded by dialogue, trust, and collaboration, is created.

Dewey (1910/1977) highlights the importance of actively involving students in learning experiences whereby they are presented with "problems to be solved by personal reflection and experimentation, and by acquiring definite bodies of knowledge leading to more specialized scientific knowledge" (p. 168). This point is underscored by Freire (1993) who contends that inquiry-based instruction void of critical reflection does not empower students or give them opportunities to serve as change agents. Instead, students simply meditate on the problem without the knowledge and/or ability to pose solutions. By coupling inquiry-based instruction with critical reflection, stakeholders are challenged to understand their role in transforming the structural barriers (e.g., teaching practices) that might serve to marginalize urban students.

Why Use It? A Need for Implementing PBL in the Science Classroom

> Students should be given problems—at all levels appropriate to their maturity—that require them to decide what evidence is relevant and to offer their own interpretations of what the evidence means. This puts a premium, just as science does, on careful observations and thoughtful analysis. Students need guidance, encouragement, and practice in collecting, sorting, and analyzing evidence, and in building arguments based on it. However, if such activities are not to be destructively boring, they must lead to some intellectually satisfying payoff that students care about. (Rutherford & Ahlgren, 1990, pp. 188–189)

Far too often, usually around the middle school grades, many urban students of color are taught to dislike science, which results in students taking the minimum required courses in science at the high school level and a disinterest in pursuing science-related careers. One of the reasons for this disinterest is that students are not encouraged to actively participate in science. For example, in many urban schools, textbooks and handouts serve as a major device for students to learn science (Darling-Hammond, 2004; Freire, 2004; Oakes, 2005). This "pedagogy of poverty" contributes very little to student learning (Haberman, 1991). Contrarily, as highlighted in the quote above, effective teachers provide students with opportunities to engage in meaningful problem-solving activities.

Because learning is culturally embedded in social constructs, developing PBL questions or scenarios that are of importance to urban students from underrepresented groups allows them to engage in relevant investigations in an authentic context. Additionally, PBL provides students with opportunities to work with their peers for the purposes of discussing, generating, and sharing information. The nature of dialogue that takes place during this interaction requires collaboration so that students may construct, deconstruct, and reconstruct knowledge (Buxton, 2003; Giroux, 2003). In addition, the differences in students' cultural backgrounds as conversations take place contribute to the transformation of perspectives. It is also through these conversations that students begin to question and recognize the subjectiveness of knowledge based on evidence, challenging the belief that science is culture-free.

For science teachers, PBL fosters an environment conducive to developing the characteristics requisite for the effective science teaching of *all* students (e.g., critical thinking, reflection, communication, collaboration). Because problems or scenarios are relevant to them and are contextually situated in their lives and the lives of those whom they will be teaching, teachers are more likely to retain this knowledge (Ball & Knobloch, 2004; Karakas, 2008). Furthermore, PBL environments promote scientific inquiry, engaging teachers in a deeper understanding of the science content, along with curriculum analyses and revisions which might lead to explorations of new instructional approaches to teaching, learning, and assessment (McConnel et al., 2008). Authentic assessment during the PBL process also allows each student to demonstrate the attainment and application of new knowledge to ongoing situations. Finally, by providing students with different avenues for acquiring science content, PBL fosters the use of differentiated instruction, an effective pedagogical tool for maximizing the learning of those who have been traditionally marginalized in the science classroom.

PBL and Science Teacher Education: How to Use It?

Teachers must be provided with opportunities to learn science content concurrently with inquiry-based teaching strategies in collaborative and discursive settings. Within this environment, teachers should also be encouraged to connect their learning to authentic contexts, such as their own, or future, classrooms. In order to facilitate the knowledge, skills, and dispositions requisite for teaching science for diversity, teachers should be presented with instructional practices that challenge their thinking and encourage them to ask questions.

Creating the Ill-Structured Problem

A vital component of PBL is that problems are ill-structured in nature (Ball & Knobloch, 2004; Karakas, 2008). That is to say, no single formula exists for conducting the investigation on how to solve the problem, the problem might change as new information is gathered, and learners can never be completely certain that they have made the "correct" decision, as more than one possible answer exists. In "traditional" classroom environments, learning begins after students are presented with a body of information. Within the context of teacher education, PBL reverses the order of learning so that learning begins after teachers are confronted with an ill-structured problem, which is indicative of the teaching profession. Just as scientists would not perform an experiment before identifying a question or problem to be investigated, teachers in a PBL classroom begin the learning process after encountering an ill-structured problem.

PBL problems are also situated in the learners' lived experiences or real world. The ideas for problems may stem from school, community, family, social issues, or learners themselves. They may also stem from current events, movies, or newspapers. A "good" problem or scenario has the following characteristics:

- Complex, ill-structured—unclear, raises questions about what is known and needs to be known; not solved easily.
- Open ended—does not result in a right or wrong answer.
- Significant—a key problem/question is identified that is authentic and meaningful to the students.
- Researchable—like science itself, it changes with the addition of new information (Gallagher, 1997; McConnel et al., 2008; Weiss, 2003).

When aimed at teachers, good problems should also scaffold their current knowledge, be authentic and relate to their future career paths, promote knowledge and skills clearly identified in the course outcomes, be appropriately challenging, facilitate peer group interaction, and require the use of science content in ways indicative of effective science teaching (Weiss, 2003). For example, preservice teachers, in small groups, might be asked what they would do if they encounter a classroom where there are students with diverse learning styles and how they would teach the concepts of heredity, forces and motions, or pollution to these diverse student groups. They could also be presented with the following dilemma:

> District personnel has asked you to consider teaching biology in a racially, ethnically, socioeconomically, and linguistically diverse classroom next year. What is your initial response to this scenario? What do you need to know to make an informed decision?

Groups could then be required to write their initial reactions to the job offer, including information about their feelings, concerns, and any other responses to the problem. For inservice teachers, teaching dilemmas might include background information on grade-specific issues drawn from their own classroom experiences such as instructional decisions, student interactions, science content, and assessment strategies (McConnel et al., 2008). Next, they might be asked to identify what they know and need to know about the scenario or problem and develop testable hypotheses that will serve as a basis for data collection.

After discussions and research utilizing available resources (e.g., library, texts, Internet, hands-on activities), groups would be asked to come up with a plan for gathering data/evidence (e.g., intervention), summarize their results, revise their hypotheses, propose and defend agreed-upon actions, and share additional questions for future research (McConnel et al., 2008). Depending of the problem or scenario, preservice teachers could implement their ideas during their field experience or student teaching semester, thereby allowing for primary data collection. Similarly, inservice teachers might collect data from their own classrooms in the form of student work or videotaped segments of science lessons. Because teachers are developing a course of action and data collection procedures using the aforementioned framework, they are applying knowledge contextually appropriate educational setting that is relevant to their lives.

Understanding by Design: A Framework for Developing PBL Experiences

When developing PBL problems or scenarios, teachers must first consider their learning goals. I advocate planning using the Understanding by Design (UBD) framework, which emphasizes the teacher as a designer and facilitator of student learning (McConnel et al., 2008; Wiggins & McTighe, 1998, 2005). UBD, which is often referred to as the backward design instructional model, begins with identifying the desired results and then working backwards to develop instruction. Wiggins and McTighe (1998) identified three main stages to the backward design framework. In stage one, the teacher should identify the desired results, which include the goals, essential questions (e.g., the problem, scenario), and enduring understandings and knowledge. In PBL, problems or scenarios should be aligned with grade-level expectations and standards. This alignment will help teachers devote time and energy to important content and concepts. The second stage, determining what constitutes acceptable evidence of competency, defines what forms of assessment will demonstrate students' acquired knowledge, understandings, and skills to answer the problem. In stage three, planning learning experiences and instruction, the teacher determines what sequence of teaching and learning experiences will equip students to develop and demonstrate the desired knowledge and understandings.

PBL problems should appeal to students and quickly interest them in pursuing the goal, problem, or question. To facilitate this pursuit, students should be familiar with the real-world issue on which the problem has been contextualized. This requires teachers to have an understanding of their students' cultural backgrounds, prior experiences, and current knowledge. This knowledge will allow teachers to create a framework for learning that encourages student engagement. Furthermore, when teaching science content, problems need to be appropriately challenging, which contributes to the construction of new science understandings (McDonald & LaLopa, 2006). In summary, PBL experiences must be carefully planned before implemented to clarify the desired results, assessment strategies, and instructional methods. The following questions may serve as a guide as teachers proceed through the stages:

Stage 1

- What is the overarching theme or "big idea?"
- What content standards will this address?
- What facts, concepts, and background knowledge will students need in order to come to an understanding?
- What are the essential and focus questions that will drive the unit?
- What inquiry and process skills are required?

Stage 2

- How and what will students do to demonstrate their understanding?
- By what criteria will performances of understanding be judged?
- Through what authentic performance tasks will students demonstrate the desired understanding?
- Through what other evidence will students demonstrate achievement of desired results?
- How will students reflect upon and self-assess their learning?

Stage 3

- What instructional activities and support will move students through the phases of the inquiry process?
- What content and skills will be scaffolded?
- What resources beyond the classroom will be required and provided?
- How will formative assessment be integrated throughout the process?
- How will the problem be tailored (differentiated) to meet the different needs, interests, and abilities of learners?

Challenges with Using PBL

While PBL can make significant contributions to student learning, there are also some challenges in using this methodology in the classroom. These include teachers' and students' shifting roles, creating a collaborative learning environment, and maintaining student engagement (Levin, Hibbard & Rock, 2002; McConnel et al., 2008; Quartaroli & Sherman, 2011). Firstly, to assist teachers in shifting from a role of "disseminator" of knowledge to that of a facilitator of knowledge, teachers should observe experienced facilitators (e.g., science teacher educators) implement PBL, practice asking open-ended questions, and work in collaborative groups to develop relevant PBL questions and scenarios (Quartaroli & Sherman). Similarly, to aid students in taking on the role of self-directed learner, teachers might consider providing scaffolding through coaching, task structuring, and hints without giving students the "correct" or final answer. These scaffolds will make learning more accessible to students by changing complex and difficult tasks to tasks that are manageable and within students' zone of proximal development. Additionally, because PBL is situated in complex, ill-structured tasks, scaffolding might be required to engage students in the sensemaking process, manage their investigation and problem-solving processes, and encourage students to communicate their thinking to teachers and their peers (Gallagher, 1997; McConnel et al., 2008; Quartaroli & Sherman, 2011).

Secondly, cooperative learning groups and collaboration are important features of PBL and science reform initiatives (NRC, 1996, 2012); teachers should provide a model or example for transitioning and working together. Students and teachers should collaboratively establish norms and rules for culturally appropriate group behaviors. Whole class debriefing sessions can also serve as a model for effective group work (Karakas, 2008; Quartaroli & Sherman, 2011). Finally, although students tend to enjoy collaborative group work, there is not always a productive utilization of time. As facilitators, teachers must carefully monitor group progress by asking students support students' claims (verbally) with the use research-based evidence. This will help students to focus on the problem or central question of the scenario while keeping an eye on the big picture.

Conclusion

> Equity in science education requires that all students are provided with equitable opportunities to learn science and become engaged in science and engineering practices; with access to quality space, equipment, and teachers to support and motivate that learning and engagement; and adequate time spent on science. In addition, the issue of connecting to students' interests and experiences is particularly important for broadening participation in science. There is increasing recognition that the diverse customs and orientations that members of different cultural communities bring both to formal and to informal science learning contexts are assets on which to build.... (NRC, 2012, p. 28)

In recent years, concern has been growing about US underperformance in the fields of science, technology, engineering, and mathematics (STEM) (NRC, 2012; Obama, 2010). This trepidation is coupled with the underrepresentation of diverse groups in STEM-related areas and the pattern of low academic performance in science among disadvantaged students. The aforementioned information depicts the fact that the United States faces a significant challenge when it comes to attracting students to actively pursue STEM careers, especially students from underrepresented groups. Despite the increased emphasis on preparing students for these fields (NRC, 2012), more must be done to ensure that all students have equitable access to STEM career pathways. Teachers must first find ways to cultivate enthusiastic, culturally and scientifically knowledgeable students.

Many researchers highlight the importance of using a real-world context for teaching science (Banks, 2006; Basu & Calabrese Barton, 2007; Bouillion & Gomez, 2001; Rivet & Krajcik, 2008). Furthermore, situating learning in the lived experiences of students has an additional value when teaching science in urban settings. In order to overcome the numerous challenges urban students face while trying to achieve academic success, opportunities must be created so that they will develop the practices and habits of mind needed for scientific literacy and to build a community in which all can participate. To promote the development of these skills, science teachers must authentically inquire into the art of teaching and learning (AAAS, 1990; NRC, 1996, 2012). Therefore, science teacher education programs have a vital challenge: to prepare science teachers to implement practices that reflect a standards-based curriculum, using students' interests and experiences as a scaffolding mechanism. In short, teacher education programs must produce reflective practitioners, capable of solving the complex and ill-structured problems they will face while teaching, one of them being how to ensure that all students are actively and authentically involved their science learning.

Teacher educators must help teachers bridge students' home cultures with the school culture, in this case, science learning (Zeichner, 1996). PBL provides an important link between students' cultural backgrounds, community issues, and science. Yet, how do we convince teachers of the value of PBL when science is often viewed as non-biased and free of cultural influences? To assist teachers in making this connection, they themselves must be given opportunities to review and create inquiry-based science lessons that are relevant to their lives and the lives of their students. In other words, teachers must be immersed in authentic, PBL contexts so that they may see value of inquiry, even as it relates to their own learning (Barnes & Barnes, 2005). These immersion experiences may provide teachers with an opportunity to learn about the brilliance and cultural capital all students bring with them into the classroom. By acknowledging, affirming, and embracing differences in cultural backgrounds, teachers can build on the prior knowledge or urban students, ensuring that they have access to equitable opportunities to learn science.

References

American Association for the Advancement of Science. (1990). *Science for all Americans*. New York: Oxford University Press.
Atwater, M. M. (1996). Social constructivism: Infusion of the multicultural science education research agenda. *Journal of Research in Science Teaching, 33*, 82–837.
Atwater, M. M. (2000). Equity for Black Americans in precollege science. *Science Education, 84*, 154–179.
Ball, A. L., & Knobloch, N. A. (2004). An exploration of the outcomes of utilizing ill-structured problems in pre-service teacher preparation. *Journal of Agricultural Education, 45*(2), 62–71.
Banks, J. A. (2006). *Cultural diversity and education* (5th ed.). Boston: Allyn & Bacon.
Barnes, A. B., & Barnes, L. W. (2005). An approach to working with prospective and current science teachers. In A. J. Rodriguez & R. S. Kitchen (Eds.), *Preparing mathematics and science teachers for diverse classrooms: Promising strategies for transformative pedagogy* (pp. 61–86). Mahwah, NJ: Lawrence Erlbaum Associates.
Basu, S. J., & Calabrese Barton, A. C. (2007). Developing a sustained interest in science among urban minority youth. *Journal of Research in Science Teaching, 44*(3), 466–489.
Bouillion, L. M., & Gomez, L. M. (2001). Connecting school and community with science learning: Real world problems and school-community partnerships as contextual scaffolds. *Journal of Research in Science Teaching, 38*(8), 878–898.
Buxton, C. (2003). *Shared responsibility: Working to reconcile "authentic" learning and high-stakes accountability in a "low-performing" urban elementary school context*. Chicago: American Educational Research Association.
Buxton, C., Lee, O., & Santau, A. (2008). Promoting science among English language learners: Professional development for today's culturally and linguistically diverse classrooms. *Journal of Science Teacher Education, 19*(5), 495–511.
Darling-Hammond, L. (2004). The color line in American education: Race, resources, and student achievement. *W E B. Du Bois Review: Social Science Research on Race, 1*(2), 213–246.

Darling-Hammond, L. (2006). Securing the right to learn: Policy and practice for powerful teaching and learning. *Educational Researcher, 35*(7), 13–24.

Darling-Hammond, L., & Bransford, J. (Eds.). (2005). *Preparing teachers for a changing world: What teachers should learn and be able to do*. Jossey Bas: National Academy of Education, Committee on Teacher Education.

Dewey, J. (1910/1997). *How we think*. Mineola, NY: Dover.

Freire, P. (1993). *Education for critical consciousness*. New York: The Continuum Publishing Company.

Freire, P. (2004). *Pedagogy of the oppressed*. New York: The Continuum Publishing Company.

Gallagher, S. A. (1997). Problem-based learning: Where did it come from, what does it do, and where is it going? *Journal for the Education of the Gifted, 20*(4), 332–362.

Gallagher, S. A., Stepien, W. J., Sher, B. T., & Workman, D. (1995). Implementing problem-based learning in science classrooms. *School Science and Mathematics, 95*(3), 136–146.

Gilbert, A., & Yerrick, R. (2001). Same school, separate worlds: A sociocultural study of identity, resistance, and negotiation in a rural, lower track science classroom. *Journal of Research in Science Teaching, 38*(5), 574–598.

Giroux, H. (2001). *Theory and resistance in education. Towards a pedagogy for the opposition*. Westport, CT: Bergin & Garvey.

Giroux, H. (2003). Critical theory and educational practice. In A. Darder, M. Baltodano, & R. Torres (Eds.), *The critical pedagogy reader* (pp. 27–56). New York: Routledge.

Goodnough, K., & Cashion, M. (2006). Exploring problem-based learning in the context of high school science: Design and implementation issues. *School Science and Mathematics, 106*, 280–295.

Gordon, P. R., Rogers, A. M., Comfort, M., Gavula, N., & McGee, B. P. (2001). A taste of problem-based learning increases achievement of urban minority middle-school students. *Educational Horizons, 79*, 171–175.

Haberman, M. (1991). The pedagogy of poverty versus good teaching. *Phi Delta Kappan, 73*(4), 290–294.

Karakas, M. (2008). Graduating reflective science teachers through problem based learning instruction. *Bulgarian Journal of Science and Education Policy, 2*(1), 59–71.

Lee, J. S., & Bowen, N. K. (2006). Parent involvement, cultural capital, and the achievement gap among elementary school children. *American Educational Research Journal, 43*, 193–218.

Lee, O. (1999). Equity implications based on the conceptions of science achievement in major reform documents. *Review of Education Research, 69*(1), 83–115.

Lee, O., & Lukyx, A. (2006). *Science education and student diversity: Synthesis and research agenda*. New York: Cambridge University Press.

Levin, B., Hibbard, K., & Rock, T. (2002). Using problem-based learning as a tool for learning to teach students with special needs. *Teacher Education and Special Education, 25*(3), 278–290.

Martin, M. O., Mullis, I. V. S., & Foy, P. (2008). *TIMSS 20007 international science report: Findings from IEA's trends in international mathematics and science study at the fourth and eighth grades*. Chestnut Hill, MA: Boston College.

McConnel, T. J., Eberhardt, J., Parker, J. M., Stanaway, J. C., Lundeberg, M. A., Koehler, M. J., et al. (2008). The PBL project for teachers: Using problem-based learning to guide K-12 science teachers' professional learning. *MSTA Journal, 53*, 16–21.

McDonald, J. T., & LaLopa, J. (2006). Problem-based learning in the science classroom. *MSTA Journal, 51*(2), 6–10.

McLaren, P. (1993). *Life in schools: An introduction to critical pedagogy in the social foundations of education*. White Plains, NY: Longman.

Moll, L. C., Amanti, C., Neff, D., & González, N. (1992). Funds of knowledge for teaching: Using a qualitative approach to connect homes and classrooms. *Theory into Practice, 31*(2), 132–141.

National Center for Educational Statistics. (2009). *National Center for Educational Statistics*. Washington, DC: U.S. Department of Education.

National Center for Educational Statistics. (2011). *The nation's report card: Science 2009*. Retrieved May 10, 2012, from http://nces.ed.gov/pubsearch/pubsinfo.asp?pubid=2011451

National Research Council. (1996). *National science education standards*. Washington, DC: National Academy Press.

National Research Council. (2012). *A framework for K-12 science education: Practices, crosscutting concepts, and core ideas*. Washington, DC: National Academy Press.

Oakes, J. (2005). *Keeping track: How schools structure inequality* (2nd ed.). New Haven, CT: Yale University Press.

Obama, B. (2010, September 16). *Remarks by the president at the announcement of the "Change the Equation" initiative*. Retrieved December 1, 2010, from www.whitehouse.gov/the-press-office/2010/09/16/remarks-president-announcement-change-equation-initiative

Parsons, E. C. (2008). Learning contexts, black cultural ethos, and the science achievement of African American students in urban middle schools. *Journal of Research in Science Teaching, 45*(6), 665–683.

Piaget, J. (1973). *To understand is to invent*. New York: Grossman.

Quartaroli, M., & Sherman, F. (2011). Problem-based learning: Valuing cultural diversity in science education with Native students. In J. Reyhner, W. S. Gilbert, & L. Lockard (Eds.), *Honoring our heritage: Culturally appropriate approaches to indigenous education* (pp. 57–74). Flagstaff, AZ: Northern Arizona University.

Rivet, A. E., & Krajcik, J. S. (2008). Contextualizing instruction: Leveraging students' prior knowledge and experiences to foster understanding of middle school science. *Journal of Research in Science Teaching, 45*, 79–100.

Rutherford, F. J., & Ahlgren, A. (1990). *Science for all Americans*. New York: Oxford University Press. http://www.project2061.org/publications/sfaa/online/sfaatoc.htm.

Sterling, D. R., Matkins, J. J., Frazier, W. M., & Logerwell, M. G. (2007). Science camp as a transformative experience for students, parents, and teachers in the urban setting. *School Science and Mathematics, 107*(4), 134–148.

Strobel, J., & van Barneveld, A. (2009). When is PBL more effective? A meta-synthesis of meta-analyses comparing PBL to conventional classrooms. *Interdisciplinary Journal of Problem-based Learning, 3*(1), 44–58.

Weiss, R. E. (2003). Designing problems to promote higher-order thinking. *New Directions for Teaching and Learning, 95*, 25–31.

Wiggins, G., & McTighe, J. (2005). *Understanding by design*. New Jersey: Merrill Prentice Hall.

Wiggins, G., & McTighe, J. (1998). *Educative assessment: Designing assessment to inform and improve student performance*. San Francisco: Jossey-Bass.

Zeichner, K. (1996). Educating teachers for cultural diversity. In K. Zeichner, S. Melnick, & M. Gomez (Eds.), *Currents in reform in preservice teacher education* (pp. 133–175). New York: Teachers College Press.

Part IV
    # Equity, Multiculturalism, and Social Justice: Diversity Issues in Science Teaching

African American and Other Traditionally Underrepresented Students in School STEM: The Historical Legacy and Strategies for Moving from Stigmatization to Motivation

Obed Norman

Introduction

There is a great need for educators to create learning environments that are welcoming and nurturing for all students but particularly for students of color in STEM education. Although the problem of underrepresentation is present in most fields, it is more pronounced in STEM fields. The issue of the underrepresentation of African American students and others is therefore of great relevance to all science educators. Chubin, May, and Babco (2005) described the situation as follows:

> The demographics are clear. Although about a third of the school-age population consists of U.S. underrepresented minority students, over three-fourths (77 percent) of the working population in science, technology, engineering, and mathematics (or STEM) occupations is predominately white, with a fair representation of Asians (about 12 percent), but only about 11 percent African American, Latino/a, and American Indian participants. While women comprise about half of the school-age population, they represent only about a fourth of the STEM workforce. (p. 74)

This chapter deals mainly with African American students in that the history cited is that of African American teachers, but the principles and problems explored in this chapter are also applicable to Latino/a and other underrepresented groups in US urban schools. The problems of students who are underrepresented in specific fields such as STEM have been identified as rooted in the structural inequality of society (Anyon, 1997; Ladson-Billings, 2006; Ladson-Billings & Tate, 1995). In this

---

The author gratefully acknowledges the support of the National Science Foundation for this research through DRK12 Contract 0732109. The study was also partly funded by the Open Society Institute. All opinions and findings are exclusively the author's and not necessarily shared by the NSF or the OSI.

O. Norman (✉)
Capstone Institute at Howard University, Holy Cross Hall,
2900 Van Ness Street, NW, Suite 427, Washington, DC 20008, USA
e-mail: onorman6@gmail.com

chapter I argue that in the specific case of classroom interactions, the teacher-student cultural disconnect is one way in which this structural inequity manifests itself. It is within this framework that this chapter explores the historical roots of this disconnect and makes recommendations for the active and deliberate creation of classroom learning communities that support learning for all students in STEM.

The Problem of Cultural Disconnect in Urban Classrooms as It Relates to STEM Education

For the sake of clarity, I make a distinction between the problems and issues of schools in general and those of classrooms in particular. Urban schools in general face the institutional and structural problems of being inadequately resourced and more often than not being located in settings of concentrated poverty (Anyon, 1997, 2005; Atwater & Butler, 2006; Ladson-Billings, 2006; Norman, Ault, Bentz, & Meskimen, 2001). Classrooms are specialized sites within schools. These are the sites set aside for curriculum enactment aimed at formal teaching and learning. Classrooms are the interactional spaces with interstices where teachers and students are "able to intellectually and relationally imbue curriculum with personal and socially relevant meaning" (Craig, 2009, p. 1042). Regarding the specific case of STEM classrooms, Norman et al. (2001) have provided this characterization:

> Our theoretical perspective on culture is informed by our focus on the role of culture in situations in which persons from diverse cultural backgrounds interact in significant and sustained ways in the pursuit of common substantive goals. We designate these sites of interaction as cultural interface zones. Urban science classrooms are particularly complex cultural interface zones in that the cultures interacting are not only those of the students and teachers but also that of science as an intellectual discipline. (p. 1103)

I argue that the problems of urban classrooms can be characterized as relational and deriving to an appreciable extent, though not exclusively, from the cultural disconnect between the predominantly White teaching corps and those students who are not White. The focus of this chapter is therefore on the relational issues of classrooms. By relational I refer to the issues pertaining to how teachers and students relate to each other and the manner in which their relating impacts how the students in turn may orient themselves to the subject matter. The terms *school* and *classroom* are used interchangeably but both will refer mainly to classroom interactions.

The characterization of the problems in urban classrooms as mainly relational is of particular relevance to science education. Merging students' home cultures in a functional relationship with the culture of science learning has been described in terms of hybridity theory (Calabrese Barton, Tan, & Rivet, 2008). Hybridity theory deals with how disparate cultural elements can be brought into harmony with each other for the accomplishment of a common goal (Kraidy, 2005). According to Moje, Collazo, Carrillo, and Marx (2001), effective science teachers merge the two cultural spaces of the students' home culture and the culture of STEM learning into an accommodating hybrid third space in which students from diverse cultural backgrounds can effectively access science learning. In this hybrid third space, different funds of knowledge and discourse modalities are brought together in a cooperative relationship. Calabrese Barton et al. (2008) have documented how young female urban students from low-SES communities cooperated with their science teachers in forging that hybrid third space in ways that enabled those girls to enhance their science learning and develop their identities as legitimate participants in science. In 2007, the National Academy of Science (2007) issued a report which concluded that the underrepresentation of women in STEM careers may be addressed by making both the STEM learning and work environments more accommodating and welcoming for females. The same has been found to be true for other underrepresented groups (Byars-Winston, Estrada, Howard, Davis, & Zalapa, 2010).

In their work on cognition and learning, Lave and Wenger (1991) characterized learning as first and foremost a "legitimate participation in communities of practice" (p. 30). Student participation can only be substantive and effective if certain conditions are met. The first condition concerns the relationship between the learner and teacher. The evidence is overwhelming that there is a "relational" problem in the education of African American students. Students from underrepresented groups and especially African American students encounter unwarranted negative perceptions about themselves not only in school but also in the general culture (Ferguson, 2001a; a Norman, Crunk, Butler, & Pinder, 2006; Barbarin & Crawford, 2006). According to Steele and Aaronson (1995) "for too many Black students, school is simply the place where more concertedly, persistently, and authoritatively than anywhere else in society, they learn how little valued they are" (p. 11). Steele cites negative school experiences as an important source of the "dis-identification" of some African American students with learning. To some, this assessment by Steele may seem extreme and even harsh, but unfortunately it is an assessment borne out by a great deal of empirical evidence. In an analysis of school data, Skiba, Michael, Nardo, and Peterson (2000) found racial disparities in reasons for student sanctions as well as referral and suspension rates after correcting for differences in student socioeconomic status. The literature is replete with anecdotes of high-performing African American students who report that teachers generally fail to recognize their superior achievements (Galletta & Cross, 2007; Gross, 1993). Moreover, high-achieving African American students consistently tell of their constant struggles to receive the same recognition and validation from their teachers that White students reportedly get (Gross; Steele & Perry, 2004). My associates and I have replicated the same results in interviews with African American students (Norman et al., 2001). African American students believe that they have to be better than White students in order to receive the same level of recognition and validation from their teachers (Andrews, 2012; Steele & Perry, 2004).

A relevant anecdote was recounted in a nationally broadcast television program titled *Shaker Heights: The Struggle for Integration*. Two African American and two White female students recounted their friendship at school which included always studying together. Every time they handed in their projects, the African American females always attained B grades and the White females always As. At one time, these four students submitted their work with the White students' names on the work done by the African American students and vice versa. The work done by the African American students received an A grade when submitted under the White students' names and the work done by the White students received a B grade when submitted under the African American students' names (See also Ogbu, 2003).

The question can reasonably be asked whether negative teacher perceptions about students really matter when it comes to learning. The consensus in the literature is that it does matter. If a classroom teacher believes, for instance, that African American students value education less than their White peers, such a belief can be expected to impact what occurs in the classroom (Lynn, Bacon, Totten, Bridges, & Jennings, 2010). Ferguson (1998) warns that if teachers "expect Black children to have less potential, teachers are likely to search with less conviction than they should for ways to help these children to improve and hence miss opportunities to reduce the Black-White test score gap" (p. 312). What Ferguson said regarding the impact of teacher beliefs about student potential is also true of teacher perceptions about student educational aspirations. Holland, Lachicotte, Skinner, and Cain (1998) expressed the same concern regarding the potent effect of teacher perceptions in general:

> Even in situations where all students are admitted to the arena of learning, learning is likely to become unevenly distributed in its specifics. Teachers will take some students' groping claims to knowledge seriously on the basis of certain signs of identity. These students they will encourage and give informative feedback. Others whom they regard as unlikely or even improper students of particular subjects... are unlikely to receive their serious responses. (p. 135)

The Special Case of Young African American Male Students

Judging from all outcome measures, boys, and especially African American boys, are the students who have the most problematic relationships with teachers. The 2010 report by the Council of the Great City Schools (CGCS) presents a troubling picture concerning the state of the educational and general progress of African American males. While students from all ethnically underrepresented groups face daunting challenges, the state of African American males is such that it warrants special attention and intervention. According to the CGCS report, African American males drop out at nearly twice the rate of White males, and their SAT scores are on average 104 points lower. Galletta and Cross (2007) have described the struggles of African American youth with academic identity development within a school context in which racial prejudice was present. In her book *Bad Boys: Public Schools in the Making of Black Masculinity*, Ferguson (2001a) carefully catalogues the disproportionally harsh treatment and judgment meted out to the African American males. She also describes how these students push back at these attempts to stigmatize and marginalize them.

Every indication is that for African American males, the interaction between teacher judgments and actual student behavior is more complex because there is empirical evidence that suggests that unwarranted stigmatization of African American males is at play in varying degrees in education, as well as in the general culture

(Barbarin & Crawford, 2006). In Delaware, an investigation was launched in 2011 into whether African American students are disproportionately singled out for disciplinary action. For males, and particularly for young African American males, this student-teacher disconnect is particularly detrimental to their academic progress as males have been identified as mainly "relational learners." In an international survey of 1,000 teachers and 1,500 K-12 male students, Reichert and Hawley (2010) found that males learn best from teachers with whom they feel supported and inspired.

Researchers, community leaders, and policymakers have all expressed the need to address the educational and social challenges faced by young African American males. Pedro Noguera's collection of thoughtful essays "The Trouble With Black Boys" (2008) makes a compelling case for why and how educators should address the pressing problems of young African American and Latino/a males. Noguera (2008) frames the problems of young African American and Latino/a males as a structural disenfranchisement. Structural disenfranchisement refers to the fact that there are societal-structural arrangements in place that prevent all students from having equal access to educational opportunities. In the section that follows, I show how others, including the eminent sociologist William Julius Wilson and the National Academy's National Research Council, have also recognized and discussed the structural disenfranchisement that Noguera identified.

Classrooms as Sites of "The Social Structure of Inequality"

In talking to both preservice and inservice teachers about the experiences of students, these teachers typically respond that they only see students and do not see students as belonging to any ethnic group (Sleeter, 1993). This is intended as a laudable sentiment, but there is no evidence that this widely held "color blindness" has transformed our classrooms into sites of equity and inclusion for all students (Andrews, 2012; Hernández Sheets, 1996). Sociologist William Julius Wilson proposes a different approach. Wilson (1998) has characterized present approaches to educational equity as predicated on an "individual-level analysis" that "focuses on the attributes of individuals in their social situations." Taking his cue from Tilly (1998), Wilson suggests that these "individual level" analyses be augmented and complemented by approaches that "consider empirically the impact of the social structure of inequality on racial group social outcomes, including the impact of relational, organizational, and collective processes" (Wilson, 1998, p. 503). The structural analysis theory suggested by Wilson (1998) is discernible in the work of many authors. Anyon (1997, 2005) and Wilson (1987) implicate class as underlying the poverty that undermines educational progress in inner cities. Massey and Denton (1993) have explored the extent to which housing segregation creates pockets of racial isolation and concentrated poverty that adversely affect education.

Recognition of the relevance of the "social structure of inequality" as articulated by Wilson (1987, 1998) is also evident in the National Research Council's (NRC) establishment of a task force to formulate an agenda and procedures to research the impact of race discrimination (Blank, Dabady, & Citro, 2004). The premise of the NRC project is that the persistence of significant disparate outcomes for racial "minorities" in the United States warrants inquiry into whether race discrimination – a structural feature of society – may at least be one of the factors causing or sustaining such disparity.

Carter (2003, 2005) explored how the distinctive cultural capital of students from nondominant groups is discounted in educational contexts. This means that to an appreciable extent, teacher perceptions about African American students are less positive than the perceptions teachers have regarding White students. It is also true that White and Black students perceive their learning environments differently (Andrews, 2012). In my own research, I have found a statistically significant difference in the way White and Black adolescents perceived their learning environments (Norman et al., 2006). White adolescents have a more positive perception of their learning environments than do Black adolescents. In urban classrooms I have also observed significant differences between the experiences of Black and White students in the same classes (Norman et al., 2001). In one class we observed there were 24 students of whom about half were African American. In observations conducted twice weekly over a 4-month period, we observed 48 instances of the teacher ignoring Black students' requests for help but not a single instance of a White student's request for help being ignored (Norman et al., 2001). The teachers involved were not acting out of malice or intentional prejudice, but they were unaware of how differently they were responding to Black and White students. Similar observations were recorded by Andrews (2012).

There is therefore empirical evidence that suggest that the "social structure of inequality" is manifested in the way teachers and the larger society use irrelevant "outward signs" of student identity to privilege or valorize certain students (mostly White) and stigmatize others (mostly Black) (Holland et al., 1998). The notion of classrooms and formal learning environments as embodiments of the "social structure of inequality" may not be as far-fetched as it may appear at first glance. Schools may after all be examples of the "social contact" settings that Loury (2002) identified as the sites of persistent vestiges of inequalities despite the commitment to equality embodied in the "social contract" codified in laws such as the *Brown* decision.

The empirical evidence for the importance of positive and supportive teacher-student relationship is fairly compelling. The pressing question now is how schools in the United States got to the place where the teacher-student relationship is one of the crucial aspects in need of reform. The answer to this question can at least in part be found by casting a historical glance backward to the circumstances surrounding the implementation of the *Brown v. Board of Education* 1954 US Supreme Court decision.

The Roots of the Classroom Cultural Disconnect: The Unintended Legacy of the *Brown* v. *Board of Education* Decision

The US Supreme Court decision in *Brown v. Board of Education* (1952) is with some justification celebrated as signaling the advent of equal education in the United States. An important fact to remember is that while the *Brown* decision may be the most well known, it is only one of a number of cases that addressed the educational interests of African Americans, Latinos/as, and other underrepresented ethnic groups. It is also unquestioned that the *Brown* decision created the legal and political space for the civil rights movement that followed. However, recent analyses have pointed out that while the *Brown* decision was a declaratory victory for equality, it was implemented in a manner destined to perpetuate unequal outcomes in education. Ladson-Billings (2004) articulated a similar analysis: "But even as *Brown I* attempted to rend us from a racially troubled past, *Brown II* worked to suture us to that history" (p. 11). *Brown* was indeed a flawed solution as it contained the seed for the perpetuation of the very ill it was designed to remedy. It was a perfect promise with an imperfect implementation. While *Brown* led to the racial integration of schools in the United States, African American students entered these integrated schools as unwelcome intruders bearing a court order and at times accompanied by armed federal marshals compelling their admission over the objections of many. All this was further compounded by the fact that *Brown* was implemented in a way that led to the mass dismissal of African American teachers from schools (see Tillman, 1994 for an extensive treatment of this issue).

Brown therefore ushered in a situation where young vulnerable African American children were attending schools where they were unwelcome and from which teachers that look like them were systematically and deliberately removed. It does not require any leap of either logic or imagination to conclude that this was at some level an institutionalizing of alienation and marginalization of underrepresented students. It is worth recalling that during the Reconstruction period following President Lincoln's Emancipation Proclamation and the Civil War, 3,000 African American candidates were educated as teachers. W.E. B. Dubois regarded this as a significant sign of African American educational advancement. In his pioneering sociological work, *The Soul of Black Folk*, W. E. B. Dubois (1903) relates the strenuous efforts made by African Americans to be educated as well as the fierce and sustained resistance that greeted all such attempts. Eventually African Americans, in collaboration with their sympathizers, established their own institutions for the education of their youth and the preparation of African American teachers. These efforts, like all other initiatives aimed at the educational advancement of African Americans had many detractors. Dubois' response to these detractors was as pointed as it was elegant:

> Above the sneers of critics at the obvious defects of this procedure must ever stand its one crushing rejoinder: in a single generation they put thirty thousand black teachers in the South; they wiped out the illiteracy of the majority of the black people of the land, and they made Tuskegee possible. (p. 4)

By contrast, the *Brown v. Board of Education* decision brought in its wake a vast reduction in the number of African American teachers – a reduction that persists to this day. I have quoted W. E. B. Dubois at length in order to draw the sharp contrast between his view of the importance of the role of African American teachers and the negative aspect of the *Brown* decision which, in the way it was implemented, resulted in a disastrous reduction in the number of African American teachers (Tillman, 1994).

Why More African American Teachers Are Needed: Role Models and a More Inclusive Iconography for STEM

A feature of the segregated schools of the pre-*Brown* era was the strong presence of African American teachers. The *Brown v. Board of Education* decision was undoubtedly a positive development in that it ended the legally sanctioned racial segregation in public schools. But the removal of African American teachers was an unfortunate consequence of how this landmark decision was implemented. This was an unfortunate course of events and part of the historical events that have shaped schooling in the United States.

In addition to the "relational" problem discussed so far, the shortage of teachers who are not White also creates a persistent problem of "iconography" for African American and other underrepresented students. Every human enterprise has representational and historical images or icons associated with that enterprise. The iconography of an enterprise sends powerful messages about who has "membership" and can legitimately participate in that enterprise. The relative absence of African American science teachers creates a problem of iconography for African American students. These students are denied the opportunity to see themselves credibly represented in the iconography of science. This "underrepresentation" of African American teachers also undermines the extent to which African American students can imagine a science identity as part of their "possible selves" (Oyserman, 1993; Oyserman, Terry, & Bybee, 2002). The issue of the exclusive historical iconography of science and possible solutions for that is a related issue (Norman, 1993) but somewhat outside the scope of this chapter.

The first thing to recognize is that the historical legacy of marginalization and exclusion will not remedy itself. The acute shortage of African American teachers, which resulted from the implementation of the *Brown* decision, has received very little attention until recently. Some observers now regard the absence of particularly African American male teachers who can act as role models for young African American males as a possible impediment to the emotional development of African American children and especially for males (Noguera, 2008). Students of all races benefit when the teachers they encounter come not only or almost exclusively from one racial group but represent the diversity of the country (Howard, 2010). Both Black and White students are disadvantaged by being taught almost exclusively by teachers of one race or ethnic group, but the empirical evidence based on academic outcomes is undoubtedly that African American students bear the brunt of this disadvantage.

The removal of African American teachers resulted in schools where African American students encountered teachers with whom they did not share a common culture and experience. This cultural gap can be seen by a closer look at the disparities reported by Skiba et al. (2000). They reported that White students receive referrals more often for "vandalism, obscene language, smoking, and leaving without permission." African American students in contrast receive referrals for "more minor and subjective reasons, such as disrespect, excessive noise, threat, or loitering" (Skiba et al., p. 16). One way to interpret these disparities is to view them as resulting from interactions between students and teachers who come from differing cultural background and as a result interpreting each other's behaviors from different cultural frames of reference.

It goes without saying that the issue of the absence of African American teachers in classrooms should be acknowledged and addressed in the medium term or at least in the long term. A shortage brought about structurally by way of deliberate policy decisions cannot rectify itself without deliberate intervention. In a North-Eastern school district with an 88 % African American student enrollment, we found the percentage of African American STEM teachers to be in the single digits. There are many deliberate interventions to address a variety of problems, but none are in place to address the specific problem of the shortage of African American teachers in general or African American science teachers in particular. Some school districts have earmarked resources and deliberate interventions to address teacher shortages by recruiting science teachers from the Philippines and other countries, but no programs or resources are targeting the shortage of African American teachers that resulted from the implementation of the *Brown* decision. This is an aspect of the problem to be tackled at the school institutional and societal structural level. Our focus for now is how the problem attendant to this underrepresentation can be addressed in the relational space of classrooms as we find them today.

Educators at all levels will have to embrace the task of developing awareness about the situation and committing to putting forth the required effort to address the problem. There has to be awareness of the fact that at least to some extent, the challenge of teaching in urban schools with high enrollments of students from ethnically underrepresented groups is one of overcoming the historical legacy that has thrown together students and teachers who come from different cultural backgrounds and have different life experiences and contexts. It is only with this awareness that teachers can embrace the challenge and responsibility of challenging and questioning their own cultural orientations and assumptions with a view to exposing and overcoming the veiled individual and institutional cultural impediments to truly caring classroom relationships.

In the next section, I discuss three interventions aimed at reforming education in ways intended to increase the extent to which learning environments are welcoming and nurturing for all students and particularly for underrepresented students. A closer look at the central tenets of all three interventions shows that all three can be implemented more effectively when implemented by a diverse teaching force adequately representing teachers from underrepresented student groups. At the same time, the interventions proposed contain principles and approaches that can be used by teachers of all backgrounds to transform classrooms into welcoming learning communities for all students.

Strategies for Creating Positive Learning Environments

A variety of approaches have been advanced to address the problem of the underrepresentation of certain groups in STEM. The list of interventions discussed here is by no means exhaustive. The Talent Quest Model (TQM) was developed by Boykin and his associates at the Capstone Institute for School Reform at Howard

University (Boykin & Ellison, 2008). The TQM model has six key interconnected principles. A closer look at two of these principles can be used to illustrate how these principles can best be implemented effectively by a diverse teaching staff in which African American and teachers from other ethnic groups are adequately represented. Co-construction is the TQM principle that speaks to the school social and cultural dynamics to be addressed in the creation of effective schools. The co-construction of knowledge or learning environments implies very clearly that the voices and the input of all the teachers, communities, and students be heard. A gross underrepresentation of Black and Latino/a teachers has the potential to result in a skewed co-construction due to the absence of the cultural perspective that such teachers bring by virtue of their membership in specific groups. In the Multiple Outcomes principle of TQM, the focus is both on the need to emphasize academic achievement as well as to help students develop positive academic identities and develop an effort optimism that hard work can result in success in school and life in general. The presence of teachers from underrepresented communities acting as role models for those students can only be a powerful means of reaching the stated goals of this principle.

The other two intervention models (developed by Dweck and Oyserman, respectively) can be regarded as more specific embodiments of the broader TQM principles. Oyserman et al. (2002) have developed an intervention based on their identification of the type of racial self-concept that is compatible with high levels of academic engagement and performance for African American and students from other underrepresented groups. This racial self-concept is characterized by a positive orientation toward one's own ethnic group as well as an optimistic engagement with respect to the dominant culture. The key premise in the identity literature is that the content and structure of racial-ethnic identity (REI) have relevance for academic achievement (Oyserman et al. 2002; Spencer, Steele, & Quinn, 1999; Steele, 1997). Of particular relevance is the REI component termed Embedded Achievement and proposed by Oyserman, Gant, and Ager (1995). "Embedded Achievement" is when people believe that their membership in a particular ethnic group imposes on them an obligation to better themselves for the sake of the larger group. An important part of motivating students is to cultivate and enhance Embedded Achievement as part of an overall strategy to help students regard academic achievement as part of their ethnic heritage and obligation. This means that teachers should help these students to start seeing school success and their identity as congruent (Ford, 1992).

The literature on equity education points to a need to identify the qualifications required by teachers who will be successful with students from traditionally underrepresented groups (see, e.g., Ladson-Billings, 1994, 1999). It is also clear from the literature that both content competency and the ability to motivate students are required for effective teaching of these students. The work of social psychologists such as Steele (1997), Dweck (1998), Oyserman (1993), and others demonstrates that enhancing teacher content knowledge is crucial but not the sole area of concern when it comes to addressing the needs of students from underrepresented groups. An equally important area of concern is to equip teachers with insights and pedagogies aimed at ensuring that schools become welcoming, nurturing, and equitable learning environments that will engender in all students high levels of "effort optimism."

The third intervention is based on the work of Dweck (2000) who states that people implicitly conceive of intelligence and ability as either static (entity theorists) or as malleable (incremental theorists). Dweck and her colleagues (Aronson, 1998; Aronson & Fried, 1998; Levy, 1998; Levy & Dweck, 1998; Levy, Freitas, & Dweck, 1998) have shown through a series of elegant experimental studies that college students' susceptibility to "stereotype threat" can be significantly reduced by instruction (Wei, 2012). The instruction involved teaching students to view intelligence and ability not as fixed or static but rather as malleable and dynamic. Students who underwent the training not only outperformed their peers but also developed a more positive attitude toward their learning environment. These studies all demonstrated that persons who hold entity theories also exhibit greater propensity to stereotype others. While the work on theories of intelligence was initially done with college students, it has subsequently been applied successfully with elementary and middle school students as well (Blackwell, Trzesniewski, & Dweck, 2007).

In her case studies of successful teachers of African American students, Ladson-Billings (1994) included both European American and African American teachers. The hopeful message from that book is that teachers from all ethnic or racial groups can be effective with students from groups other than their own. In the rest of this chapter, we will discuss some of the strategies that teachers can use to transform classrooms into learning communities that are welcoming and nurturing for all learners, especially for those from communities that have hitherto been marginalized and excluded. According to Ferguson (2001b), "Black children do not require exotic instructional strategies that allegedly suit Black children better than Whites" (p. 367). What is needed is for all teachers to make the effort to relate to all students with equal empathy and acceptance. This will require effort on the teacher's part.

A simple practical solution to avoid unintended discriminatory treatment of students will be for teachers to regularly elicit feedback from all students as to how the students are perceiving their treatment by the teachers. Teachers can then correct their classroom responses as indicated by the students' feedback. I make this suggestion based on my work analyzing student responses on large national databases such as the National Longitudinal Study of Youth (NLSY, 1979, 1997). We conducted a statistical analysis of NLSY data to determine how a large random and reasonably representative sample of African American and White students responded to survey items that addressed their educational aspirations and also their perceptions about their learning environments. We analyzed the responses of students to relevant items on the National Longitudinal Study of Youth (NLSY) 79 and NLSY 97 (Norman et al., 2006). The NLSY 97 consists of a nationally representative sample of approximately 9,000 youths who were 12–16 years when they were first surveyed in 1997. The NLSY 79 is a nationally representative sample of 12,686 young men and women who were 14–22 years old when they were first surveyed in 1979. Each cohort continues to be surveyed annually or on a biennial basis. We also used data from a smaller student sample from a National Science Foundation-funded research project called Performance Enhancement for African

American Students in Science (PEASS). Our statistical analysis of variance (ANOVA) found no significant difference between the educational aspirations expressed by White and African American adolescents. We did, however, find a statistically significant difference in the way White and African American adolescents perceived their learning environments. White adolescents have a more positive perception of their learning environments than do Black adolescents. It is on the basis of this empirical finding that we can regard as an unwarranted stigmatization the persistent and pervasive belief that Black students have lower academic aspirations than White students. By conducting their own class-wide surveys from time to time, teachers can develop a realistic sense of students' perceptions and beliefs about their classroom experiences. This can go a long way in helping teachers change their strategies and approaches in ways intended to create more welcoming, nurturing, and equitable learning experiences for all students.

The Need for a Substantive Discourse on the Stigmatization and Valorization of Students

Teacher education seems to be the appropriate place to start thinking about recommending specific courses of action to address the issue of creating positive learning environments for all students. With respect to this issue, there is a clear need to initiate a substantive discourse with a robust conceptual framework and clearly defined constructs. The empirical basis for the framework and the associated constructs should be clearly visible. Without being exhaustive I want to offer here the beginnings of the outline for such a discourse for use in teacher education.

The consistent stream of empirical data that point to the very disparate school experiences of US White and Black students points to a need for teacher education to explore in some depth the extent to which stigmatization and its converse, valorization, may be part of the social structure of school and schooling. Valorization is the unwarranted attribution of positive qualities or perceptions to a particular group of people. Stigmatization, by contrast, is the unwarranted attribution of negative qualities or perceptions to another group of people. These two constructs are useful for framing a conversation on the sometimes very different experiences of Black and White students. The relationship between teachers and students can be complicated by the extent to which teachers may use irrelevant "outward signs" of student identity to privilege or valorize certain students and stigmatize others (Holland et al., 1998).

African American students are routinely stigmatized, while European American students can be said to be mostly valorized. That is how we can view the experiences of the African American students in the story we related earlier of the four girls who submitted their work under each other's names. These Black students' work earned an A grade for the first time when their work was submitted under their White colleague's names. Similarly the work done by the White students received a B grade for the very first time when submitted under the African American students' names. This anecdotal case of Black stigmatization and White valorization is confirmed empirically by the studies and reports cited earlier. To these reports we can also add the 2012 report of the US Department of Education's Civil Rights Data Collection (CRDC, 2012) survey. This national survey of 72,000 schools shows that racial disparities in school discipline, including suspensions, expulsions, and arrests, remain alarmingly high in districts and states across the United States.

These surveys and research reports of racial disparities in student experiences should be brought to the attention of preservice teachers in a systematic way and in the context of a discourse such as the one proposed here. Teachers and others working with students from underrepresented groups should recognize the need to have a solid empirical basis for perceptions about students so as to avoid unwarranted stigmatization of such students.

It is important to consider that stigmatization may not only undermine student performance by engendering stereotype threat. Stigmatization may also undermine student academic engagement by compromising the development of the type of student self-concept that social psychologists have identified as associated with high academic achievement.

In the foregoing paragraphs, I have already made reference to the work by Oyserman, Terry, and Bybee (2002) who have identified the two types of racial self-concepts that are compatible with high levels of academic engagement and performance for African American students. The first one is called "dual identity" and refers to a racial self-concept characterized by a positive orientation toward one's own ethnic group, as well as an optimistic engagement with respect to the dominant culture. The second type is called the "minority" identity and involves a positive attitude toward one's own ethnic group coupled with skepticism and a defensive vigilance toward the mainstream culture. If schools are to seriously take on the task of eliminating the achievement gap, schools must create learning environments that promote the development of the productive kind of racial self-concepts among African American adolescents.

The negative school experiences of discrimination and stigmatization undermine the ability of African American students to develop the optimistic orientation and engagement required for the formation of a dual-identity racial self-concept. Such negative experiences and perceptions may also exacerbate the defensive vigilance and skepticism of students with "minority" racial self-concept and thus compromise the ability of these students to maintain a pragmatic and productive engagement with the larger society.

There is a strong suggestion that the school experiences and perceptions of African American students are more geared toward fostering the development of exactly the types of racial self-concepts that Oyserman et al. (2002) have identified as associated with low academic performance and engagement (see also Norman et al., 2001). An example of such a self-concept would be the "in-group-focused" identity which is the one most prevalent among African American adolescents. Students who have this type of racial self-concept have retreated and disengaged from the wider culture to the ostensible safety of group-centered solidarity. Unwarranted stigmatization and negative perceptions about students can only exacerbate the test score gap.

When teacher education students participate in discussions such as proposed above, they are more likely to develop the appropriate empathy for their own students. Empathy for students means that teachers need to be mindful of the realities of different students' lives inside and outside the classroom. Teachers should recognize that students from underrepresented groups experience impediments to their learning that are largely attributable to structural features of society. The schools and school personnel should find ways to orient themselves toward the students and their communities in ways that create social capital for students. Students can then draw on that social capital to thrive. In an extensive ethnographic study in schools and communities in St. Louis and Atlanta, Morris (2009) described how African American students can thrive when schools and communities can come together and create accessible social capital for these students. Morris' work also shows that the ability of schools to forge these social capital-enhancing bonds with communities will be strengthened if the community is adequately represented in the ethnic composition of the school personnel.

Motivating Students

The central thrust of this chapter is that science teacher educators should communicate to teachers that classrooms should be places where students are motivated rather than stigmatized. So far we have discussed stigmatization. We can conceptualize Steele's (2004) construct of "stereotype threat" as essentially a response to stigmatization. Motivation would then be the counterpoint to stereotype threat. The insight from social psychology is that the cultivation of "possible selves" (images of the self one would like to attain) is critical for motivating action (Strauman & Higgins, 1988). Studies have also found that adolescents with school-focused possible selves are at reduced risk of involvement in delinquent activities, do better at school, and feel more connected to school (Oyserman, 1993; Oyserman & Harrison, 1998). Teacher education should equip teachers with the skills and dispositions required to provide students with the school experiences aimed at cultivating in African American students the perceptions that foster the development of the types of racial self-concepts or "possible selves" that Oyserman et al. (1998) have identified as associated with high academic performance and engagement. These authors have also provided an empirical basis for the idea that academic possible selves are rooted in part in racial identity.

Adolescence is the time when youth grapple with the critical identity development questions of who they are and whom they can become (Chavous, Harris, Rivas, Helaire, & Green, 2004). For youth from groups outside the majority, the challenge of identity development is further compounded by the fact that it takes place within a context in which they encounter discrimination and negative stereotyping of their identities as discussed in this chapter.

Conclusion

A vast body of empirical data on school outcomes for African American students points to a problem of teacher-student cultural disconnect. To an appreciable extent, this disconnect can be seen as a continuing legacy of the implementation of the *Brown v. Board of Education* (1954) decision which led simultaneously to the integration of students as well as the large-scale removal of African American teachers from schools. This has resulted in urban classrooms having to deal, among other things, with the relational problems arising from the cultural disconnect between the large numbers of African American students and the predominantly White teaching corps. While the focus here is exclusively on the relational issues of classrooms, it is important to at least mention the other problem arising out of the absence of African American teachers, particularly in STEM subjects. African American students in classes where African American teachers are grossly underrepresented are deprived of role models that they can follow into those fields. As adolescents grapple with the critical identity development questions of who they are and whom they can become (Chavous et al., 2004), their vision of the "possible selves" (Oyserman, 1993) they can attain is constrained if they are deprived of role models from their own ethnic or racial group. This alone is sufficient reason for why the acute shortage of African American and teachers from other underrepresented groups should be addressed at least in the long term.

The issue as discussed here suggests a number of specific recommendations to help educators create learning environments that are welcoming and nurturing for all students. At the policy level, the underrepresentation of black STEM teachers should be addressed. The present underrepresentation of Black teachers is the result of systematic and deliberate action and can only be reversed by equally systematic and deliberate policy initiatives. Beyond resolutions and mission statements, there is little evidence that this problem is being addressed in any substantive manner even as the percentage of such teachers is trending downward.

In the interim teacher education programs should take seriously the task of equipping teachers with both the skills and attitudes required for the relational challenges posed by culturally diverse students. In the short term, the relational problems in classrooms can be addressed and significantly reduced if teachers embrace the challenge and transform their classrooms into learning communities where all students are motivated instead of stigmatized.

I close this chapter with some of the most insightful words on the need for creating nurturing classroom environments. These are the words of a teacher at one of the schools where we conducted our research. This teacher wrote:

> The school-to-prison pipeline doesn't just begin with cops in the hallways and zero tolerance discipline policies. It begins when we fail to create a curriculum and a pedagogy that connects with students, that takes them seriously as intellectuals, that lets students know we care about them, that gives them the chance to channel their pain and defiance in productive ways. Making sure that we opt out of the classroom-to-prison pipeline will look and feel different in every subject and with every group of students. But the classroom will share certain features: It will take the time to build relationships, and it will say, "You matter. Your culture matters. You belong here." (Christensen, 2012)

Christensen advocates a critically engaged and culturally conscious approach that is very different from the well-meaning but wholly inadequate color-blind approach that says "I do not see color. I just see kids."

References

Andrews, D. (2012). Black achievers' experiences with racial spotlighting and ignoring in a predominantly White High School. *Teachers College Record, 114*(10), 46.
Anyon, J. (1997). *Ghetto schooling: A political economy of urban educational reform.* New York: Teachers College.
Anyon, J. (2005). What counts as educational policy? Notes toward a new paradigm. *Harvard Educational Review, 75*(1), 65–88.
Aronson, J. (1998). *The effects of convincing ability as fixed or improvable on responses to stereotype threat.* Unpublished manuscript.
Aronson, J., & Fried, C. (1998). *Reducing stereotype threat and boosting academic achievement of African Americans: The role of conceptions of intelligence.* Unpublished manuscript.
Atwater, M. M., & Butler, M. (2006). Professional development of teachers of science in urban schools. In J. Kincheloe, K. Hayes, K. Rose, & P. M. Anderson (Eds.), *Urban education: An encyclopedia* (pp. 153–160). Wesport, CT: Greenwood Press.
Barbarin, O., & Crawford, G. (2006). Acknowledging and reducing stigmatization of African American boys. *Young Children, 61*(6), 79–86.
Blackwell, L., Trzesniewski, K., & Dweck, C. (2007). Implicit theories of intelligence predict achievement across an adolescent transition: A longitudinal study and an intervention. *Society for Research in Child Development, Inc, 78*(1), 246–263.
Blank, R., Dabady, M., & Citro, C. (Eds.) (2004). *Measuring racial discrimination. Panel on methods for assessing discrimination.* Committee on National Statistics Division of Behavioral and Social Sciences and Education. Washington, DC: National Academies Press. National Research Council of the National Academies.
Boykin, W., & Ellison, C. (2008). The talent quest model and the education of African American children. In H. Neville, B. Tynes, & S. Utsey (Eds.), *Handbook of African American psychology.* Thousand Oaks: Sage.
Brown v. Board of Education, 347 U.S. 483 (1954).
Byars-Winston, A., Estrada, Y., Howard, C., Davis, D., & Zalapa, J. (2010). Influence of social cognitive and ethnic variables on academic goals of underrepresented students in science and engineering: A multiple-groups analysis. *Journal of Counseling Psychology, 57*(2), 205–218.
Calabrese Barton, A., Tan, E., & Rivet, A. (2008). Creating hybrid spaces for engaging school science among urban middle school girls. *American Educational Research Journal, 45*(1), 68–103.
Carter, P. (2003). "Black" cultural capital, status positioning, and schooling conflicts for low-income African American youth. *Social Problems, 50*(1), 136–155.
Carter, P. (2005). *Keepin' it real: School success beyond black and white.* Oxford, NY: Oxford University Press.
Chavous, T., Harris, A., Rivas, D., Helaire, L., & Green, L. (2004). Racial stereotypes and gender in context: African Americans at predominantly Black and predominantly White colleges. *Sex Roles, 51*, 1–16.
Christensen, L. (2012). The classroom to prison pipeline. *Rethinking Schools, 26*(2), 2011–2012.
Chubin, D., May, G., & Babco, E. (2005). Diversifying the engineering workforce. *Journal of Engineering Education, 94*, 73–86.
Council of the Great City Schools (CGCS). (2010). *Report on the education of Black males.* Retrieved from: http://cgcs.org/newsroom/Black_Male_Achievement.pdf
Craig, C. (2009). The contested classroom space: A decade of lived educational policy in Texas schools. *American Educational Research Journal, 46*(4), 1034–1059.

Dubois, W. E. B. (1903). *Souls of Black folks*. New York: Penguin Books.
Dweck, C. (1998). The development of early self-conceptions: Their relevance for motivational processes. In J. Heckhauser & C. Dweck (Eds.), *Motivation and self – regulation across the life span* (pp. 257–280). Cambridge, NY: Cambridge University Press.
Dweck, C. (2000). *Self-theories: Their role in motivation, personality, and development*. Philadelphia, PA: Psychology Press.
Ferguson, R. (1998). Teacher perceptions and expectations and the Black-white score gap. In C. Jencks (Ed.), *The black-white test score gap*. Washington, DC: Brookings Institution Press.
Ferguson, A. (2001a). *Bad boys: Public schools in the making of black masculinity (law, meaning, and violence)*. Ann Arbor, MI: University of Michigan Press.
Ferguson, R. (2001b). Test-score trends along racial lines, 1971 to 1996: Popular culture and community academic standards. In N. Smelser, W. Wilson, & F. Mitchell (Eds.), *American becoming: Racial trends and their consequences* (pp. 348–390). Washington, DC: National Academy of Science Press.
Ford, M. (1992). *Motivating humans: Goals, emotions, and personal agency beliefs*. Newbury Park, CA: Sage.
Galletta, A., & Cross, W. (2007). Past as present, present as past: Historicizing Black education and interrogating 'integration'. In A. Fuligni (Ed.), *Contesting stereotypes and creating identities: Social categories, social identities, and educational participation* (pp. 15–41). New York: Russell Sage.
Gross, S. (1993). Early mathematics performance and achievement: Results of a study within a large suburban school system. *Journal of Negro Education, 62*, 269–287.
Hernández Sheets, R. (1996). Urban classroom conflict: Student-teacher perception: ethnic integrity, solidarity, and resistance. *The Urban Review, 28*(2), 165–183.
Holland, D., Lachicotte, W., Skinner, D., & Cain, C. (1998). *Identity and agency in cultural worlds*. Cambridge, MA: Harvard University Press.
Howard, J. (2010). The value of ethnic diversity in the teaching profession: A New Zealand case study. *International Journal of Education, 2*(1), 1.
Kraidy, M. (2005). *Hybridity: Or the cultural logic of globalization*. Philadelphia: Temple.
Ladson-Billings, G. (1994). *The dreamkeepers: Successful teachers of African American children*. San Francisco: Jossey-Bass.
Ladson-Billings, G. (2004). Landing on the wrong note: The price we paid for Brown. *Educational Researcher, 33*(7), 3–13.
Ladson-Billings, G. (2006). From the achievement gap to the education debt: Understanding achievement in U.S. schools. *Educational Researcher, 35*, 73–12.
Ladson-Billings, G. J. (1999). Preparing teachers for diverse student populations: A critical race theory perspective. *Review of Research in Education, 24*, 211–247.
Ladson-Billings, G., & Tate, W. (1995). Toward a critical theory of education. *Teachers College Record, 97*, 47–68.
Lave, J., & Wenger, E. (1991). *Situated learning: Legitimate peripheral participation*. Cambridge, MA: Cambridge University Press.
Levy, S. (1998). *Children's static versus dynamic conceptions of people: Their impact on intergroup attitudes*. Doctoral dissertation, Columbia University, New York.
Levy, S., & Dweck, C. (1998). Trait-focused and process-focused social judgment. *Social Cognition, 16*, 151–172.
Levy, S., Freitas, A., & Dweck, C. (1998). *Acting on stereotypes*. Unpublished manuscript.
Loury, G. (2002). *The anatomy of racial inequality*. Cambridge, MA: Harvard University Press.
Lynn, M., Bacon, J., Totten, T., Bridges, T., & Jennings, M. (2010). Examining teachers' beliefs about African American male students in a low-performing High School in an African American School District. *Teachers College Record, 112*(1), 289–330.
Massey, D., & Denton, N. (1993). *American apartheid: Segregation and the making of the underclass*. Cambridge, MA: Harvard University Press.
Moje, E. B., Collazo, T., Carrillo, R., & Marx, R. W. (2001). "Maestro, what is 'quality'?": Language, literacy, and discourse in project-based science. *Journal of Research in Science Teaching, 38*(4), 469–498.

Morris, J. (2009). *Troubling the waters: Fulfilling the promise of quality public schooling for Black children*. New York: Teachers College Press.

National Academies [National Academy of Sciences, National Academy of Engineering, and Institute of Medicine]. (2007). *Beyond bias and barriers: Fulfilling the potential of women in academic science and engineering*. Washington, DC: National Academies Press.

Noguera, P. (2008). *The trouble with Black boys: And other reflections on race, equity, and the future of public education*. New York: Teachers College Press.

Norman, O. (1993). Beyond Kuhn: The historiography and epistemology of an inclusive sociocultural discourse on science. Proceedings of the 3rd International Conference on History, Philosophy and Science Teaching. University of Minnesota.

Norman, O., Ault, C., Bentz, B., & Meskimen, L. (2001). The black-white 'achievement gap' as a perennial challenge of urban science education: A sociocultural and historical overview with implications for research and practice. *Journal of Research in Science Teaching, 38*(10), 1101–1114.

Norman, O., Crunk, S., Butler, B., & Pinder, P. (2006). *Do Black adolescents value education less than White peers? An empirical and conceptual attempt at putting a thorny question in perspective*. Paper presented at the annual meeting of the American Educational Research Association Conference, San Francisco.

Ogbu, J. (2003). *Black American students in an affluent suburb: A study of academic disengagement*. Mahwah, NJ: Lawrence Erlbaum Associates.

Oyserman, D. (1993). Who influences identity: Adolescent identity and delinquency in interpersonal context. *Child Psychiatry and Human Development, 23*(3), 203–214.

Oyserman, D., & Harrison, K. (1998). Implications of ethnic identity: African American identity and possible selves. In J. Swim & C. Stangor (Eds.), *Prejudice: The target's perspective* (pp. 281–300). San Diego: Academic.

Oyserman, D., Gant, L., & Ager, J. (1995). A socially contextualized model of African American identity: Possible selves and school persistence. *Journal of Personality and Social Psychology, 69*, 1216–1232.

Oyserman, D., Terry, K., & Bybee, D. (2002). A possible selves, intervention to enhance school involvement. *Journal of Adolescence, 24*, 313–326.

Reichert, M., & Hawley, R. (2010). *Reaching boys, strategies that work – and why*. San Francisco: Wiley.

Skiba, R., Michael, R., Nardo, A. C., & Peterson, R. (2000). *The color of discipline: Sources of racial and gender disproportionality in school punishment*. Retrieved from http://www.indiana.edu/~iepc/

Sleeter, C. (1993). How white teachers construct race. In C. McCarty & W. Crichlow (Eds.), *Race, identity and representation in education*. New York: Routledge.

Spencer, S. J., Steele, C. M., & Quinn, D. M. (1999). Stereotype threat and women's math performance. *Journal of Experimental Social Psychology, 35*, 4–28.

Steele, C., & Aronson, J. (1995). Stereotype threat and the intellectual test performance of African-Americans. *Journal of Personality and Social Psychology, 68*, 797–811.

Steele, C. (1997). A threat in the air: How stereotypes shape intellectual identity and performance. *American Psychologist, 52*(6), 613–629.

Steele, C., & Perry, T. (2004). *Young, gifted, and Black: Promoting high achievement among African-American students*. Boston: Beacon Press.

Strauman, T., & Higgins, T. (1988). Self-discrepancies as predictors of vulnerability to distinct syndromes of chronic emotional distress. *Journal of Personality, 56*, 685–707.

Tillman, L. C. (1994). African American principals and the legacy of Brown. *Review of Research in Education, 28*, 101–146.

Tilly, C. (1998). *Durable inequality*. Berkeley, CA: University of California Press.

U.S. Department of Education [ED]. (2012). *The civil rights data collection (CRDC)*. Washington, DC: Office for Civil Rights, U.S. Dept. of Education.

Wei, T. (2012). Sticks, stones, words, and broken bones: New field and lab evidence on stereotype threat. *Education Evaluation and Policy Analysis, 34*, 465–488.

Wilson, W. (1987). *The truly disadvantaged: The inner city, the underclass, and public policy*. Chicago: The University of Chicago Press.

Wilson, W. (1998). The role of the environment in the black-white test score gap. In C. Jencks & M. Phillips (Eds.), *The black-white test score gap* (pp. 501–509). Washington, DC: Brookings Institution Press.

Preparing Science Teachers for Diversity: Integrating the Contributions of Scientists from Underrepresented Groups in the Middle School Science Curriculum

Rose M. Pringle and Cheryl A. McLaughlin

The students watched intensely as we flashed images on the projector screen of individuals from a variety of ethnic backgrounds, some dressed in their typical work gear while others wore regular outfits. Who are the scientists in this presentation? Why do you think they are scientists? One student, Alexis, selected the image of a White male dressed in a black pants and white shirt and jacket with a black bow tie, who was in fact a waiter at a five-star restaurant in downtown Los Angeles. "Why do you think he is a scientist," we asked curiously. "Well, because he looks so distinguished." Another student selected an image of an older White male bespectacled with unkempt hair, dressed in similar garb as the waiter. "Why?" "Well, you know, he looks like the typical scientist…crazy hair, safety goggles, bow tie. The only thing missing is the test tube and beaker." The other students laughed while another chimed in; "Yea, he has the 'Einstein' look alright." We were not amused, but we were curious. "And what about the Chinese and the well-dressed Black lady in the suit?" we continued. In response, Jan shared, "He is possibly an engineer or a computer specialist and she, a manager in an office or a receptionist." When informed that the Chinese was a member of his country's elite gymnastic team and the lady an engineer, the class broke into much laughter. We probed; "Where did your image of a typical scientist come from?"

Profiling Scientists

The preservice teachers' responses shared in the opening vignette illustrate the taken-for-granted misconceptions and stereotypes that are pervasive about the images of a scientist. Profiling is the use of specific characteristics such as age or race to make generalization about a person's way of life. The preservice teachers in the vignette above created profiles of scientists based on stereotypes that have become

R.M. Pringle (✉) • C.A. McLaughlin
College of Education, School of Teaching and Learning, University of Florida,
2423 Norman Hall PO Box 117048, Gainesville, FL 32611, USA
e-mail: rpringle@coe.ufl.edu; chermac72@ufl.edu

embedded in their consciousness over a given period of time. Stereotypes are oversimplified mental images that are often regarded as embodying a group or class of people. These images contribute to the beliefs and perceptions that are typically reflected in the way science lessons are implemented in the classroom. Given the inextricable link between teachers' belief and instructional practices (Bandura, 2000; Tschannen-Moran & Woolfolk Hoy, 2007), we believed it was important to tease out the stereotypes that preservice teachers have about scientists with the understanding that the process of learning to teach begins with making explicit one's beliefs about the issues related to teaching and learning. Embracing the proposal by Banks et al. (2001), our goal as we engaged the preservice teachers in multicultural pedagogy was to make science more representative and inclusive of the nation's diversity while reshaping the frames of references, perspectives, and concepts that make up school science knowledge. This activity provided opportunities for the preservice teachers to confront their own beliefs and stereotypes about scientists while increasing their awareness of the impact of these perceptions. Our whole group discussion was framed within two profound questions: What do scientists do? What images come to mind when you hear the word scientist? The following ideas were offered as to what scientists do: Scientists "look for explanation of how the world works," "seek ways of making our lives better by finding cures for illnesses," "do experiments and make new things," and "are involved in research and exploration." In reviewing the profiles constructed by our students, the recurring theme suggested that scientists were "well educated about science" as evidenced by the level of discussions, articulation, and command of the science content knowledge. One preservice teacher insisted that scientists were "just bright people and it is easier to recognize someone who is a non-scientist based on how they project themselves."

Further examination and analyses of the features identified in the profiles constructed by our students reveal that the image of the White male overwhelmingly persists as the classical representation of a scientist. Some students had argued that while more females are becoming scientists, they were far outnumbered by their male counterparts and that ethnic groups do not readily come to mind when they think about scientists. When asked about the contributions of ethnic groups to what we know as science, the following were some of the responses: "Well, I am not sure," "maybe, but not much in the real sense of what we learn as science or will eventually teach."

A summative task of this activity required the students to complete a journal entry describing possible experiences that may have contributed to their views about the profiles of scientists. Several of these entries suggested that popular media played an essential role in defining their images of scientists:

> The cartoons I watched when I was younger always depicted boys as scientists: Johnny Quest, Jimmy Neutron…Also most of the movies that are created typically portray scientists as White men. There is the doctor in Jurassic Park, the engineer in Independence Day, and the archaeologist in Stargate. I could go on and on…
>
> The images of mad scientists are everywhere. Albert Einstein is typically the inspiration for these depictions with his unkempt hair and lab coat…
>
> …it is just recently that they have depicted women, let alone African Americans, as scientists on TV.
>
> Bill Nye the science guy is the image that first comes to my mind…

The importance of popular media, including television, Internet, and magazines, in shaping stereotypes and cultural identities within a given population cannot be underscored here. Cinema, radio, and television have become the primary source of information for a growing number of individuals in our population and, as such, play a critical role in shaping the public perceptions of a variety of issues. Other entries indicated the extent to which school activities played a role in their images of scientists:

> In high school, we would visit the scientists working in their labs at the university. We would watch them and they would show us the cool things they were working on. Usually they were all White dudes and they had on their white lab coats and goggles…
> In middle school mainly we would do biographies of famous people and usually they were people like Newton and Galileo and others.
> In elementary school, we had posters in the library of scientists, I do not remember who they were but I know they were all men … and I think they were all White.

Scholars and researchers involved with the development of curriculum that embraces multicultural education cannot assume that students' perceptions of scientists have transcended the White male stereotype. We contend that preservice teachers should be given opportunities to develop images of scientists beyond the monoculture of White male dominance in order to effectively implement science curriculum that acknowledges diversity. We also support recent calls for changes in approaches to multicultural education in teacher education to include programs that move beyond the focus on curriculum and toward a framework where prospective teachers are encouraged to challenge their existing beliefs about equity, diversity, teaching, and learning (Luft, Bragg, & Peters, 1999; Monhardt, 2000) because teachers explicitly and implicitly impart their beliefs and expectations of students—some doing more harm than good for struggling students (Yerrick, Schiller, & Reisfeld, 2011). Teacher candidates have well-established beliefs about students from diverse backgrounds and their capabilities before entering their teacher education programs (Darling-Hammond, 2002), and one of their commonly held beliefs is about the academic capabilities of students from diverse backgrounds particularly for students from traditionally underrepresented groups. Embracing the notion that these students are less capable of academic success than others, Song and Christiansen (2001) posit that preservice teachers will continue to tailor instructions and implement science curricula in middle schools that foster low expectations unless their misconceptions about who can do science is challenged. This recipe for continuing low achievement and marginalization has, for years, contributed to underrepresentation of traditionally underrepresented and marginalized groups in science.

Multicultural Science Education

Prompted by the resounding call for the inclusion of all students regardless of gender, culture, or ethnicity (American Association for the Advancement of Science [AAAS], 1989, 1993; National Research Council [NRC], 2000, 2012), educators today feel hard-pressed to increase participation of underrepresented groups in the

STEM fields. In order to achieve this goal, teachers should consider the variety of ways in which they can promote equity in STEM for their culturally and ethnically diverse student population. Very often, the cultural beliefs and perceptions of students from traditionally underrepresented groups are overlooked during the implementation of science curriculum, thus portraying science as an abstract discipline far beyond their scope of participation. Multicultural science education recognizes the limitations of the traditional science curriculum in this regard and as such promotes multiple cultural views of all students while challenging stereotypes that for a long time have contributed to prejudice, racism, and inequality. According to Atwater (1993), multicultural science education is "a field of inquiry with constructs, methodologies, and processes aimed at providing equitable opportunities for all students to learn quality science in schools, colleges, and universities" (p. 3). Furthermore, students are engaged in a knowledge construction process that allows them to understand, investigate, and determine how implicit cultural assumptions, frames of references, perspectives, and biases of textbook writers influence the ways in which knowledge is constructed (Banks, 1993, 2009). In a multicultural science classroom, the cultural contexts and traditions of all students are therefore recognized and respected, the scientific contributions of all cultural groups are appreciated and valued, and science is depicted as a discipline that is open to all students from different cultural and ethnic backgrounds.

Not many preservice teachers are prepared to teach science in a multicultural classroom and, therefore, struggle to engage in culturally responsive pedagogy and cultural relevant teaching strategies (Aikenhead & Jegede, 1999; Calabrese Barton, Tan, & Rivet, 2008). Culturally responsive teaching as an approach acknowledges, values, and integrates the cultural identities and experiences of students in ways that enhances the quality of their learning environments (Banks et al., 2005; Villegas & Lucas, 2002). The onus falls on teachers to engage in culturally responsive teaching practices using curricula that in many cases do not give substantial attention to the culture and contribution of underrepresented groups in science. Building on the cultural resources students bring to school, culturally relevant teaching encourages students to critically examine educational content and process while constantly questioning how this knowledge contributes to a truly democratic and multicultural society (Howard, 2003; Ladson-Billings, 1992). According to Tate (1995), such culturally relevant teaching also requires students to maintain cultural integrity as well as strive for academic excellence. Teachers therefore need to be prepared to be able to empower their students intellectually while bridging the gap between marginalized and mainstream cultures (Davis, 2006).

Preparing educators to teach science to diverse student population is, therefore, a daunting task (Moore, 2006), and many inservice teachers acknowledge that teaching science in contemporary and equitable ways is an equally complex and challenging endeavor. Teaching science to diverse populations is further compounded by the fact that middle school teachers are largely female and White (Henke, Peter, Li, & Geis, 2005) while the student population consists of an increasing number of students of color (Ladson-Billings, 1997; Lee & Fradd, 1998; Villegas & Lucas, 2002) with a projection that this trend will continue beyond the twenty-first century. Underrepresented students' failure to identify with the culture of science and their seemingly active resistance to learning the subject may be attributed to the differences between their cultural frames and those of their science teachers. Furthermore, differences in the cultural frames of references affect teachers' ability to provide meaningful learning experiences that connect with students' prior knowledge or have the potential to reshape the frames of references, perspectives, and concepts that make up school science knowledge.

The success of underrepresented students in STEM areas depends on the ability of teachers to heighten their academic performance in science by engaging in both culturally responsive and relevant teaching practices. Preparing science teachers to promote social justice and equity in the science classroom is, however, a necessary task if we are to promote the academic achievement of students from traditionally underrepresented groups in the areas of STEM.

Researchers contend that many of the challenges associated with the preparation of preservice science teacher for the multicultural classroom are tied to issues of race and class, the inequitable distribution of resources, the rich diversity of school-age children, and stereotypical views and expectations (Villegas & Lucas, 2002). In order to understand the complexities of teaching science to diverse student populations, preservice teachers should be encouraged to confront their own stereotypes as they attempt to understand their own views of diversity, science, and teaching diverse learners. Thus, the inclusion of multicultural education in preservice teacher education programs continues to be strongly advocated by educators as empirical studies repeatedly reveal discrepancies in the educational achievement levels of culturally diverse students and their counterparts in mainstream society (Murrell, 2002). As a result, there is disproportionate participation of some ethnic groups in science and science-related careers. While this disproportionately low number undoubtedly reflects a complex interplay of numerous and well-known forces, some have argued that such underrepresentation in science is the result of socioeconomic and environmental issues (Hill, Corbett, & St Rose, 2010). Others have suggested that the content of science is divorced from the students' lived experiences resulting in recurring marginalization. Only a few have viewed this disparity within the larger multicultural context of science teaching and learning (Hogan & Corey, 2001; Lee & Fradd, 1998). In response to this achievement gap, several school districts are insisting on additional educational opportunities, accreditation agencies are mandating multicultural coursework and field experiences, and teachers are continuing to seek strategies for teaching culturally diverse students (Banks et al., 2001).

We believe that the onus is on science teacher educators to begin the process of addressing the issue of equitable science during teacher preparation program in general, and specifically in their science education courses. In this chapter, we describe our efforts to provide pedagogical opportunities for preservice teachers to broaden their concept of multicultural science education and ways to engage the personal and cultural identities of their learners into their science lessons. Some of these strategies specifically involve the infusion of the contributions to science made by scientists from underrepresented groups in the sciences such as African Americans, Latinos/as, Native Americans, and Asian Americans. In addition, we will discuss the impact and reactions of the teacher candidates to these teaching and learning activities along with implications for science teacher education. Using Banks (1993) five dimensions of multicultural education, we framed our advanced science education course on principles including (a) content integration, (b) the knowledge construction process, (c) prejudice reduction, (d) an equity pedagogy, and (e) an empowering school culture and social structure. We acknowledge that the dimensions, though conceptually distinct, overlap in practice and are interrelated. Therefore, the objective of the two teaching activities described in our chapter is for our preservice science teachers to acquire the knowledge and skills needed for effectively enacting the processes of multicultural science education while providing equitable opportunities for all students to learn science. During the process of their learning, preservice teachers will understand and confront the implicit cultural assumptions, frames of references and biases of science curriculum, and textbook writers that influence the ways in which scientific knowledge is presented. In addition, they will conduct research to explore the contributions of non-mainstream scientists and identify specific points in the middle school science curriculum for integrating such knowledge.

Challenging Stereotypes

Concerns about the continued lack of success and participation by underrepresented groups in science prompted a focus on equity in major science education reforms in the 1990s (AAAS, 1989; NRC, 2000) and a call for rethinking multicultural education. Attempts to bring about school reforms to respond to the academic needs of a multicultural nation include the rise of the ethnic studies movement in the 1960s whose primary goal was that of challenging the negative images and stereotypes of African American prevalent in mainstream scholarship by depicting accurate descriptions of the life, history, and contributions of underrepresented groups. The growing body of literature reflecting the strong correlation between teachers' beliefs and instructional practices contributed to our consensus that teachers need to first learn to recognize their attitudes toward the owners and contributors of science knowledge and skills before they are able to direct their students and adequately present the contributions made to the discipline by underrepresented groups.

Consistent with the premise of multicultural education that cultural diversity enhances the effectiveness of science learning (Atwater, 1993; Howard, 2003), we explored with our preservice teachers the contributions to the discipline of science by ethnic groups. We challenged the preservice teachers to adopt the practice of making the science curriculum relevant to their diverse student population by integrating, where necessary and appropriate, the contributions of underrepresented groups in the development of science knowledge. To achieve such, we first analyzed the mandated science curriculum and the corresponding textbooks adopted by the school district over the last three adoption periods to identify how scientists were represented in the texts. The criteria used in the analysis included counting the number of times the science knowledge was connected to the scientists credited with its development, indications of race or ethnicity, gender, place of origin, and the presence of biographies. While some improvements have been made over time, our collective analyses of the science textbooks revealed that to a large extent, images of scientists were still portrayed as predominantly White males interspersed with pictures of White females and people of other ethnicity. Such images give credence to the historical notion of science knowledge being owned by the dominant group (Melear, 1995; Parsons, 1997). In addition, some texts incorporated short biographies of prominent scientists in the respective fields, but our preservice teachers noted that because the focus in the science courses is on the content knowledge, the information contained in the blocked sections of the texts is usually overlooked during the teaching process.

A science curriculum and its associated pedagogy that attend to the needs of a diverse nation should recognize and celebrate the contributions made to the discipline by all members of the scientific community. If preservice teachers do not recognize racial and ethnic diversity among the community of scientists, they will not be able to provide examples and instruction for children to confront their beliefs about themselves as science learners and to seek to move from the periphery of science learning toward the pursuit of scientific fields of study or careers. By all indications, many science teachers rely heavily on science texts during their teaching. Information in these texts and the mandated curriculum as presented to teachers do not always recognize the contributions of individuals outside of the mainstream culture. In addition, many of the middle school science teachers without the knowledge of the contributions made will have difficulty making the necessary adjustments in their teaching to reflect the contributions of scientists from diverse ethnic groups.

Literature suggests a strong connection between teachers' knowledge and practice and explains that what teachers know and believe become evident in their practice (Cochran-Smith & Lytle, 1999; Little, 2003). It is, therefore, likely that teachers who are unaware of the contributions of scientists from underrepresented groups will not be able to integrate such in the enactment of their middle school science curriculum. This is compounded by the fact that many of the middle school teachers are from the dominant group portrayed as the developers of science and whose ways of thinking have been shaped by the dominant culture. We contend that there is the need for specific learning tasks during teacher education classes to allow preservice teachers to broaden their understanding and appreciation of the various contributions made to science beyond that presented in school texts. This, however, is a first step toward achieving the tenets of multicultural science education as embraced by Atwater (2010) and Banks et al. (2005). Teacher educators are, therefore, encouraged to assist preservice teachers in recognizing the contributions made to science by all people and then create experiences for them to learn how to make relevant changes in the content and process (Atwater, 2010) of their enacted curriculum. This task according to Banks (2009) is the first and basic approach toward integration of multicultural content into the curriculum. While this approach does not fundamentally change the curricular content, it provides a framework from which to engage students in conversations that include the struggles for inclusion in and recognition by mainstream science community.

In the following activity, our preservice teachers explored the contributions to scientific knowledge of ethnic groups. Many of our students, even though they indicated that they had prepared extensive biographies on famous scientists during their middle school years, never had the opportunity to explore contributions made to science by individuals outside of mainstream society. In pairs, preservice teachers were instructed to use any available resources to identify the contributions made to science by a member of an ethnic group. The assignment was submitted as a poster presentation and included a brief biography of the scientists' work including the description of the life history and contributions made to the discipline. Additionally, their presentation required them to identify specific areas in the science curriculum that the life history and achievement of the scientist could be incorporated. Recognizing that continuous discussion and reflection about issues of ethnicity can cause dissonance, we further required that the preservice teachers maintain a log of their reflections and reactions to the activity. The goal was to record the specific triggers and personal reactions that would reemerge in later class discussions. Despite the discomfort, Villegas and Lucas (2002) contend, "preservice teachers must be challenged and helped to recognize ways in which taken-for-granted notions regarding the legitimacy of the social order area are flawed" (p. 23). Thus, the dissonance becomes a catalyst for change among the preservice teachers when their beliefs as espoused seem inconsistent with appropriate actions. The Internet provided the preservice teachers with a wealth of information on the life history and contributions of their selected scientists ranging from post-slavery era to the twenty-first century. Needless to say they did a remarkable presentation of the works and life history of their selected scientists. The table below reveals some of the scientists the preservice teachers presented in response to the assignment (Table 1).

The journaled responses to this activity highlighted a "wow factor" indicating that students were pleasantly surprised by the extent to which many of our current understandings in science are made possible through the contributions of members of various ethnic groups. For instance, impressed with Carver, one student who expressed a love for botany wrote, "Carver's contributions should be front and center when such topics as botany, or biotechnology are being taught in the science lessons." Additionally several students lamented the fact that this information was not brought to their attention at an earlier point in their learning experiences. One student indicated, "This research certainly forced me to reconsider some of my conceptions of scientific knowledge and how the contributions of marginalized individuals have been excluded from regular textbooks." Although some students were able to appreciate the objective of the assignment, some journals reflected levels of tensions as some preservice teachers grappled with the notion of whose science curriculum should be developed in their classrooms.

It was evident that some students were comfortable with the status quo and had difficulty recognizing the biases and discriminatory practices reflected by the exclusions in the texts and the science curriculum. Some questioned the widely accepted notion that science as a subject was free of cultural influence now that they have

Table 1 Some scientists from underrepresented groups and their accomplishments

Name	Race/ethnicity	Occupation	Inventions/accomplishments
Carver, George Washington	African American	Botanist	Discovered hundreds of uses for previously useless vegetables and fruits, principally the peanut
Boykins, Otis	African American	Inventor, engineer	Invented the control unit for the artificial pacemaker
Walker, C. J	African American	Inventor	Invented conditioning system to straighten Black hair
Jemison, Mae	African American	Astronaut	First female African American astronaut in history of NASA
Bose, Amar	Asian American	Physicist	Designed the Bose speaker systems
Ho, David	Asian American	AIDS Researcher	Assisted in research leading to development of antiviral drug for AIDS
Kalpana, Chawla	Asian American	Astronaut	First Asian American woman to go into space
Chandrasekhar, Subrahmanyan	Asian American	Astrophysicist	Studied physical processes of importance to the structure and evolution of the stars
Alvarez, Luis	Latino/a	Physicist	Helped design ground-controlled radar system for aircraft landings
Ochoa, Ellen	Latino/a	Astronaut	First Hispanic American woman astronaut
Chang-Diaz, Franklyn	Latino/a	Astronaut	First Costa Rican astronaut
Molina, Mario	Latino/a	Chemist	Did extensive research on chlorofluorocarbons (CFCs)

discovered that the discipline reflect and perpetuate the cultural and hegemonic norms proliferated in the textbooks. It was clear that even though some were becoming aware of their own biases and beliefs, they felt constrained by the curriculum and the need to prepare children for the state's standardized tests in science. Perhaps the preservice teachers were using the tests and the mandated standards as avoidances toward recognizing and acknowledging their stereotypes and prejudices. They, according to Weinstein, Tomlinson-Clarke, and Curran (2004) consider their own cultural norms to be neutral and universal, accepting as normal and without question the programs and discourse of schools.

As a group, they did not understand the privileges they were afforded because of their identity as European American. At this stage in their preparation, they did not understand themselves in relation to the many cultural influences over their lifetimes and as products of an education system bent on reproducing the established status quo. This can be problematic because preservice teachers need to understand the genesis of their beliefs in order to recognize the connections between their experiences and their actions as teachers.

Cultural Awareness in the Diverse Classroom

As we grapple with the issue of increasing and broadening participation in science, what is missing are deliberate efforts to lead teachers and students to a greater sensitivity of cultural awareness and less cultural stereotyping within the educational system. Effective pedagogy and enhanced cultural awareness are important in the process of providing equitable opportunities for learning science, but preservice teachers need to understand the privileges afforded to the dominant group when the contributions of other ethnic groups are disregarded. Teachers in effect need to understand their students' cultural backgrounds and experiences by upholding the notion that "cultural diversity is appreciated in science classrooms because it enhances rather than detracts from the richness and effectiveness of science learning" (Atwater, 1993, p. 3). This richness in diversity is then extended when the contributions made by people of color in science are not merely acknowledged but actively legitimized, and integrated into science teaching.

Contemporary research on "how people learn" and what constitutes major themes for science learning have provided a knowledge base for national science framework and standards of curriculum and instruction (NRC, 2012). These provide the guiding principles for developing K-12 science curriculum. Notably, the contributions to science by people of color have been given scant regard as evidenced by the omission in mainstream curricular frameworks. In fact, many students including our preservice teachers have gone through K-12 science education with a monocultural view regarding the ownership of science and who are the generators of science knowledge. Our attempts at sensitizing the preservice teachers to the contributions of scientists from underrepresented groups and affording them deliberate opportunities to address the issue of their exclusion from the curriculum were met with mixed feelings. Our goal was to provide our preservice teachers with opportunities to incorporate the contributors to the vast pool of knowledge known as science. Science teachers who recognize and value that science is not culture free are more likely to believe and internalize that students from marginalized groups are capable of learning science. They are also more likely to integrate science lessons that promote equity into their science classrooms where appropriate, conveying to their students explicitly that individuals from all backgrounds are capable of achieving success and pursuing degrees and careers in STEM areas. Giving credence to the scientific contributions of all peoples and with equal respect could lead to greater sensitivity of cultural awareness and less cultural stereotyping, thus allowing for broader participation in science by diverse students. According to Banks and Banks (1995), such curricular transformation not only brings the contributions from the margin to the center of the curriculum but also allows the students from underrepresented groups to understand the nation's common heritage and traditions.

We can no longer be satisfied with engaging learners in a curriculum that does not accurately present the contributions made to science by members of all ethnic groups. We believe that all teachers should acquire the knowledge and skills required to develop and implement an equity pedagogy (Banks & Banks, 1995) that provides students from diverse ethnic and cultural groups with an equal opportunity to attain academic and social access in school. Thus, as teacher educators we have to attend to the nature and quality of activities afforded to our preservice teachers during teacher preparation.

We have presented two of the activities used in our advanced science education methods course to highlight the notion that no single culture has a monopoly on the generation of science knowledge. We wanted to reinforce the idea that "education that legitimizes one culture within a pluralistic society robs students from other cultural backgrounds of self-esteem and contributes to discrimination" (Atwater, 1993, p. 3). By sensitizing preservice teachers to the contributions made to the development and advancement of science by people of color and then requiring them to develop a repertoire of activities that specifically seeks to integrate such knowledge into their practice, we are equipping them with the knowledge and skills required to teach science to "all" students. Recognizing the contributions to science by "all" people may be an initial step toward the development of cultural competence among preservice teachers. With such level of cultural competence, preservice teachers will be able to extend participation of underrepresented groups in science by engaging students in discussions that build on their cultural and linguistic resources (Cochran-Smith, 2004). However, achieving the goal of "science for all" will require a reconceptualization of fundamental issues of pedagogy, science content, and introduction to the culture of science within the context of day-to-day teaching beginning in preservice teacher education.

The Role of the Science Teacher Educator

The degree to which teachers are able to accomplish the task of teaching science to all children in ways that would encourage learning and their ultimate participation in the culture of science is affected by factors such as their level of preparedness to teach their particular subject area (Michaels, Shouse, & Schweingruber, 2008) and the preparation and experience teaching science to diverse populations. Bryan and Atwater (2002) posed the following question: What is the science teacher educators' role in facilitating preservice teachers learning to teach science in equitably ways? In other words, how can we effectively prepare a population of teachers equipped with the necessary content knowledge, skills, attitudes, and values to impact reform efforts in the ever-growing multicultural learning environments? Preparing teachers to enact a science curriculum is challenged by the need for teacher education programs to move beyond the traditional courses that seek to introduce teacher candidates to specific issues of diversity. In typical teacher education programs, preservice teachers are introduced to issues of diversity and multicultural education in general education courses. Although such courses play an important role in diversity, the superficial mention of general issues regarding diversity does not go far enough when confronting the more salient issues of the contribution of various cultures to the discipline of science.

The curriculum and pedagogy that teacher educators use in teaching preservice teachers about diversity include a range of pedagogical and programmatic strategies. These are important in helping to foster positive dispositions among them and to further their understanding of diversity. However, as we think about the roles of science teacher educators, we believe that the ideas addressed in the general education or specially designed multicultural courses should be expanded with continued reinforcement in content-specific pedagogical courses. Rather than viewing diverse student populations through a cultural deficit model, proponents of multicultural science education advocate that all students can learn in contexts where different ways of knowing and different constructions of science are brought into the science classroom (Atwater, 2011; Calabrese Barton, & Tan, 2009; Emdin, 2011).

The goal of teacher education is to educate future teachers and equip them with the relevant skills that will allow them to effectively modify prescribed curriculum and pedagogy in ways that attend to the equitable recognition of diverse cultures in science. Their knowledge base should, therefore, include the contributions of underrepresented populations to the developing science knowledge. Unfortunately, even at this juncture in our history, differences in race, culture, ethnicity, language, and class are still perceived as barriers to effective science instruction due, in part, to stereotypical ideas associated with the capabilities and capacities of diverse students. Teacher education programs that fail to respond to issues of diversity and that do not provide adequate preparation for preservice teachers to teach for diversity are indirectly thwarting the achievement of one of the major goals in the current educational reform efforts.

Multicultural education continues to be strongly supported by teacher educators who are all too familiar with current classroom rituals, routines, and curricular practices that exclude or distort the life experiences, histories, and contributions of scientists from underrepresented groups. We are aware that these practices inadequately address the complexities of a diverse student population resulting in marginalization of underrepresented groups. To successfully move the preservice teachers beyond the level of awareness and treatment of the science curriculum in a diverse society, as we have presented in this chapter, teacher educators must also articulate a vision of teaching and learning that shifts preservice teachers toward transformative practices and a recognition of their roles as effective agents for social change. We embrace the call made by Banks and Banks (1995) for an interrogation and reconstruction of existing school structures that foster inequities. They posit that curricular implementation within the context of existing assumptions and structures are insufficient to result in the kind of transformation in which underrepresented groups challenge the status quo. Preservice teachers must therefore critically examine their own beliefs and, as agents of change, employ practices that affirm the views of students from underrepresented groups and empower school culture and social structure toward increasing underrepresented students in STEM areas. As advocates for the systematic infusion of multicultural education and transformative practices, we envision a burgeoning research agenda that seeks to explore and challenge the taken-for-granted practices in science education classrooms in which our preservice teachers are being prepared.

Expanding the Research Agenda

One of the goals of science education reform is to encourage broader participation in science. This is no easy task because of the range of systemic issues that plague K-12 education in general and specifically as it relates to the science instruction of students from diverse race, ethnic, cultural, and socioeconomic backgrounds. Teachers have a role in the promotion of the deficiencies or in adjusting the trajectories of all learners. Therefore, as science teacher educators and in our quest to effect changes, we need to further adjust our actions within multicultural science education to pointedly deal with specific issues such as challenging the implicit messages about ownership of science knowledge. It is time for students from traditionally underrepresented groups in STEM areas to experience a sense of place in their science classes. We acknowledge that changes have to occur at a number of places within the bureaucratic education system. Some changes have begun in teacher education programs as many of the courses offered now address preservice teachers' beliefs about diversity and multicultural education and their impact on teaching and learning. In addition to pedagogical practices and classroom norms available to guide the enactment of multicultural science education, we believe that the images emanating from the science curriculum being implemented in our middle schools do not position students from traditionally underrepresented groups as participants or achievers in science. Rather, the subliminal messages about the mainstream development of science knowledge promote the traditional, exclusionary views of the subject.

As teacher educators and research advocates, we propose a research agenda that involves the investigation of preservice teachers' beliefs about the ownership of science knowledge and the role that race and ethnicity plays in one's entry into the community of science. Teacher candidates bring both past experiences and beliefs to their teacher education programs. This intersection of experience and beliefs creates a powerful combination that can impact their reactions to the activities in their programs and ultimately their own decisions about teaching and learning. We offer the following questions as triggers to initiate conversations among preservice teachers and to bring to the fore some of their strongly held beliefs:

- Who owns the discipline, language, and culture of science?
- Who are the participants in the discipline, language, and culture of science?
- Who or what determines who can and cannot do science?

The preservice teachers' responses to these questions will offer much insight into their beliefs and the expectations they hold for students from underrepresented groups. Unraveling their beliefs and rebuilding a process for giving credence to the contributions of people of color in science may be an effective strategy for demystifying diversity in a multicultural nation. Integrating the contributions of scientists from underrepresented groups in the middle school science curriculum will certainly offer the challenge that no single group has a monopoly on science knowledge; hence, the doors should open for all learners to become potential learners of science.

Our preservice teachers were comfortable with the kind of teaching that offered privileges to mainstream students—those who most often identify with the dominant groups. They were of the dominant group and, without hesitation, embraced the stereotypes presented in the texts, media, and their own schooling experiences regarding the contributors of science knowledge and ultimately those who can learn science. They were, however, unaware of how such schooling practices influence the achievement of students from other groups. Much of the literature on teacher expectations of student achievement helps us understand when teachers believe in students' abilities, the students are likely to be successful (Ladson-Billings, 1999; Ogunleye, 2009; Rodriguez, 2001). The two activities discussed here along with other pedagogical strategies sought to challenge the beliefs of the preservice teachers and to expose the taken-for-granted biases that exist within the education system. While the post-activity deliberations reveal a level of awareness of the issues among the preservice teachers, all were able to identify specific curricular connections to the scientists they selected and whose life histories they explored. The following vignette extracted from one student's journal offers some hope for the future:

> I wished I had studied about Carver's contribution to science in Middle school.
> I wished not only because he has done so much but his story has given me a new perspective that all people can do science.
> I wish my students will (sic) be able to learn not only science for the test but also learn about the humans behind the science in the curriculum.

The implication of the story communicated in the student's vignette provides some assurance that issues in multicultural science education reform can be propelled from the margin to the center of the curriculum. The onus is therefore on science teacher educators to provide those critical experiences to challenge the preservice teachers beyond awareness to transformative curricular practices. We should therefore explore the utility of social justice pedagogy in the science teacher education program, which will legitimize the culture of underrepresented groups in science (Atwater & Suriel, 2010). We should empower preservice teachers to challenge and confront the stereotypes embedded within existing secondary science curriculum. We should help preservice teachers to develop a belief in the ability and worth of underrepresented students and to hold them to high academic standards. With a social justice perspective or frame of reference, preservice teachers will be more committed to implementing science lessons that embody multiculturalism.

References

Aikenhead, G. S., & Jegede, O. J. (1999). Cross cultural science education: A cognitive explanation of cultural phenomena. *Journal of Research in Science Teaching, 36*(3), 269–287.

American Association for the Advancement of Science. (1989). *Science for all Americans*. Washington, DC: Oxford University Press.

American Association for the Advancement of Science. (1993). *Benchmarks for science literacy*. Washington, DC: Oxford University Press.

Atwater, M. (1993). Multicultural science education: Assumptions and alternative views. In *Science for all cultures: A collection of articles from NSTA's journals* (pp. 1–5). Arlington, VA: NSTA.

Atwater, M. (2010). Multicultural science education and curriculum materials. *Science Activities, 47*, 103–108.

Atwater, M. (2011). Significant science education research on multicultural science education, equity, and social justice. *Journal of Research in Science Teaching, 49*(1), 1–5.

Atwater, M., & Suriel, R. L. (2010). Science curricular materials through the lens of social justice: Research findings. In T. Chapman & N. Hobbel (Eds.), *Social justice pedagogy across the curriculum: The practice of freedom* (pp. 273–282). New York: Taylor & Francis.

Bandura, A. (2000). Self-efficacy: Foundation of agency. In W. Perrig & A. Gorb (Eds.), *Control of human behavior, mental processes, and consciousness* (pp. 17–33). Mahwah, NJ: Lawrence Erlbaum.

Banks, J. A. (1993). Multicultural education: Historical development, dimensions, and practice. *Review of Research in Education, 19*, 3–49.

Banks, J. A. (2009). Multicultural education: Dimensions and paradigms. In J. A. Banks (Ed.), *Routledge international companion to multicultural education* (pp. 9–32). New York, NY: Routledge/Taylor & Francis.

Banks, C. A., & Banks, J. A. (1995). Equity pedagogy: An essential component of multicultural education. *Theory into Practice, 34*(3), 152–158.

Banks, J., Cochran-Smith, M., Moll, L., Richert, A., Zeichner, K., LePage, P., et al. (2005). Teaching diverse learners. In L. Darling-Hammond & J. Bransford (Eds.), *Preparing teachers for a changing world: What teachers should learn and be able to do* (pp. 232–274). San Francisco: Jossey-Bass.

Banks, J. A., Cookson, P., Gay, G., Hawley, W. D., Irvine, J. J., Neito, S., et al. (2001). Diversity within unity: Essential principles for teaching and learning in a multicultural society. *Phi Delta Kappan, 83*(3), 196–203.

Bryan, L. A., & Atwater, M. M. (2002). Teacher beliefs and cultural models: A challenge for science teacher preparation programs. *Science Teacher Education, 86*(6), 821–839.

Calabrese Barton, A., & Tan, E. (2009). Funds of knowledge and discourses and hybrid spaces. *Journal of Research in Science Teaching, 46*(1), 50–73.

Calabrese Barton, A., Tan, E., & Rivet, A. (2008). Creating hybrid spaces for engaging school science among urban middle school girls. *American Educational Research Journal, 45*(1), 68–103.

Cochran-Smith, M. (2004). *Walking the road; Race, diversity, and social justice in teacher education*. New York: Teacher's College Press.

Cochran-Smith, M., & Lytle, S. L. (1999). Relationships of knowledge and practice: Teacher learning in communities. *Review of Research in Education, 24*, 249–305.

Darling-Hammond, L. (2002). Learning to teach for social justice. In L. Darling-Hammond, J. French, & S. P. Garcia-Lopes (Eds.), *Learning to teach for social justice*. New York: Teacher's College Press.

Davis, B. M. (2006). *How to teach students who don't look like you: Culturally relevant teaching strategies*. Thousand Oaks, CA: Corwin.

Emdin, C. (2011). Dimensions of communication in urban science education: Interactions and transactions. *Science Education, 95*, 1–20.

Henke, R. R., Peter, K., Li, X., & Geis, S. (2005). *Elementary/secondary school teaching among recent college graduates: 1994 and 2001* (NCES 2005-161) U. S. Department of Education, National Center for Education Statistics. Washington, DC: U. S. Government Printing Office.

Hill, C., Corbett, C., & St Rose, A. (2010). *Why so few? Women in science, technology, engineering, and mathematics*. Washington, DC: American Association of University Women.

Hogan, K., & Corey, C. (2001). Viewing classrooms as cultural contexts for fostering scientific literacy. *Anthropology and Education Quarterly, 32*(4), 214–243.

Howard, T. C. (2003). Culturally relevant pedagogy: Ingredients for critical teacher reflection. *Theory into Practice, 42*(3), 195–202.

Ladson-Billings, G. J. (1992). Culturally relevant teaching: The key to making multicultural education work. In C. A. Grant (Ed.), *Research in multicultural education: From the margins to the mainstream*. Bristol, PA: Falmers Press.

Ladson-Billings, G. (1997). It doesn't add up: African American students' mathematics achievement. *Journal for Research in Mathematics Education, 28*, 697–708.

Ladson-Billings, G. J. (1999). Preparing teachers for diverse student populations: A critical race perspective. In I.-N. Asghar & P. D. Pearson (Eds.), *Review of research in education* (Vol. 24, pp. 211–247). Washington, DC: American Educational Research Association.

Lee, O., & Fradd, S. H. (1998). Science for all, including students from non-English language backgrounds. *Educational Researcher, 27*, 12–21.

Little, J. W. (2003). Inside teacher community: Representations of classroom practice. *Teachers College Records, 105*(6), 913–945.

Luft, J. A., Bragg, J., & Peters, C. (1999). Learning to teach in a diverse setting: A case study of a multicultural science education enthusiast. *Science Education, 83*(1), 100–118.

Melear, C. (1995). Multiculturalism in science education. *The American Biology Teachers, 57*(1), 21–26.

Michaels, S., Shouse, A. W., & Schweingruber, H. A. (2008). *Ready, set, science: Putting research to work in K-8 science classrooms*. Washington, DC: National Academy.

Monhardt, R. M. (2000). Fair play in science education: Equal opportunities for minority students. *The Clearing House, 74*(1), 18–22.

Moore, F. (2006). Multicultural preservice teachers' views of diversity and science teaching. *Research and Practice in Social Sciences, 1*(2), 98–131.

Murrell, P. C. (2002). *African-centered pedagogy: developing schools of achievement for African American children*. Albany: State University of New York Press.

National Research Council [NRC]. (2000). *Inquiry and the national science education standards: A guide for teaching and learning*. Washington, DC: National Academy Press.

National Research Council [NRC]. (2012). *A framework for K-12 science education: practices, crosscutting concepts, and core ideas*. Washington, DC: The National Academies Press.

Ogunleye, A. O. (2009). Defining science from multicultural and universal perspectives: A review of research and its implications for science education in Africa. *Journal of College Teaching and Learning, 6*(5), 57–71.

Parsons, E. C. (1997). Black high school females' images of scientist: Expression of culture. *Journal of Research in Science Teaching, 34*(7), 745–768.

Rodriguez, A. (2001). From gap gazing to promising cases: Moving toward equity in urban education reform. *Journal of Research in Science Teaching, 38*(9), 1115–1129.

Song, K., & Christiansen, F. (2001). *Achievement gap in preservice teachers in urban settings*. East Lansing, MI: National Center for Research on Teacher Learning. (ERIC Document Reproduction Service No. ED456187)

Tate, W. F. (1995). Returning to the root: A culturally relevant approach to mathematics pedagogy. *Theory into Practice, 34*(3), 166–173.

Tschannen-Moran, M., & Woolfolk Hoy, A. (2007). The differential antecedents of self-efficacy beliefs of novice and experienced teachers. *Teaching and Teacher Education, 23*, 944–956.

Villegas, A. M., & Lucas, T. (2002). Preparing culturally responsive teachers: Rethinking the curriculum. *Journal of Teacher Education, 53*(1), 20–32.

Weinstein, C. S., Tomlinson-Clarke, S., & Curran, M. (2004). Toward a conception of culturally responsive classroom management. *Journal of Teacher Education, 55*(1), 25–38.

Yerrick, R., Schiller, J., & Reisfeld, J. (2011). "Who are you callin' expert?": Using student narratives to redefine expertise and advocacy lower track science. *Journal of Research in Science Teaching, 1*(48), 13–36.

The Triangulation of the Science, English, and Spanish Languages and Cultures in the Classroom: Challenges for Science Teachers of English Language Learners

Regina L. Suriel

Latinos/as comprise the largest ethnic grouping in the US population (U. S. Census Bureau, 2011), with Latino/a school age population steadily increasing. As of 2006, one in five students attending US public schools is of Latino/a origin (Fry & Gonzales, 2008). It is projected that the Latino/a school enrollment will increase by 25 % by the year 2020 (Hussar & Bailey, 2011). Furthermore, the number of Spanish-speaking English language learners (ELLs) is also expected to rise (Fry & Gonzales, 2008). Currently, 80 % of all ELLs speak Spanish as their first language (Gándara & Rumberger, 2009). A persistent challenge for educators teaching Latinos/as and Latino/a ELLs, science educators in particular, is the Latino/a high dropout rate (Brown & Rodríguez, 2009; Kohler & Lazarin, 2007; U.S. Department of Education & National Center for Education Statistics, 2011), substandard performance in standardized science achievement examinations, low college enrollment, and low enrollment in postsecondary science and science-related fields (Levine, Gonzalez, Cole, Fuhrman, & Floch, 2007; National Academy of the Sciences [NAS], 2010; U.S. Department of Education & National Center for Education Statistics, 2007, 2012). If we are to diversify ideas, contributions, and perspectives in the sciences and if we are to subscribe to values and norms espoused by equity, emancipation, and social justice movements, then the educational needs of Latinos/as must be addressed.

It is critical to promote scientifically literate citizens while expanding enrollment in science courses for Latino/a students. If this goal is to be realized, educators need to better appreciate the linguistic demands imposed upon Latino/a bilinguals and need to understand pedagogical interventions required for effectively educating this (McInstosh, 2011; Quinn, Lee, & Valdés, 2012). In fact, mandates from the *No Child Left Behind* (2001) require these interventions and hold every state

---

R.L. Suriel (✉)
Valdosta State University, Middle, Secondary, Reading, & Deaf Education, Dewar College of Education & Human Services, 1500 N. Patterson Street, Valdosta, GA 31698, USA
e-mail: rlsuriel@valdosta.edu

accountable for helping limited English proficient children achieve academically and be held to the same academic standards as their English-speaking peers (NCLB. Part A, Subpart 1). Thus, this chapter presents issues and recommendations regarding the education of Latinos/as. First, this chapter begins with an examination of learning: sociocultural, second language acquisition, and concept formation. From a pedagogical perspective, a teaching scenario serves as the platform for understanding the linguistic barriers and challenges experienced by Latino/a children while in science classrooms. Essential to the discussion of Latino/a students is the impact that various cultural barriers have on Latino/a science learners. This chapter presents a number of approaches for nurturing culturally congruent "science" classroom environments. Lastly, this chapter offers a synopsis of challenges science educators face relative to how Latino/a students experience science as they transition into colleges and science-related majors, highlighting how teachers can facilitate this transition.

Learning Theories

Central tenets of sociocultural theory posit that knowledge is socially constructed and that learning and development occurs through the interactions between learners and teachers (Vygotsky, 1978). Such contexts are recognized as having social, cultural, and political elements (Freire, 1993; Nieto, 2002). Vygotskian theory suggests that knowledge construction occurs as a series of transformations. First, the learner is presented with an external activity or cultural tool (e.g., scientific concepts, language forms, and functions) as an interpersonal experience, which then the learner transforms into an *intrapersonal* event. For some functions, internalization occurs gradually and in a short period of time; for other functions, external activities remain unresolved. Learning advances ahead of intellectual development and thinking is challenged beyond present understandings. Learning and development are self-regulated in an active process. As a learner interacts with more knowledgeable social agents—teachers and peers—he or she constructs and internalizes knowledge. In this process, learners are provided with cultural tools (i.e., symbolic language and its expressions) that provoke qualitatively improved thinking and reasoning (Vygotsky, 1978).

Cummins (1981) proposed a language learning theory paralleling Vygotsky's theory of learning. Cummins proposed that emergent bilinguals first develop basic interpersonal communicative skills (BICS) and subsequently develop cognitive academic language proficiency (CALP). Basic interpersonal communicative skills (BICS) refer to the informal language skills used to navigate everyday situations and develop within the first or second year of exposure to the second or host language. BICS help learners to increase their speech capacity with the aid of contextual clues. On the other hand CALP combines language proficiency and cognitive processes, often requiring four to 10 years to develop (Ramirez, Pasta, Yuen, Billings, & Ramey, 1991; Slama, 2012). It is also important to note that conceptual understanding and curricular assessments require the use of CALP.

Similar to BICS, in learning social language (including the language of science), children first learn by speaking through the voice and actions of a more competent other (Hay & Fielding-Barnsley, 2012; Mercer, 2008; Vygotsky, 1978; Wertsch, 1991) until they internalize the language and use it for their own purposes. Once language is internalized, thoughts are organized and actions are regulated. Thus, for Cummins and Vygotsky, learning is first a social, then an individual process.

Science learning involves knowledge construction mediated by social interactions and the use of cultural tools, i.e., science language and science ways of thinking and doing (Beeth & Hewson, 1997; Kirch, 2010; Wertsch, 1991). Science language here refers to specific use of symbols, such as science words, phrases, and modes of expressions, with meanings and sets of rules, prescribed by the scientific community (Lemke, 1990). A learner's interaction with the environment, human and nonhuman alike, involves the use of senses to process information and make meaning. To learn science means using, organizing, and synthesizing factual information within a conceptual framework to make meaning. Preconceptions are formed first, as curiosity-driven individuals explore and interact with the world and gather factual knowledge. Then, through socially mediated interactions with more knowledgeable agents (parents, siblings, teachers, and peers), conceptual frameworks begin to evolve. Elaborate conceptual frameworks continue to evolve with increasing masses of factual information and experiences (National Research Council [NRC], 2005). Conceptual understandings are influenced by society. Interpretations of the world, including natural phenomena, are informed by the cultural values, norms, and belief systems of communities and are transmitted in the discourse (Aukrust, 2011; Chalmers, 1999). Thus, how scientists conceive science, how educators teach science, and how students learn science depend on cultural belief systems, including those espoused by the scientific community.

Instructional goals aim to advance learning by building on prior experiences and understandings (Aukrust, 2011, p. 7). Learning is advanced by constructing or "building on" existing understandings and experiences (NRC, 2005, p. 4). The entry point of effective pedagogical approaches, regarding learners from different ethnic, cultural, and linguistic backgrounds, is to build on prior knowledge (Lee, 2003; Mason & Hedin, 2011). This crucial pedagogical approach becomes particularly important when educating Latino/a science learners because of the heterogeneity in cultural perspectives and linguistic abilities endemic to the Latino/a community. To compound matters, these learners are also subjected to the language of science. Thus, science classrooms become the crucible for the triangulation of cultures and languages: Spanish (L1), English (L2), and the language of science (L3). Science teachers face the challenge of effectively triangulating three distinct languages and cultures. If this trio is not adequately addressed, science becomes incomprehensible and unattainable for the learner, impeding knowledge construction for Latino/a students acquiring the English language.

Who Are the Latino/a English Language Learners?

The Latino/a Community Possesses a Plethora of Cultural Identities

Latinos/as, unified by a common Spanish ancestry, are a culturally heterogeneous group residing in the United States and varying in social, political, and religious orientations as well as educational and socioeconomic (SES) backgrounds (Torres-Saillant, 2005; Wallesrstein, 2005). Some Latinos/as are native to the United States, others are recent immigrants, and many are second-, third-, and fourth-generation descendants of immigrants. While some Latinos/as identify exclusively with their home culture, e.g., as Ecuatorianos, Salvadoreños, Bonrinqueños,[1] or Quisqueyanos[2] (Garcia-Preto, 2005, p. 154), many Latinos/as identify exclusively with hegemonic groups, mainly Whites, i.e., historical Anglo-Saxon and other Northern European populations in the United States (Schleef & Cavalcanti, 2010; Tafoya, 2005). Some Latinos/as identify with both parent culture and hegemonic cultures (e.g., Mexican-American), while others identify with local or regional cultures, such as the Dominican York[3] and Tejanos.[4]

Linguistically, Latinos/as Encompass a Spectrum of Bilingualism

Latinos/as express a wide range of linguistic heterogeneity. Hammers and Blanc (2000) present a continuum of linguistic identities that best describe (Latino/a) bilinguals. On both ends of the spectrum, there exist the dominant bilinguals who are more competent in one language (parent language or L1) than another (host language or L2). For the purposes of this chapter, an emergent dominant Latino/a bilingual[5] is defined as one with either complete or incomplete Spanish language (L1) fluency (reading, writing, listening, and speaking) and acquiring the English language (L2). Often, these types of bilinguals are referred to as English language learners (ELLs) or dual-language learners (DLLs). In between the extremes of the spectrum exist a wide range of bilinguals only differing in the degree of their L1 and L2 competencies, with balanced bilinguals indicating equal levels of competence in L1 and L2. Albeit distinct, linguistic identities follow similar patterns to cultural identities with some bilinguals identifying monoculturally (strong ties to a parent culture or dominant cultures) while others identifying biculturally (with L1 and L2 cultures) or multiculturally (L1, L2, and regional cultures, e.g., Nuyoricans).[6]

---

[1] Borinqueño/a is an indigenous term used to identify natives of the island of Puerto Rico.
[2] Quisqueyano/a is an indigenous term used to identify natives of the Dominican Republic.
[3] Dominican York is a term used to identify Dominicans residing in New York City.
[4] Tejano/a refers to individuals of Mexican descent who were born/raised in Texas.
[5] For an excellent discussion on emergent bilinguals, refer to Garcia, Kleifgen, and Falchi (2008).
[6] Nuyoricans are individuals of Puerto Rican descent residing in the United States.

Linguistic Challenges in Educational Settings

In this section, various linguistic challenges experienced by Latino/a bilinguals are examined, particularly within the context of a physical science lesson. Then, issues regarding language interference for Latino/a science learners are explored and discussed. Finally, instructional approaches for the science classroom are presented with the goal of assisting science teachers in developing and supporting language acquisition in emergent dominant Latino/a bilingual.

Emergent Bilinguals Face a Number of Linguistic Challenges

Bilingualism increases cognitive abilities, particularly the ability to remain focused on tasks and knowing how language is structured and used (Bialystok, Craik, & Luk, 2012; Garbin et al., 2010). For bilinguals, strong linguistic foundations in the native language transfer[7] to the second language, and L2 acquisition is facilitated in this process (Cummins, 1988; Farrell, 2011; Hammers & Blanc, 2000).

However, younger and adolescent bilinguals experience different challenges that affect the rate and quality of linguistic transfer. For instance, young emergent bilinguals (EBs) face the challenge of learning a second language while still acquiring the structure of their first language (Chomsky, 1969). Yet, without continued development of L1, the ability to communicate in L1 is lost, and the cognitive benefits of bilingualism is hampered (Francis, 2005; Menken & Kleyn, 2010; Wright & Taylor, 2000). In contrast, older EBs have an advantage over younger bilinguals with linguistic transfer because their first language is further developed and they can better negotiate meaning between the two languages (though explicit instruction is necessary to construct meaning). However, middle-/secondary-level EBs endure harder academic and linguistic demands for successfully achieving, especially in exit and graduation exams, while still mastering CALP in L2 (Center for Education Policy, 2008). Also, because of identity development and socialization practices, adolescent EBs are more likely to feel embarrassed about their limited knowledge of L2 (Gándara, Gutierrez, & O'Hara, 2001; Manavathu & Zhou, 2012) and shy away from sharing knowledge and ideas, behaviors that affect their motivation to succeed academically (Suárez-Orozco & Suárez-Orozco, 2001). Another linguistic challenge that both young and adult Latino/a EBs face is linguistic interference, particularly in science classrooms. The following section further elaborates on the linguistic challenges Latino/a encounters in science learning environments.

---

[7] For an excellent discussion on language transfer, see Treffers-Daller and Sakel (2012).

L2 Interferes with the Acquisition of L3

Some linguists argue that understanding is hampered and confusion is likely to occur when learners switch from one language to another, e.g., Spanish to English (Allard, Bourdeau, & Mizoguchi, 2011; Sandoval, Gollan, Ferreira, & Salmon, 2010). Piggybacking and complicating such a notion, the language of science as practiced in science classrooms, increases the difficulties threefold. Therefore, the commonly held idea of language interference is plausible, especially when teachers receive delayed, mistaken, or no responses from their bilingual learners. The following example of a knowledge construction "moment" illustrates language interference in context (Suriel, 2011). The scenario occurs between two Spanish-English bilingual students and a Latina bilingual teacher in a secondary-level bilingual physical science enrichment activity.

Gabby's Misunderstanding About Electrical Circuits

During a lesson on electrical circuits, students were supplied with a worksheet. Accompanying the Spanish translation of the problem statement were three drawings depicting simple electrical circuits. Students were asked to respond to the following problem statement regarding electrical circuits:

Instructor R: A 5.0 Ω lamp requires 0.20 A of current to operate. In which circuit (s) would the lamp operate when the switch S is closed? *Una lampara de 5.0 Ω require 0.20 A de corriente para operar. ¿En cúal circuito prenderá la lampara si cerramos el switch S (interruptor S)?*
Student Gabby (pseudonym): I say is B.
R: Why?
Gabby: Because even if you turn off the switch, energy…
R: Well, what is the difference between B and C? If you close the switch?
Gabby: That the switch is right next to the battery.
R: Oh, you are saying that the location of the battery and the switch has something to do with the current?
Gabby: Yes, because, right here. If you turn it off, it will still go like that [tracing a longer electron flow path in diagram B].
Student Pablo: It takes longer.
R: Take a look at diagram C.
Gabby: But [diagram] C, it goes like that and it doesn't close.
R: Gabby…Gabby, you were doing good. What happens in [diagram] C? If I close the switch…what will be the path of the electron flow?
Gabby: [diagram] C? It doesn't matter. Wait, what?
[Gabby reads the statement in Spanish to herself.]
Gabby: Oh, cuando el circuito prende, da la luz. (Oh, when the circuit turns on, it gives light.) Oh sorry, I thought you meant which will turn off.
R: Oh, you read it in Spanish, and then it made sense?
Gabby: Yeah
R: Oh. Interesting.
Gabby: Oh, umm, I don't know.

In her attempt to answer this statement, Gabby illustrates two sources of linguistic confusion. The terms "closed/turn off" and "open/turn on" were interpreted incorrectly. Like Pablo, Gabby interprets the term "close" in the phrase "when the switch S is closed?" as indicating the location of the switch on the circuit as in "close or nearby" to the battery. When prompted by the instructor to compare and trace electron flow in both diagrams, Gabby then realizes that it is not the location of the switch that matters. Gabby then proceeds to read the statement in Spanish and fixes on the term "turn on" that is clearly recognizable in the Spanish translation "prender." The terms to "turn on" and "turn off" a switch may be a source of confusion for many monolingual and bilingual novice physical science learners alike. To "turn on" or "open" a light refers to the process of closing the switch, and to "close/turn off" a light refers to the process of opening the switch to prevent electron flow. Here, the English terms used to describe the process of electricity (lighting) and the words used to describe the actual physical process of electrical circuits are at odds, possibly causing language interference between L2 and L3, particularly for Gabby.

In this example, we can see that language, lexical semantics in particular, confounds interpreting scientific information. In effect, the ambiguities associated with L2 interfere with the students' sense-making efforts about science.

L1 Assists with the Comprehension of L3

When Gabby read the statement in Spanish, she comprehended what was being asked by the question. Her sense-making efforts were not simply a matter of her being comfortable reading in Spanish, her mother tongue. Instead, the terms used to describe the electrical process are obvious to her. For instance, in the context of electrical circuits, the term "prender" means to give light, and the term switch translates to "interruptor" in Spanish (or to interrupt as in "to interrupt" the electron flow). The less-ambiguous Spanish sequence of words provides clearer and better meaning for Gabby compared to the less concrete English descriptions when understanding electrical circuits. Furthermore, these particular terms and phrases may have been embedded in Gabby's personal experiences with light switches. The use of L1 in the Spanish translation of the statement provided Gabby with the opportunity to access L3. In contrast, L2 acted as the source of language interference and misunderstanding.

This scenario illustrates the linguistic demands bilingual learners can experience when learning science. Science teachers are thus challenged to develop the knowledge and skills to effectively address the linguistic demands of the multilingual classroom if they are to assist knowledge construction with Latino/a bilinguals (National Council for Accreditation of Teacher Education, 2011). To achieve this end, an efficacious multilingual teacher is challenged to first triangulate L1, L2, and L3, then identify lexical and conceptual semantics within each language, and finally triangulate lexical and conceptual semantics for every lesson.

Numerous instructional approaches have been suggested to support teachers in their efforts to triangulate L1, L2, and L3 (see Diversity and Equity in Next

Generation Science Standards).[8] While some teaching strategies presented here aim to develop basic literacy in general, others are specific to science. For example, journaling can be considered a general strategy supporting academic and scientific writing in students. On the other hand, emphasizing knowledge of Latin-based roots, prefixes, and suffixes in the science language is specifically geared to developing vocabulary awareness within the context of science. In the following section, various teaching approaches are described that further develop basic literacy skills (reading, writing, listening, and speaking) and support bilingualism for both primary- and secondary-level emergent bilinguals in the science classroom. Although some of the listed strategies are specific to teaching emergent Latino/a bilinguals, others have been proven effective for teaching non-Latino/a emergent bilinguals and English speakers developing linguistic competencies.

Approaches Supporting and Developing Vocabulary Understanding in L1, L2, and L3

Development and Understanding of Word Cognates

In this strategy, translations of words that look and sound similar in English and Spanish (crosslinguistic cognates) are used to draw upon prior knowledge. For example, Spanish-speaking students may recognize and identify the term filtrate as *filtrar* or accelerate as *acelerar*, terms that they may be familiar with in Spanish.[9] However, it is important to note here that students may be familiar with the Spanish equivalent of the English words/terms and not necessarily with the concept behind them, i.e., the concept of filtration or acceleration. Thus, science teachers should develop conceptual understanding along with vocabulary development. When used as such, the use of cognates in the science classroom helps develop vocabulary in both Spanish and English, assist with comprehension, and support bilingualism and biculturalism in emergent bilinguals (Carlo et al., 2004; Dressler, Carlo, Snow, August, & White, 2011; Kelley & Kohnert, 2012).

Unlocking Academic Vocabulary

In building vocabulary with students, it is important to recognize the basic levels of vocabulary. Beck, McKeown, and Kucan (2002) propose three levels or tiers of vocabulary:

- Tier One (basic everyday vocabulary; high-frequency words)

---

[8] For an extended discussion on the mechanics of language acquisition, see Fillmore and Snow (2000); Echevarria, Vogt, and Short (2004); and Hammers and Blanc (2000).

[9] E-lists of Spanish-English cognates specific to science are readily available.

- Tier Two (general-purpose academic vocabulary; vocabulary used in school curricula)
- Tier Three (content-specific academic vocabulary; terms used in the science classrooms)

Emergent bilinguals struggle most with Tier Two and Tier Three vocabularies. Two strategies science teachers can implement in their teaching to help increase understanding of Tier Three vocabulary are to deconstruct and to break down science terms and concepts (Gabby's misunderstanding; Language-Rich Inquiry Science for English Language Learners (LISELL) project [LISELL], 2012) and to interchange Tier One with Tier Two and Tier Three terms and phrases (DeLuca, 2010). DeLuca (2010) provides an excellent example of this interchanging technique that combines both the use of Spanish-English cognates and the deconstruction of a science statement (Tier Three). DeLuca's example is presented as follows: "When a volcano is dormant, we may say that it is asleep." The first phrase includes both Tier Two and Tier Three terms and the continuing statement "we may say that it is asleep" uses Tier One terms facilitating comprehension. Furthermore, this example uses the term dormant, a Spanish-English cognate that translates in Spanish into the term *dormido* and, by extension, inactive. Thus, the emergent Latino/a bilingual could potentially translate the statement into Spanish as *El vocán esta dormido o inactivo* (the volcano is dormant or inactive) and correctly identify the scientific concept at hand.

For individual teaching units, and particularly when working with Spanish dominant emergent bilinguals, it is also important that teachers identify all Tiers One, Two, and Three vocabularies, including nouns particular to each unit (Walqui & Heritage, 2012). For instance, on a lesson on light reflection and refraction, teachers should identify and define Tier One vocabulary such as nouns (e.g., mirrors, light, light source, lamp, water), verbs (bend, curve, bounce), adjectives (straight, wavy, dissimilar), prepositions (e.g., on, under, through), and words and phrases such as same as and similar to. Tier Two words for this lesson could include terms such as trajectory, transparent, disperse, and unite. Tier Three terms for this lesson could include terms such as reflection, refraction, translucent, medium, dense, solid, liquid, gas, atoms, and photons (Walqui & Heritage, 2012). Unlocking vocabulary as it is used in scientific expressions (both oral and written) helps with developing and supporting comprehension of scientific phenomena and metalinguistic competency.

Using Pictures and Realia to Build Vocabulary

Due in part to the various dialects of Spanish, teachers who teach science in Spanish in US classrooms also experience linguistic interference, Spanish language interference in this case.[10] The use of visual tools such as pictures, illustrations, or realia is

---

[10] Spanish-speaking nations, as well as the United States, teach either Castilian or Latin American Spanish (Garcia & Torres-Guevara, 2008).

a highly recommended strategy for all multilingual/multicultural classrooms (Ashton, 1996; Ellis, 1994; Manavathu & Zhou, 2012; Pray & Monhardt, 2009; Short, Vogt, & Echevarria, 2011).

Approaches Supporting and Developing Writing Skills in L1, L2, and L3

Writing and Journaling

In learning language, it is important to develop the mechanics of the host language, e.g., its grammar and morphology including appropriate modes of expressions. Science journaling provides writing opportunities to articulate scientific inquiry, that is, making sense of observations, questioning, hypothesizing, and concluding by describing, explaining, and inferring (Akerson & Young, 2005; Huerta & Jackson, 2010; Lee, Penfield, & Buxton, 2011; Winsor, 2008). Studies conducted on ELLs and science learning indicate that in conjunction with hands-on collaborative activities and journaling or "writing like scientists," students' writing abilities improved (Bravo & Garcia, 2004; Rivard, 2004). With journaling, teachers have an opportunity to provide appropriate and constructive feedback, clarifying grammatical and morphological errors such as verb tenses, plural and possessive forms of nouns, and the use of articles (Holmes & Moulton, 1997; Peyton & Reed, 1990). In a study conducted on effective teaching and learning strategies among secondary-level science Latino/a emergent bilinguals, students identified journaling as an excellent learning technique but suggested that journaling be highly structured and innovative in order to be effective. Students in this study suggested the following strategies:

- The use of highly structured journaling that included summarizing key aspects of the lessons including reviewing questions and statements and using drawings, pictures, and diagrams to convey meaning
- The freedom to write in English or Spanish or both
- Ejournaling—or the ability to use technology to document learning (Suriel, 2011)

In an ongoing effort to support writing in different genres, in addition to journaling, science teachers should also (a) assign laboratory and research reports and (b) require elaborative written responses in formal assessments (Buxton et al., 2013). However, when assessing grammar in writing, science teachers should remain flexible and support learning (CALP requires four to 10 years to develop). The uses of semantic maps to connect ideas and synthesize key concepts also help front-loading information before elaborating ideas (Jackson, Tripp, & Cox, 2011).

Approaches Supporting and Developing Oral Competency in L2 and Scientific Discourse (L3)

Speaking

Out of the four basic language skill competencies, oral competency is the quickest to develop for individuals learning a second language (Cummins, 1988). This is due in part to the need to navigate and negotiate linguistically in everyday situations and in social interactions (see BICS). Collaborative group work, as those that occur in science classrooms, provides the context for supporting oral communication, in addition to supporting scientific discourse. In fact, one of the most recommended approaches to developing oral competency in L2 among emergent bilinguals is group work embedded in science discourse (Lee et al., 2011; Short et al., 2011). In essence, science discourse (L3) enables emergent bilinguals to discuss observations, hypothesis, questions, inferences and conclusions in the pretext of understanding and explaining scientific phenomena (Howe, et al., 2007). Furthermore, and compared to other disciplines, the context of science often affords the use of different senses (e.g., touching, seeing, smelling). Though emergent bilinguals are challenged with communicating in L2, they are still able to learn and experience science (L3). When purposely grouped, emergent bilinguals will interact with more balanced bilinguals. These social interactions not only provide the context to co-construct and further develop language (L1, L2, and L3) but also serve to encourage new/different cultural perspectives as those espoused by the new host culture or the culture of science.

It is important to note that as learners gain basic literacy and academic language, the demand for the triangulation of languages diminishes (Harper & de Jong, 2004). Thus, the semantically savvy science educator can concentrate on constructing meaning bilingually between L2 and L3.

Cultural Challenges

In this section, literature that support and develop cultural congruence between Latinos/as and the culture of science is presented first. Then, a discussion of the cultural incongruence between US school culture and Latino/a culture is presented to highlight challenges Latinos/as face in their early academic careers and that may impede them from realizing careers in the science fields. Finally, a systemic plan of action is proposed to help nurture Latino/a participation in science and science-related careers.

The Culture of Science can Be Congruent with Latinos/as Ways of Knowing and Doing

The culture of science need not be discordant with Latino/a culture and ways of understanding and doing science (Settlage & Southerland, 2011). Latinos/as' interactional learning styles are congruent with ways of doing science. Latinos/as learn in collectivistic ways, highly valuing cooperation, attentiveness, and intergroup articulations of keen observations (Greenfield et al., 2006). Similar interactional patterns are exercised as scientific practices (Next Generation Science Standards (NGSS)). As a commonly accepted practice in science education, reform-based science instruction immerses students in the culture of science and provides emergent bilinguals with opportunities and the context for developing, supporting, and triangulating L1, L2, and L3.[11]

Another highly recommended pedagogical approach to bridging the varying cultures and beliefs systems of learners and the culture of science is nurturing culturally congruent classrooms. Cultural congruence refers to various curricular approaches that specifically incorporate the culture (language, values, and norms) of students (Ladson-Billings, 1995). Among these culturally congruent approaches are creating, supporting, and maintaining culturally inclusive science classrooms, curricula, and pedagogy (Key, 2003; Lee, 2003; Lee & Buxton, 2010). To achieve this, science teachers can act as knowledge transmitters and facilitators, rather than knowledge dispensers, drawing upon students' culture (e.g., language, different worldviews) to initiate, encourage, and/or extend discussions. For example, when teaching a lesson on plants, teachers may include samples and discussions of plants students are culturally familiar with, including their historical perspectives and uses (Suriel, 2010). To this end, teachers have to transform curricula to include knowledge typically omitted from textbooks and school curricula (Banks, 2004; Suriel & Atwater, 2012). Science teachers can also construct hybrid spaces or "third spaces" in science classrooms (physical or intellectual) that provide emergent bilingual opportunities to fuse different types of discourses (i.e., L1 and L2 discourses) with scientific discourse (Aikenhead, 2001; Calabrese-Barton & Tan, 2008; Moje, Collazo, Carrillo, & Marx, 2001; Quigley, 2011). Hybrid spaces are important for the Latino/a science classroom because they serve as safe and flexible environments to negotiate meaning between L1, L2, and L3 and their respective cultures.

Science classrooms provide rich and unique contexts for supporting various cultural worldviews. With this purpose in mind, it is important to accommodate students' cultural knowledge, knowledge that is informed by their cultural beliefs. As a starting and welcoming platform, science teaching should first draw on students' prior knowledge, then incorporate their understandings, and further elaborate and clarify their ideas. Teachers can then help co-construct meaning, in culturally

---

[11] For an extended discussion on the impacts of language development through science inquiry activities, see Abrams and Ferguson (2004/2005); Bergman (2011); Bravo and Garcia (2004); Hampton and Rodriguez (2001); Lynch, Kuipers, Pyke, and Szesze (2005).

congruent ways, about the way "Western" science explains scientific phenomena. In this way, science teachers serve as trilingual cultural ambassadors, supporting the triangulation and development of L1, L2, and L3.

Culture of the School Need Not Impede Latinos/as/Latinas from Enrolling and Participating in Science and Science-Oriented Programs

Science learning is not devoid of cultural, social, and political agendas especially those espoused by mainstream schooling. Educators are charged with the task of imparting the competencies required for an individual to function successfully within the larger collective. However, the expectations and value systems that teachers and institutions of education at large typically embrace are different than those of Latinos/as and are often informed by White middle-class values (Banks, 2004). For example, meritocracy and individual achievement are fundamental to the US educational system and hold that all students, regardless of their class, race, ethnicity, age, gender, and physical and mental capabilities, have equal access to opportunities for high-quality educational outcomes and, by implication, upward social mobility.

In terms of educational achievement, a number of differences exist in the responses and activities related to learning between Latino/a culture and the culture of US schools. These modes of action are visible in a number of situations including a dualistic individualistic vs. collectivistic orientation, the level of value placed on "cultures of learning" in the household, and attitudes about the possibility of achievement. For instance, hegemonic groups in the United States and other developed Western societies emphasize individualistic ways of learning and development. Western culture encourages self-expression and autonomy. Child-initiated questions are often encouraged and even praised. Collectivistic ways of learning, for example, among immigrant Mexican Americans, often center upon group, family, and community membership (Bryan & McLaughlin, 2005; Greenfield et al., 2006), which can be in direct conflict with Western views. Specifically, activities that support cooperation, attentiveness, and keen observation are highly valued, while questioning and assertiveness are discouraged (Greenfield et al., 2006), though both are essential to scientific inquiry and inquiry-based instruction (Bergman, 2011).

What Is the Nature of Differential Educational Outcomes Between Latinos/as/Latinas and Their White Counterparts?

When addressing the differential outcomes between Latino/a students and European American or White students, the question of "is it simply a matter of literacy level (i.e., the degree of symbol and tool mastery)" often comes up. High-quality educational outcomes, being tied to high-quality SES outcomes, are associated with excellence (Crissy & Bauman, 2010). High-quality educational outcomes are those that reflect the imparting of high levels of both basic and higher-order skills and literacies; by implication, a student that has acquired a high-quality educational outcome is worthy of positions requiring technical aptitude, leadership, initiative, community involvement, and other personal qualities also associated with excellence.

In a meritocratic society, high-quality educational outcomes are often based on access and social capital. The components of high-quality educational outcomes are the result of access to the meritocracy. For example, a Latina student may be mature and academically gifted, but because she does not have access to culturally sensitive and sympathetic gatekeepers, i.e., academic counselors, teachers, to advise her about career choices, college destinations, and science-related careers, or because the school did not provide a rigorous curriculum (AP courses), the student may be excluded from educational outcomes that befit that student's capabilities and interests (Ceglie, 2009; Young, 2005). Students belonging to a hegemonic or assimilated group are more likely to (a) receive seamless access to information about best colleges and their admission requirements, (b) opportunities to network or intern with professionals in fields of interest, (c) participate in extra curricular academic oriented activities, and (d) participate in preparatory courses to enhance their ability to do well on standardized assessments. All these may be seen as gates to high-quality educational outcomes; those that mediate access to these gates are, in effect, the gatekeepers to these outcomes. True equity, then, would include interventions throughout the academic life of a student to expose the student to opportunities for high-quality educational outcomes.

When schools provide empowering experiences to Latinos/as, students can become more motivated and develop the ability and confidence to succeed (Cummins, 2001). Educators and theorists have consistently pointed to the importance of the student's self-agency or feeling that she or he is involved in the learning process; science literacy is dependent in large part on a properly motivated student's partnership and contribution to the literacy acquisition process (Bandura, 1977; Bruner, 1961; Bryan, Glynn, & Kittleson, 2011; Dewey, 1929; Pintrich, 2000; Zimmerman, 2000). It is in this context that student exposure to gatekeepers, through strategically placed interventions and other exposures, can be seen as an important component in any plan to increase Latino/a involvement in the sciences (Levine et al., 2007).

A more comprehensive cohort of Latino/a advocates such as counselors, science administrators, and teachers would be the cornerstone of a program that includes familiar strategies such as mentorship and involvement through science-oriented after-school programs. Similarly, the education of science teachers would include self-regulation and self-reflection regarding academic and science-related goals for Latinos/as, the redesign of science curricula to reflect the linguistic and cultural triangulation of L1, L2, and L3, as well as to reflect classrooms where high expectations of students are uniform across the board. The keys to developing scientific literacy and skills involve (a) extended, sustained, and meaningful interactions with gatekeepers providing affective and academic support (mentors, counselors, teachers, coaches, peers); (b) exposure to highly rigorous academic programs that are inquiry based—after-school, mentoring, tutoring, science-related and competitive learning opportunities, advanced placement courses, and internships—(c) culturally relevant curricula; and (d) focused policy that specifically addresses the eradication of the achievement gap as it relates to Latino/a science learners.

Summary/Concluding Thoughts

Knowledge is gained and shared through social interactions, and learning is optimized when co-constructed between our peers and us. Language in all its forms (informative, expressive, and directive) serves as a cultural tool and allows us to participate in knowledge construction. When language impedes understanding, learning is hindered. Science classrooms are often multilingual environments; thus, it is very likely that science students experience language interference. For instance, and in addition to language interference due to learning a second language, language interference can occur between English and the language of science or between the language of science and the language of mathematics. The complexities of these languages and their combinations are more compelling for emergent bilingual learners in science classrooms. Navigating the linguistic demands of the science classroom can be confusing and may, at the very least, delay comprehension and at the worst prevent knowledge construction. Such experiences have the potential to disenchant students about future careers in the sciences. Thus, it is imperative that teachers and teacher educators develop the knowledge and skills to address language as it is used in the science classroom and as it pertains to the acquisition of English and the language of science.

Furthermore, attention must be placed on the triangulation of the English, Spanish, and the language of science when educating Latino/a emergent bilinguals. Creating hybrid spaces or a classroom environment where emergent bilinguals can negotiate, interchange, and co-construct meaning has been a highly recommended and effective practice for bilingual science classrooms (Calabrese-Barton & Tan, 2008; Moje et al., 2001; Quigley, 2011; see approaches in SIOP, LISELL, ESTELL, CALLA).

Cultural congruence is also necessary to nurture academic success in Latino/a students. Hegemonic ideas, values, and norms are often embedded in educational institutions and are often expressed, poignantly, in curricula. Advocates for multicultural learning environments, science classrooms in particular, challenge such worldviews and propose new ways to accommodate different cultural perspectives. The inclusion, elaboration, and discussion of cultural histories, especially those regarding non-Western science ways of knowing, are excellent pedagogical approaches supporting culturally relevant and harmonious teaching and learning.

Science educators play an important role in (a) developing the knowledge, skills, and attitudes in science teachers for effectively teaching emergent bilinguals, especially Latino/a bilinguals (Harper & de Jong, 2009), and (b) supporting science teachers' transition from content teachers to content English as second language teachers (Welsh & Newman, 2010). In this effort, science educators in teacher education should explicitly address the linguistic challenges faced by emergent bilinguals, science EBs in particular. Furthermore, science educators should require all science teachers to incorporate curricular strategies in their curricula development and teaching practices that are specifically geared to further develop and support English language acquisition in all students, particularly in emergent bilinguals (Quinn et al., 2012; Santos, Darling-Hammond, & Cheuk, 2012). Science educators should also provide opportunities for science teachers to examine their own cultural and belief systems and the impact of culture on individual behavior so that they may gain a greater appreciation of the struggles faced by EBs (Keengwe, 2010). The goal here is to support a sense of advocacy for EBs in science teachers, potentially leading to curricula transformation inclusive of linguistic and multicultural approaches (Suriel & Atwater, 2012). Social justice for Latino/a science learners is enacted when we support, nurture, and guide them in knowledge construction, particularly regarding science.

References

Abrams, J., & Ferguson, J. (2004/2005). Teaching students from many nations. *Educational Leadership, 62*(4), 64–67.
Aikenhead, G. S. (2001). Students' ease in crossing cultural borders into school science. *Science Education, 85*, 180–188.
Akerson, V. L., & Young, T. A. (2005). Science the "write" way. *Science and Children, 43*(3), 38–41.
Allard, D., Bourdeau, J., & Mizoguchi, R. (2011). Addressing cultural and native language interference in second language acquisition. *CALICO Journal, 28*(3), 677–698.
Ashton, P. (1996). The concept of activity. In L. Dixon-Krauss (Ed.), *Vygotsky in the classroom: Mediated literacy instruction and assessment*. White Plains, NY: Longman.
Aukrust, V. G. (2011). *Learning and cognition in education*. Oxford, UK: Elsevier.
Bandura, A. (1977). Self-efficacy: Toward a unifying theory of behavioral change. *Psychological Review, 84*, 191–215.
Banks, J. A. (2004). Multicultural education: Historical development, dimensions and practice. In J. A. Banks & C. A. M. Banks (Eds.), *Handbook of research on multicultural education* (pp. 3–29). San Francisco: Jossey-Bass.
Beck, I., McKeown, M., & Kucan, L. (2002). *Bringing words to life: Robust vocabulary instruction*. New York: Guilford Press.
Beeth, M. E., & Hewson, P. W. (1997). *Learning to learn science: Instruction that supports conceptual change*. Paper presented at the Annual Meeting of the European Science Education Research Association, September 2–6, 1997, Rome, Italy.
Bergman, D. (2011). Synergistic strategies: Science for ELLs is science for all. *Science Scope, 35*(3), 40–44.
Bialystok, E., Craik, F. I. M., & Luk, G. (2012). Bilingualism: Consequences for mind and brain. *Trends in Cognitive Sciences, 16*(4), 240–250.
Bravo, M., & Garcia, E. (2004). *Learning to write like scientists: English language learners' science inquiry & writing understandings in responsive learning contexts*. Paper presented at the American Educational Researchers Association. Retrieved from http://www.ncela.gwu.edu/files/rcd/BE022124/Learning_to_Write_Like_Scientists.pdf

Brown, T. M., & Rodríguez, L. F. (2009). Empirical research study: School and the co-construction of dropout. *International Journal of Qualitative Studies in Education, 22*, 221–242.

Bruner, J. S. (1961). The act of discovery. *Harvard Educational Review, 31*(1), 20–32.

Bryan, R. R., Glynn, S. M., & Kittleson, J. M. (2011). Motivation, achievement, and advanced placement intent of high school students learning science. *Science Education, 95*(6), 1049–1065.

Bryan, L. A., & McLaughlin, H. J. (2005). Teaching and learning in rural Mexico: A portrait of student responsibility in everyday school life. *Teaching and Teacher Education: An International Journal of Research and Studies, 21*(1), 33–48.

Buxton, C., & Allexsaht-Snider, M., co-PIs. (2012). *Language-rich inquiry science for English language learners (LISELL) project*. University of Georgia. http://www.coe.uga.edu/lisell/whats- new/

Buxton, C., Allexsaht-Snider, M., Suriel, R., Kayumova, S., Choi, Y., Bouton, B., et al. (2013). Using educative assessments to support science teaching for middle school English language learners. *Journal of Science Teacher Education, 24*(2), 347–366.

Calabrese-Barton, A., & Tan, E. (2008). Funds of knowledge and discourses and hybrid spaces. *Journal of Research in Science Teaching, 46*(1), 50–73.

Carlo, M. S., August, D., McLaughlin, B., Snow, C. E., Dressler, C., Lippman, D., et al. (2004). Closing the gap: Addressing the vocabulary needs of English-language learners in bilingual and mainstream classrooms. *Reading Research Quarterly, 39*(2), 188–215.

Ceglie, R. J. (2009). Science from the periphery: Identity, persistence, and participation by women of color pursuing science degrees. (January 1, 2009). *Dissertations Collection for University of Connecticut.* Paper AAI3361001.

Chalmers, A. F. (1999). *What is this thing called science* (3rd ed.). Indianapolis, IN: Hackett Publishing Company.

Chomsky, N. (1969). *The acquisition of syntax in children from 5 to 10*. Cambridge, MA: MIT Press.

Crissy, S. R., & Bauman, K. B. (2010). *Between a diploma and a bachelor's degree: The effects of sub-baccalaureate postsecondary educational attainment and filed of training on earnings*. Paper presented at the annual meeting of the Population Association of America. Dallas, TX, April 15–17, 2010. Retrieved from http://www.edweek.org/media/censusdiplomas-34jobs.pdf

Cummins, J. (1981). The role of primary language development in promoting educational success for language minority students. In *Schooling and language minority students: A theoretical framework*. Los Angeles: California State University; Evaluation, Dissemination, and Assessment Center.

Cummins, J. (1988). Language proficiency, bilingualism and academic achievement. In J. Cummins (Ed.), *Bilingualism and special education: Issues in assessment and pedagogy* (pp. 130–151). San Diego, CA: College-Hill Press.

Cummins, J. (2001). Empowering minority students: A framework for intervention. *Harvard Educational Review, 71*(4), 656–675.

DeLuca, E. (2010). Unlocking academic vocabulary. *Science Teacher, 77*(3), 27–32.

Dewey, J. (1929). *The sources of a science of education*. New York: Horace Liveright.

Dressler, C., Carlo, M. S., Snow, C. E., August, D., & White, C. E. (2011). Spanish-speaking students' use of cognate knowledge to infer the meaning of English words. *Bilingualism: Language and cognition, 14*, 243–255.

Echevarria, J., Vogt, M., & Short, D. (2004). *Making content comprehensible for English language learners: The SIOP model* (2nd ed.). Boston: Pearson Education.

Ellis, E. S. (1994). Integrating writing strategy instruction with content-area instruction: Part 1– Orienting students to organizational devices. *Intervention in School and Clinic, 29*(3), 169.

Farrell, M. (2011). Bilingual competence and students' achievement in physics and mathematics. *International Journal of Bilingual Education and Bilingualism, 14*(3), 335–345.

Fillmore, L. W., Snow, C. E. (2000). *What teachers need to know about language*. Report of the Clearinghouse on Languages and linguistics. Retrieved from Clearinghouse on Languages and linguistics website: http://www.cal.org/resources/digest/0007bredekamp.html

Francis, N. (2005). Research findings on early first language attrition: Implications for the discussion on critical periods in language acquisition. *Language Learning, 55*(3), 491–531.

Freire, P. (1993). *Pedagogy of the oppressed: New revised 20th-anniversary edition.* New York: Continuum.

Fry, R., & Gonzales, F. (2008). *One-in-five- and growing fast: A profile of Hispanic public school students.* Retrieved from PEW Hispanic Center website: http://pewhispanic.org/reports/report.php?ReportID=92

Gándara, P., Gutierrez, D., & O'Hara, S. (2001). Planning for the future in rural and urban high schools. *Journal of Education for Students Placed at Risk (JESPAR), 6*(1&2), 73–93.

Gándara, P., & Rumberger, R. (2009). Immigration, language and education: How does language policy structure opportunity? *Teachers College Record, 111,* 6–27.

Garbin, G., Sanjuan, A., Forn, C., Bustamante, J. C., Rodriguez-Pujadas, A., Belloch, V., et al. (2010). Bridging language and attention: Brain basis of the impact of bilingualism on cognitive control. *NeuroImage, 53*(4), 1272–1278.

Garcia, O., Kleifgen, J. A., & Falchi, L. (2008). *From English language learners to emergent bilinguals* (A research initiative of the campaign for educational equity, equity matters: Research, review, Vol. 1). New York: Teachers College, Columbia University.

Garcia, O., & Torres-Guevara, R. (2008). Monoglossic ideologies and language policies in the education of U. S. Latinas/os. In E. Murillo Jr., S. A. Villenas, R. T. Galván, J. Sánchez Munoz, C. Martínez, & M. Machado-Casas (Eds.), *Handbook of latinos/as and education: Theory, research and practice* (pp. 182–192). New York: Routledge.

Garcia-Preto, N. (2005). Latino/a families: An overview. In M. McGoldrick, J. Giordano, & N. Garcia-Preto (Eds.), *Ethnicity and family therapy* (3rd ed., pp. 153–165). New York, NY: Guilford Press.

Greenfield, P., Trumbull, E., Keller, H., Rothstein-Fisch, C., Suzuki, L. K., & Quiroz, B. (2006). Cultural conceptions of learning and development. In P. Alexander & P. H. Winnie (Eds.), *Handbook of educational psychology* (2nd ed., pp. 675–692). Mahwah, NJ: Lawrence Erlbaum Associates.

Hammers, J. F., & Blanc, H. A. (Eds.). (2000). *Bilinguality and bilingualism* (2nd ed.). New York: Cambridge University Press.

Hampton, E., & Rodriguez, R. (2001). Inquiry science in bilingual classrooms. *Bilingual Research Journal, 25*(4), 461–478.

Harper, C., & de Jong, E. (2004). Misconceptions about teaching English-language learners. *Journal of Adolescent and Adult Literacy, 48*(2), 152–162.

Harper, C. A., & de Jong, E. J. (2009). English language teacher expertise: The elephant in the room. *Language and Education: An International Journal, 23*(2), 137–151.

Hay, I., & Fielding-Barnsley, R. (2012). Social learning, language and literacy. *Australasian Journal of Early Childhood, 37*(1), 24–29.

Holmes, V. L., & Moulton, M. R. (1997). Dialogue journals as an ESL learning strategy. *Journal of Adolescent and Adult Literacy, 40*(8), 616–621.

Howe, C. J., Tolmie, A., Thurston, A., Topping, K., Christie, D., Livingston, K., et al. (2007). Group work in elementary science: Towards organizational principles for supporting pupil learning. *Learning and Instruction, 17,* 549–563.

Huerta, M., & Jackson, J. (2010). Connecting literacy and science to increase achievement for English language learners. *Early Childhood Education Journal, 38*(3), 205–211.

Hussar, W. J. & Bailey, T. M. (2011). *National center for education statistics, 2011.* Projections of Education Statistics to 2020 from http://nces.ed.gov/pubsearch/pubsinfo.asp?pubid=2011026

Jackson, J., Tripp, S., & Cox, K. (2011). Interactive word walls: Transforming content vocabulary instruction. *Science Scope, 35*(3), 45–49.

Keengwe, J. (2010). Fostering cross cultural competence in preservice teachers through multicultural education experiences. *Early Childhood Education Journal, 38*(3), 197–204.

Kelley, A., & Kohnert, K. (2012). Is there a cognate advantage for typically developing Spanish-speaking English-language learners? *Language Speech and Hearing Services in Schools, 43*(2), 191–204.

Key, S. G. (2003). Enhancing the science interest of African American students using cultural inclusion. In S. M. Hines (Ed.), *Multicultural science education*. New York: Peter Lang.

Kirch, S. A. (2010). Identifying and resolving uncertainty as a mediated action in science: A comparative analysis of the cultural tools used by scientists and elementary science students at work. *Science Education, 94*(2), 308–335.

Kohler, A. D., & Lazarin, M. (2007). *Hispanic education in the United States*. Retrieved from. http://www.nclr.org/index.php/site/pub_types/issue_briefs/P20/

Ladson-Billings, G. (1995). Toward a theory of culturally relevant pedagogy. *American Educational Research Journal, 32*(3), 465–491.

Lee, O. (2003). Equity for linguistically and culturally diverse students in science education: A research agenda. *Teachers College Record, 105*(3), 465–489.

Lee, O., & Buxton, C. (2010). *Diversity and equity in science education: Theory, research, and practice*. New York: Teachers College Press.

Lee, O., Penfield, R. D., & Buxton, C. (2011). Relationship between "form" and "content" in science writing among English language learners. *Teachers College Record, 113*(7), 1401–1434.

Lemke, J. L. (1990). *Talking science: Language, learning and values*. Norwood, NJ: Ablex.

Levine, R., Gonzalez, R., Cole, S., Fuhrman, M., & Floch, K. C. L. (2007). The geoscience pipeline: A conceptual framework. *Journal of Geoscience Education, 55*(6), 458–468.

Lynch, S., Kuipers, J., Pyke, C., & Szesze, M. (2005). Examining the effects of a highly rated science curriculum on diverse students: Results from a planning grant. *Journal of Research in Science Teaching, 42*(8), 1–35.

Manavathu, M., & Zhou, G. (2012). The impact of differentiated instructional materials on English language learner (ELL) students' comprehension of science laboratory tasks. *Canadian Journal of Science, Mathematics and Technology Education, 12*(4), 334–349.

Mason, L. H., & Hedin, L. R. (2011). Reading science text: Challenges for students with learning disabilities and considerations for teachers. *Learning Disabilities Research and Practice, 26*(4), 214–222.

McInstosh, S. (2011). *State high school tests: Changes in state policies and the impact of the college and career readiness movement. Center for Education Policy*. From: http://www.cep-dc.org/displayDocument.cfm?DocumentID=385

Menken, K., & Kleyn, T. (2010). The long-term impact of subtractive schooling in the educational experiences of secondary English language learners. *International Journal of Bilingual Education and Bilingualism, 13*(4), 399–417.

Mercer, N. (2008). Talk and the development of reasoning and understanding. *Human Development, 51*, 90–100.

Moje, E. B., Collazo, T., Carrillo, R., & Marx, R. W. (2001). "Maestro, what is 'quality'?": Language, literacy, and discourse in project based science. *Journal of Research in Science Teaching, 38*(4), 469–498.

National Academy of the Sciences [NAS]. (2010). *Expanding underrepresented minority participation: America's science and technology talent at the crossroads*. (A research report co-sponsored by the Committee on Underrepresented Groups and the Expansion of Science and Engineering Workforce Pipeline; Committee on Science, Engineering, and Public Policy; Policy and Global Affairs; National Academy of Sciences, National Academy of Engineering, Institute of Medicine). Washington, DC: National Academies Press. Available from http://www.nap.edu/catalog/12984.html

National Council for Accreditation of Teacher Education. (2011). *Unit Standards in effect 2008, Standard4: Diversity*. Retrieved from. http://www.ncate.org/Standards/NCATEUnitStandards/UnitStandardsinEffect2008/tabid/476/Default.aspx#stnd4

National Research Council. (2005). *How students learn: Science in the classroom*. (Committee on How People Learn: A Targeted Report for Teachers.). In M. S. Donovan, & J. D. Bransford, (Eds.), Washington, DC: The National Academies Press.

Nieto, S. (2002). *Language, culture, and teaching: Critical perspectives for a new century*. Mahwah, NJ: Lawrence Erlbaum Associates.

No Child Left Behind Act, U. S. C. §107 (2001).
Peyton, J. K., & Reed, J. (1990). *Dialogue journal writing with nonnative english speakers: A handbook for teachers*. Alexandria, VA: TESOL.
Pintrich, P. R. (2000). The role of goal orientation in self-regulated learning. In M. Boekaerts, P. R. Pintrich, & M. Zeidner (Eds.), *Handbook of self-regulation* (pp. 451–502). San Diego, CA: Academic.
Pray, L., & Monhardt, R. (2009). Sheltered instruction techniques for ELLs. *Science and Children, 46*(7), 34–38.
Quigley, C. (2011). Pushing the boundaries of cultural congruence pedagogy in science education towards a third space. *Cultural Studies of Science Education, 6*(3), 549–557.
Quinn, H., Lee, O., & Valdés, G. (2012). *Language demands and opportunities in relation to next generation science standards for English language learners: What teachers need to know. Understanding language: Language, literacy, and learning in the content areas*. http://ell.stanford.edu
Ramirez, J. D., Pasta, D. J., Yuen, S., Billings, D. K., & Ramey, D. R. (1991). *Final report: Longitudinal study of structural immersion strategy early-exit, and late-exit transitional bilingual education programs for language minority children*. San Mateo, CA: Aguirre International.
Rivard, L. P. (2004). Are language-based activities in science effective for all students, including low achievers? *Science Education, 88*(3), 420–442.
Sandoval, T. C., Gollan, T. H., Ferreira, V. S., & Salmon, D. P. (2010). What causes the bilingual disadvantage in verbal fluency? the dual-task analogy. *Bilingualism: Language and Cognition, 13*(2), 231–252.
Santos, M., Darling-Hammond, L. & Cheuk, T. (2012). *Teacher development to support english language learners in the context of common core state standards. Understanding language: Language, literacy, and learning in the content areas*. http://ell.stanford.edu
Schleef, D., & Cavalcanti, H. B. (2010). *Latinos/as in Dixie: Class and assimilation in Richmond*. Virginia Albany, NY: State University of New York Press.
Settlage, J., & Southerland, S. A. (2011). *Teaching science to every child: Using culture as a starting point* (2nd ed.). New York, NY: Routledge.
Short, D. J., Vogt, M. J., & Echevarria, J. J. (2011). *The science model for teaching science to english learners*. Boston: Pearson.
Slama, R. B. (2012). A longitudinal analysis of academic english proficiency outcomes for adolescent English language learners in the United States. *Journal of Educational Psychology, 104*(2), 265. Academic One File. Web. 10 July 2012.
Suárez-Orozco, C., & Suárez-Orozco, M. M. (2001). *Children of immigration*. Cambridge, MA: Harvard University Press.
Suriel, R. (2010). Spanish moss: Not just hanging in there. *Science Activities: Classroom projects and curriculum ideas, 47*(4), 133–140.
Suriel, R. (2011). *The trilingual science teaching ambassador: Exploring the triangulation of Spanish, English, and the language of science*. Unpublished dissertation, University of Georgia, Athens, GA.
Suriel, R., & Atwater, M. M. (2012). From the contribution to the action approach: White teachers' experiences influencing the development of multicultural science curricula. *Journal of Research in Science Teaching, 49*(10), 1271–1295.
Tafoya, S. M. (2005). Shades of belonging: Latinos/as and racial identity. *Harvard Journal of Hispanic Policy, 17*, 58–78.
Torres-Saillant, S. (2005). Latino/a. In S. Oboler & D. J. González (Eds.), *The Oxford encyclopedia of Latinos/as and Latinas in the United States* (Vol. 2, pp. 507–510). East Los Angeles, CA: Oxford University Press.
Treffers-Daller, J., & Sakel, J. (2012). Why transfer is a key aspect of language use and processing in bilinguals and L2-users. *International Journal of Bilingualism, 16*(1), 3–10.
U. S. Census Bureau. (2011). *Profile America facts for features: Hispanic heritage month 2011*. September 15–October 15 from: http://www.census.gov/newsroom/releases/archives/facts_for_features_special_editions/cb11-ff18.html

U. S. Department of Education, National Center for Education Statistics. (2012). *Postsecondary expectations of 12th-graders*. Retrieved from http://nces.ed.gov/programs/coe/figures/figure-ect-1.asp

U.S. Department of Education, National Center for Education Statistics. (2007). *Science performance of students in grades 4, 8, and 12*. Retrieved from http://www.ed.gov

U.S. Department of Education, National Center for Education Statistics. (2011). *The condition of education 2011 (NCES 2011–033)*. Retrieved from http://nces.ed.gov/fastfacts/display.asp?id=16

Vygotsky, L. S. (1978). *Mind in society: The development of higher psychological processes*. Cambridge, MA: Harvard University Press.

Wallesrstein, I. (2005). Latin@s: What is in a name? In R. Grosfoguel, N. Maldonado-Torres, & J. D. Saldivar (Eds.), *Latino/as in the world-system: Decolonization struggles in the 21st century U. S. empire* (pp. 31–39). Hendon, VA: Paradigm Publishers.

Walqui, A. & Heritage, M. (2012). *Instruction for diverse groups of english language learners. Understanding language: Language, literacy, and learning in the content areas.* http://ell.stanford.edu

Welsh, L. C., & Newman, K. L. (2010). Becoming a content-ESL teacher: A dialogic journey of a science teacher and teacher educator. *Theory into Practice, 49*(2), 137–144.

Wertsch, J. V. (1991). *Voices of the mind: A sociocultural approach to mediated action*. Cambridge, MA: Harvard University Press.

Winsor, M. S. (2008). Bridging the language barrier in mathematics. *Mathematics Teacher, 101*(5), 372–378.

Wright, S. C., & Taylor, D. M. (2000). Subtractive bilingualism and the survival of the Inuit language: Heritage- versus second-language education. *Journal of Educational Psychology, 92*(1), 63–84.

Young, H. (2005). Secondary education systemic issues: Addressing possible contributors to a leak in the science education pipeline and potential solutions. *Journal of Science Education and Technology, 14*(2), 205–216.

Zimmerman, B. J. (2000). Self-efficacy: An essential motive to learn. *Contemporary Educational Psychology, 25*, 82–91.

Part V
Policy Reform for Science Teacher Education

STEM-Based Professional Development and Policy: Key Factors Worth Considering

Celestine H. Pea

Any opinions, findings, conclusions, or recommendations expressed in this chapter are those of the author and does not reflect the views of the National Science Foundation in any way.

Introduction

This chapter looks at science, technology, engineering, and mathematics (STEM)-based professional development and policy, as well as related implications for the curriculum and cultural pedagogy, as factors worth considering for improving science education for all 6–12 public school students. These factors are of national significance as the United States strives to remain globally competitive, economically sound, and generally secure. However, maintaining preeminence in the world will require a workforce that is highly skilled and competent in STEM education. To that end, the need is to move from traditional ineffective practices to newer paradigms in science education (Cuban, 2012; Fullan, 2010).

The aim of this move is to develop highly effective STEM teachers, great STEM school leaders, and highly qualified STEM graduates to help keep the United States strong as described in reports from the President's Council of Advisors for Science and Technology (PCAST, 2010, 2012), the National Research Council (NRC, 2012b, 2013), the Carnegie Foundation of New York (2009), and the Business-Higher Education Forum (BHEF, 2007, 2010). Hence, it is imperative that science teachers and educators and school leaders engage in STEM-based professional development to ensure that the quality of teaching and learning in science inspires and prepares students for the future workforce (BHEF, 2010; PCAST, 2012; Stephens & Richey, 2011).

C.H. Pea (✉)
National Science Foundation, 4201 Wilson Blvd, Arlington, VA 22230, USA
e-mail: cpea@nsf.gov

STEM-based professional development offers a viable option for helping to build, sustain, and move the nation forward in acquiring such a workforce. It is characterized by twenty-first-century skills (NRC, 2012b) and the use of state and/or district level plans informed by rich data sets of information from schools, principals, teachers, students, and the learning environment (Bausmith & Barry, 2011; Supovitz, Foley, & Mishook, 2012). Such information is used to develop individual, group, and collaborative team activities that focus on deeper learning, specific content, and pedagogical content knowledge and skills as part of a whole school reform effort.

This information is also used to promote active learning for teachers that reflects what they are expected to do with their students in their own classrooms. It also supports coaching, modeling, observation, and feedback for teachers and involves the use of multiple assessments and ongoing analysis of student work. STEM-based professional development promotes coherence, is of longer duration, and is linked to an analysis of teacher and student learning (*Learning Forward*, 2010; Loucks-Horsley, Love, Stiles, Mundry, & Hewson, 2003; NRC, 2012a, b).

As evidenced-based evaluation, research, and practice proclaim, this type of professional development calls for a sharp departure from traditional approaches, yet it recognizes that it is the teacher who remains the determinant of success for students who will make up the future workforce (BHEF, 2007; Fullan, 2007a; Porter, Birman, Garat, Desimone, & Yoon, 2004). Therefore, teachers are viewed as the target and agent (Metz, 2009) for reforming science education, and professional development is identified as the primary venue for reaching all teachers. Hence, a closer look at STEM-based professional development and policy is extremely important in developing an exemplary future STEM workforce.

STEM-based professional development also reflects the spirit of the *Blueprint* (2010) of the reauthorized Elementary and Secondary Education Act (ESEA); *A Framework for K-12 Science Education: Practices, Crosscutting Concepts, and Core Ideas* (*NAS*, 2010); and the Next Generation Science Standards (NGSS). These documents reflect real-world changes in science education and will likely require world-class policy support for STEM-based professional development.

Consistent with STEM-based professional development and policies, the curriculum and cultural pedagogy remain a challenge at 6–12 grades in science education, while at the same time, NGSS call for the addition of new dimensions (e.g., crosscutting variables, designed-based engineering, and learning progression). Consequently, insights about the critical role these factors play in science education need to be researched more soundly to gain a better understanding of the implications for teaching and learning. It is imperative that states and school districts address STEM-based professional development, policy, curriculum, and cultural pedagogy to help improve the performance of teachers and leaders, who in turn will be better positioned to improve the performance of all students (Mizell, 2008; NRC, 2012b).

Professional Development: A Necessary Strategy

Great Teachers/Great Leaders. For more than a half century, professional development has been the "go to" strategy for reforming science education. However, large- and small-scaled evaluations of professional development activities show that despite great intentions, results have been tepid with little or no sustainability (Darling-Hammond, Wei, Andree, Richardson, & Orphanos, 2010; Porter et al., 2004; Wei, Darling-Hammond, Richardson, Andree, & Orphanos, 2010).

In addition, most offerings remained traditional in form and duration and focused on too many topics in a single setting (NRC, 2007; Schwartz, Sadler, Sonnert, & Tai, 2008). Despite this knowledge about the ineffectiveness of traditional approaches, many states and districts continue to use them. However, the science education community is engaging in more research about STEM-based professional development models that use a full range of competencies to promote deeper learning and knowledge transfer between students and teachers (NRC, 2012b).

One such model involves the use of learning progressions (Corcoran, 2007; Duncan & Rivet, 2013; Elmesky et al., 2012; Next Generation Science Standards [NGSS], 2013; Rogat et al., 2011; Smith, Wiser, Anderson, & Krajcik, 2006; Stevens, Delgado, & Krajcik, 2009) to inform practice and present pathways to deeper learning in science. For example, Furtak (2012) is midway into a 5-year Faculty Early Career (CAREER) grant award that is working on an "existence proof" of how an educative learning progression designed for high school biology teachers and their students can be used to improve instruction and student outcomes. The learning progression serves as a scaffold for teachers in inquiry-based settings by helping them to not only anticipate the ideas that students have but also suggest instructional strategies tailored to these ideas. Further, the aim is to establish baseline data for teachers' instructional practice and to measure student learning with a pre-post assessment of students' understanding of natural selection.

Moreover, the study tracks and measures change in teachers' classroom instructional practices during the teaching of natural selection to see whether these changes in teachers are associated with changes in student learning about evolution. Subsequently, this study can inform the field about learning, how students learn natural selection, and how teachers may learn to teach it more effectively. Although learning progressions in science have not been proven to be effective in large-scale reform, lessons learned from this study can contribute to the design of focused, sophisticated, and effective STEM-based professional development (Furtak, 2012).

Another model by Maskiewicz and Winters (2012) examines how teachers facilitate scientific inquiry while at the same time documenting teachers' and students' learning progressions. In this multi-year study, teachers participated in STEM-based professional development that encouraged them to pose questions about and explore scientific phenomena in a single domain while pursuing ideas and questions brought up by students. Teachers were encouraged to consider the strengths students brought to the classroom from both intellectual and epistemological viewpoints. However, initial results showed that "a focus only on the teacher can lead to inappropriate assumptions about a teachers' change in practice." (Maskiewicz & Winters, p. 460). Thus, going forward, emphasis must be placed on change in both students and teachers participating in the learning progression.

Building on results from the 2010 study by Maskiewicz and Winters, in 2012, these researchers initiated a second study that involved an in-depth investigation of the interplay between teacher and students learning, highlighting how the two co-constructed classroom inquiry practices. The hypothesis was that as teacher practices increase in sophistication, teacher and student learning progressions must be inextricably linked. To that end, "the ultimate measure and description of where a teacher is on a learning progression should include what the students are doing in the class, along with the intellectual and epistemological resources of these students, and how the teacher responds and adapts to those." (Maskiewicz and Winters, p. 460).

As noted from the evaluation results by Porter et al. (2004), reform efforts must focus on both students and teachers. As such, studies by Furtak (2012) and Maskiewicz and Winters (2012) reinforced that position that studies should examine learning for both students and teachers, investigate transition points between teacher and student learning, and document the transfer of knowledge between the two (Maskiewicz & Winters).

In addition to studying various STEM-based professional development models that focus on learning progressions, research by Fields, Levy, Tzur, Martinez-Gudapakkam, and Jablonski (2012) illustrates that 7th through 12th grade students scored better on science subject tests when their teachers participate in STEM-based professional development in the same science subject. The greatest difference was evident in biology where 70 % of students taught by teachers who participated in long-term professional development scored higher than the 55 % of students whose teachers did not participate in the professional development activities. Again, evaluations by Porter et al. (2004), Wei et al. (2009), and Darling-Hammond et al. (2010) demonstrated that STEM-based professional development focused on specific curriculum content and pedagogies was more effective for teachers relative to implementation of strategies.

Whereas, it is generally agreed that a significant number of science teachers need to participate in long-term STEM-based professional development like the studies described above, research shows that teachers learn from each other as well. For example, research shows that teachers learn from other teachers who are effective and that teachers are better at raising student achievement when they are connected to other teachers who raise student achievement (Early & Shagoury, 2010; Jackson & Bruegmann, 2009). This exchange in knowledge and skills leading to improvements in student achievement often occurs in settings where there is strong peer-to-peer support within and/or across grade levels as well as great leadership at the school level (Early & Shagoury, 2010; Pea, 2012).

Great Leaders. To bring about even greater success and change, research finds that "effective principals are critical to strengthening teaching and learning, but there has been an insufficient investment in recruiting, preparing, and supporting great principals, particularly for high-poverty schools" (*U. S. Department of Education*, 2010, p. 7). Reports from studies and programs show that other than the teacher, the most important school-based variable affecting student achievement is school leadership (BHEF, 2007; U. S. Department of Education, 2010; Elmore, 2000; Goldring, Spillane, Huff, Barnes, & Supovitz, 2006; Spillane, 2000). Notably, research by Early and Shagoury (2012) showed that teachers cited the local administrators as a major factor in how they view their school. However, not all teachers in the study felt that way. Almost half of the teachers saw local administrators as a source of encouragement or discouragement. Taken collectively, more teachers than not feel good when their local administrators know their names, provide positive feedback, support their requests, and stop by their classrooms to see how they are doing (Early & Shagoury, 2012).

Therefore, research confirms that highly effective teachers and leaders are the nation's best resources for increasing the interest in and selection of STEM careers by a greater number of students. *Learning Forward* (2010) included principals in their description of effective professional development and notably the *Blueprint* (2010) shifted STEM-based professional development activities from a single focus on teachers to include principals and other educators. Specifically, ESEA stated that:

> We will elevate the teaching profession to focus on recognizing, encouraging, and rewarding excellence. We are calling on states and districts to develop and implement systems of teacher and principal evaluation and support, and to identify effective and highly effective teachers and principals on the basis of student growth and other factors. (p. 4)

Effective teachers and leaders connect and collaborate on ensuring higher achievement by all students and closing of the achievement gap in STEM fields.

Today, many districts recognize the need for great leaders and have established leadership academies and/or programs to help principals become better STEM leaders (Corcoran, Schwartz, & Weinstein, 2009; Drago-Severson, 2012; Goldring, Huff, May, & Camburn, 2008). In 2009, the Education Development Center (EDC) released an evaluation report conducted on districts, states, and higher education institutions funded by the Wallace Foundation to improve principal preparations programs. EDC reported that the impact of these programs is evident in the quality of instruction, hiring decisions/placement, professional development offerings, and support for teachers in schools led by principal-participants. Many teachers agreed that effective principals contributed heavily to their decisions to remain in teaching or at a particular school (Ingersoll & Merrill, 2010; Pea, 2012; Wallace Foundation, 2011a).

This evaluation showed that principals who participated in programs that focused on educating great leaders, in turn, provided more access to STEM-based professional development for their teachers. However, evaluations from large-scaled professional development paint a different picture about what contributes to poor leadership at the school level. For example, Darling-Hammond et al. (2010) showed that "opportunities for sustained, collegial professional development of the kind that produces changes in teaching practice and student outcomes are much more limited in the United States than in most high-achieving nations abroad (p. v)." The evaluation also showed that teachers in the United States have less time in their regular work schedules for professional development through cooperative work with colleagues than other nations do. Additionally, the little time teachers do have rarely allowed for deep engagement in joint efforts to improve teaching and learning (Garet, Porter, Desimone, Birman, & Yoon, 2004; Muijs & Lindsay, 2008).

To change these outcomes, more efforts must be implemented to help principals become better leaders in science education through the development and assessment of quality programs (Corcoran et al., 2009; Drafo-Severson, 2007; King, Levinger, & Schoener, 2006). Only when school leaders and policymakers take on the hard and necessary steps to engage teachers in real change will the nation began to benefit from the type of leadership and STEM-based professional learning described in the *Blueprint* (2010) and suggested by evaluation and research (BHEF, 2007; Darling-Hammond et al., 2010; Davis, 2002; Drago-Severson, 2012; Fullan, 2007a; Guskey & Yoon, 2009; Porter et al., 2004).

Change. The need for meaningful change in leadership and teaching is evident in the number of students who continue to score at or below basic on state, national, and international tests. While across the nation there are pockets of success relative to lessons learned from emerging research on professional development and the impact on student achievement, recent assessment results (Fleming, 2012) are showing that most students are not scoring at higher levels in science. The National Assessment of Educational Progress (NAEP, 2009, 2011) and other sources reveal that students continue to have a shallow grasp of science when required to use laboratory equipment to perform science experiments.

Results from these tests also revealed that students are not doing well using new interactive computer tasks to solve scientific problems. Specifically, the results showed that:

> Both the hands-on and computer tests asked students to predict what might happen in a particular scientific scenario, make observations about what occurred in the scenarios, and explain the findings of the experiments or investigations they launched. These questions examined how well students could conduct and reason through "real life" science situations and grasp the scientific concepts of what occurred in their investigations according to the report from the National Center on Education Statistics, the U. S. Department of Education division that administers NAEP. (Fleming, 2012, p. 1)

Even though students were able to report what was happening in some activities, in other situations, students had trouble manipulating variables and running experiments. Likewise, more students could draw the right conclusion than could provide an explanation for their answers using data from their findings. These findings signal the need for districts and states to ensure that the key resources for improvement in student achievement—teachers and leaders—have the support and policy needed to bring about changes in practices at the classroom level.

More of the same will not likely provide students and teachers with the knowledge and skills to do better (Fleming, 2012; NRC, 2012b). Spillane (2012) noted that decades of policy shifts have seemingly led to the beliefs that "…the more things have changed, the more things seem to have stayed the same," (p. 123). One of the underlying reasons is the lack of knowledge of and attention to what it actually requires for change to occur in educational settings (Davis, 2002; Fullan, 2007a; Spillane & Callahan, 2000). Researchers are quite clear that for students to use twenty-first-century knowledge and skills, significant changes must occur nationwide along the K-16 continuum (Elmore, 2004; Fullan, 2007a, 2007b; Mizell, 2008; Spillane & Callahan, 2000; Wallace Foundation, 2011b).

Even with considerable support for reform since National Defense Education Act (P. L. 85–864) of 1957, Davis (2002); Stafford & Bales (2011); and Stylinski, Parker, and McAuliffe (2011) argue that planners and providers of professional development have not attended to certain conditions that need to be met for people to change the theories that guide their teaching and learning. Researchers contend that for STEM-based professional development to be effective, there are a few conditions about change that must be met (Feldman, 2000; Fullan, 2007a, 2007b).

First, teachers must become discontented with a practical theory because they believe that the current methods they are using are not being effective with most students (Annetta et al., 2013; Davis, 2002). On the contrary, teachers must see the new practical theory as reasonable and beneficial to improving learning (Davis). Hence, STEM-based professional development should provide teachers (and principals) with opportunities to address preexisting knowledge and beliefs about teaching and learning (Stafford & Bales, 2011; Stylinski, Parker, & McAuliffe, 2011). Second, it is recommended that teachers receive continued opportunities to deepen and expand their subject matter and pedagogical knowledge and pedagogical content knowledge; without such, it is difficult for teachers to accommodate the diverse cultures of their students (Feldman, 2000; Stafford & Bales, 2011).

Many researchers (Davis, 2002; Stafford & Bales, 2011) agree that to facilitate change in science teaching practices, principals and other relevant education stakeholders need to focus on ways to observe teachers' actions in context. Davis and Stafford and Bales confirm that professional development embedded in or aligned with the implementation of a new program or innovation, related to the desired change, can provide multiple opportunities for teachers to engage in the process of change. Such opportunities allow teachers to interact, pilot or observe implementation, reflect and share in feedback sessions, as well as avail themselves of other support activities (Feldman, 2000; Loucks-Horsley et al., 2003).

Such broadly defined models of STEM-based professional development should include not only cognitive and achievement indexes but also a wide range of affective variables as well (Bykerk-Kauffman, 2010). Affective variables may include students' motivation, interest, teacher expectation, attitudes and beliefs, and self-efficacy (Ford, 1992; R. Elmore, personal communication, March 24, 2003; H. Gardner, personal communication, April 21, 2003; Maltase & Tai, 2010).

The idea that affective factors might be included in STEM-based professional development was the topic of recent research by Swarat, Ortony, and Revelle (2012). Since research (PCAST, 2010; Swarat, Ortony, & Revelle, 2012) shows that students are losing interest in STEM and switching out of STEM courses at an alarming rate during their college and high school years, Swarat et al. (2012) ". . . sought to identify sources of student interest or ways of fostering interest" (p. 515) in STEM. Keenly aware of the lack of agreement in the literature about how interest can be defined as well as the numerous elements that can influence students' interest, the researchers used a questionnaire and follow-up interviews to capture students' interest in science at the middle school level where students' interest in STEM drops significantly (Swarat et al., 2012, Wells, Sanchez, & Attridge, 2007).

The results showed that "when thinking about how interesting an instructional episode was, students seemed to be mostly concerned with the form of the activity, and not so much with the topic and learning goal." (Swarat et al., 2012, p. 530). Moreover, the results revealed that students were most interested in lessons that actively engaged them in learning, used scientific tools and technologies, and included few teacher-centered practices (Swarat et al., 2012).

These outcomes align with evaluation results (Porter et al., 2004; Wei, Darling-Hammond, & Adamson, 2009) and outcomes from research by Stafford and Bales (2011) and Stylinski et al. (2011). Both groups of researchers suggested that providing teachers with a deeper understanding of the theories behind newly proposed ideas, prior to the actual implementation, buffers fears and doubts as well as shore up confidence levels about engaging instructional change (Fullan, 2007a, 2007b). For example, Johnson and Marx (2009) used a transformative professional development (TPD) model to help change teachers' practices and beliefs in urban schools. For Johnson and Marx, "TPD is based on the premise that through effective, sustained, collaborative professional development, climates of schools, as well as beliefs and practices of teachers can be positively transformed over time" (p. 118). The researchers showed that this can be done by allowing students to experience effective science instruction by changing their learning environments into positive climates. For this to occur, teachers must be given opportunities in non-confrontational sessions to let go of ineffective practices and replace them with new effective ones.

Spillane and Callahan (2000) showed that principals and other school leaders do not intentionally sabotage reform efforts that call for significant changes. Instead, these leaders sometimes misconstrue the intent of the reform agendas and often misuse terminology, such as hands-on, to represent reform changes because that is what they actually understood. However, their superficial understanding of reform rarely meets the true meaning of policy mandates.

In 2011, Peled, Kali, and Dori completed a 7-year retrospective analysis of principals' influence on science teachers' implementation of technology into the science curriculum. The results showed that long-term support for or against technology implementation by the principals was significant toward the teachers' ability and motivation to integrate technologies into their teaching. Furthermore over time teachers began to need less and less principal support. The study further showed that teachers were classified as initiators, followers, evaders, and objectors based on the mode and extent to which they integrated technologies into their teaching. Similarly, principals were classified as initiating, empowering, permitting yet preventing, and resisting (Peled, Kali, & Dori, 2011). The major lessons learned was that while principals reacted relatively true to how they were classified, teachers who were initiators, even when working in an environment where they have little or no support, "find ways to lead a process of change in an education system and research setting" (Peled et al., 2011, p. 243).

Whereas research by Fields et al. (2012) Guskey (2000) demonstrated that principals do not always make wise decisions when reassigning teachers annually to schools and subjects, the new emphasis on building great leaders places special attention on school and subject assignments. These types of improvements in policy decisions will go a long way in creating leaders more knowledgeable about and supportive of science education reforms.

Professional Development and Policy: Policy Matters

To ramp up science reform following more than a decade of losing ground will necessitate the need to revisit local, state, and/or national policies to ensure proper aligning of STEM-based professional development with newer reform mandates (U. S. Department of Education, 2010; Guskey, 2000; PCAST, 2010). As such, major policy recommendations will likely focus on the urgent need to develop strategies and approaches for significantly increasing the number of effective STEM teachers, great STEM leaders, and highly qualified STEM graduates (Corcoran, 2007; PCAST, 2010, 2012). To achieve the highest levels of success, policies aimed at increasing and maintaining system-wide capacity need to be at the forefront of developing the internal STEM expertise needed to ensure that the system's capacity remains stable (Spillane, 2012; Yin, 2006).

With the universal beliefs that teachers are the targets and agents of change, some policymakers promote professional development as a reform strategy on the theory that the mere existence of policies will automatically bring about immediate buy-in by teachers. The expectation is that buy-in by teachers will automatically lead to changes in teacher participation, changes in teacher practice, and improvement in student achievement (Spillane, 2012; Weinbaum, Weiss, & Beaver, 2012). In reality, that is not the case.

Largely because on the one hand, policymakers rarely look beyond teachers in building capacity to meet policy mandates, while on the other, policymakers frequently overlook teachers in the decision-making process. Researchers (Grobe & McCall, 2004) showed that many teachers feel virtually ignored by all levels of school administration and that their voice is literally silent in decision-making about professional development or any other proposed school or district change. In instances where teachers feel ignored, many do not embrace reform change. Of those that do, over time, some revert back to their old way of teaching or graft pieces of the reform vision onto what is familiar to them without actually changing their everyday practices significantly (Fullan, 2007a, 2007b; Haney & McArthur, 2002).

Shaver, Cuevas, Lee, and Avalos (2007) surveyed elementary teachers about how educational policies affected their science instruction. The results indicated that as science accountability increases, it is important to understand teachers' perception of the influence of policies on their classroom practices. Therefore, policymakers need to include teachers' input in all policy mandates before policy changes are made and prepare teachers for implementing policy mandates (U. S. Department of Education, 2010). By carefully planning for policy changes, from inception to implementation, policymakers can increase teachers' reception of policies while easing the burden of the demands of STEM-based professional development on the classroom instruction and practices (Church, Bland, & Church, 2010; U. S. Department of Education, 2010).

Beyond involving teachers in decision-making about proposed changes, district and state leaders usually do not fund or support policies as written (Guskey, 2010). Even though nearly 90 % of funding for educational purposes is in state and local budgets, many rely on external funding to meet local reform needs. While competing and/or seeking external funds to support school improvements is important, Fullan (2007a) and Yin (2006) argue that external support are rarely powerful enough to change or sustain reform once the support ends. Moreover, stipulations associated with external funding, uncertainty of funding availability, political influences, and economic downturns often negatively impact science reforms supported largely by external funding (Yin).

Furthermore, when policy calls for a relevant school-related changes that could involve extensive STEM-based professional development (e.g., teacher release time, teacher tuition, incentives, new curricular products, technology tools), policymakers should embrace the fact that such changes will require strong leadership, significant time, and substantial funding (NRC, 2012a; Time Act, 2009; Wallace Foundation, 2011a, 2011b). Collectively, policies should be viewed from a perspective that fully endorses what it will actually take to bring change to scale (Darling-Hammond, LaPointe, Meyerson, & Orr, 2007; Elmore, 2004; Spillane & Callahan, 2000; Yin, 2006).

To address these challenges, researchers contend that policymakers should first draft policies based on relevant school, teacher, and student data and education research; develop a plan for implementation; allow for more flexibility in the use of state and federal funds; become cognizant of factors that impact time and change; and include the voice of teachers from the beginning (Davis, 2002, 2011; Fullan, 2007a, 2007b; E. Massey, Personal Communication, June 29, 2012; Penuel, McWilliams McAuliffe et al., 2009). Such policies should impact classroom practice directly in agreement with but not driven by large-scale testing and only increase requirements for teachers and/or students—with concomitant changes in other relevant domains.

Professional Development and Policy: Implications for the Curriculum

For decades, there have been serious debates about what the science curriculum should be (Cuban, 2012; DeBoer, 1991, 2000; Schwartz et al., 2008). In the early years, the focus was mostly on what to teach and, to some degree, at what levels. Later on, uniform curriculum standards, common assessments, and higher accountability, the space gap; attention to the best and brightest; social conditions; and space explorations dominated the curriculum (Cuban, 2012; Science and Engineering Indicators, 2000). Today, innovation and discovery, global competitiveness, international tests, broadening participation, and culture and diversity drive curricular offering in one way or another (Loveless, 2011).

Achieve (2010) is helping to establish the curriculum for what K-12 students should accomplish by the end of 12th grade. The NGSS were released April 2013. NGSS will require rethinking of professional development and policy to ensure that states and school districts are preparing all students to choose, engage in, and complete a more rigorous curriculum in STEM.

An idea being proposed by leading stakeholders (e.g., federal, states, business, educators, and policymakers) is that a rigorous curriculum is needed to inspire students to be creative and resourceful in their studies, such that what they learn and do will lead to new innovations and discoveries in STEM (PCAST 2012). Already leaders of the NGSS (Fleming, 2012) are comparing and contrasting the new standards with the findings from the 2009 and 2011 NAEP assessment tests.

The dismal outcomes notwithstanding, the results show that NGSS will address those areas where students did not perform well on the 2009 and 2011 NAEP tests. More importantly, NGSS aim to improve depth over breadth regarding student understanding about and practical application of scientific principles (Schwartz et al., 2008). Even further, NGSS intend to shift the focus from describe, distinguish, and explain to construct explanations, plan new investigations, integrate engineering designs, and design methods for conducting activities in novel situations (S. Pruitt, Personal Communication, June 28, 2012; Wayne et al., 2008).

A change in focus by NGSS will also transform the epistemological model that will undergird the new curricular shift toward more effective practices. Education researcher (Metz, 2009) argues that this will have a major effect on teachers because "the epistemological model underlying reform curricula is typically far removed from teachers' experience, from both curriculum they have used before and their own schooling" (p. 2). For example, science reform efforts will require teachers to implement new curricula that often have newer pedagogies (e.g., learning progression, cultural pedagogy), sometimes embedded assessment and frequently integrated learning technologies (Schneider, Krajcik, & Blumenfeld, 2005).

Again, literatures across these domains ". . . have repeatedly found that reform curriculum context alone, divorced from teacher professional development, is insufficient to support change in the classroom" (Metz, 2009, p. 2). In this regard, during the decision-making process, policymakers should be deliberate in their plans for ensuring that STEM-based professional development includes the enduring qualities needed to help teachers make the transition from current to future practice.

Beyond the impact of the *Framework*/NGSS, researchers are looking at other curriculum-related factors that policymakers might consider as they mandate curriculum changes through STEM-based professional development. For example, Lynch, Pyke, and Grafton (2012) are conducting a retrospective view of scaling up middle school science curriculum materials in a changing policy climate, while Penuel, Fishman, Yamaguchi, and Gallagher (2007) explored strategies that best fostered curriculum implementation. Davis (2011) is exploring methods to add information to existing curricula materials to test whether or not educative scaffolds provide experiences for teachers and students to learn core disciplinary ideas and crosscutting concepts through engaging in scientific practices. The scaffolds are being designed to support teachers in their move toward a more student-centered science teaching process with STEM-based professional development embedded in the educative supplements (Davis and Krajcik, 2005).

Heller, Daehler, Wong, Shinohara, and Miratrix (2012) conducted research that identified links along the teacher improvement student achievement continuum. The results suggested that deeper content knowledge for teachers had a powerful effect on students test scores. However, students' ability to explain why their answers were correct did not improve until professional development for teachers included an analysis of student conceptual understanding. This study suggests that professional development alone is not as effective in impacting student achievement in science as well as professional development that integrates content analysis of student learning (Yoon et al., 2007).

Schneider et al. (2005) investigated how reform innovations linked to curricular materials might help inservice teachers enact the reforms embedded in the materials. Even though, half of the teachers enacted the curriculum as designed, they were less successful in contexts that challenged their thinking and actions, presenting challenging science ideas, responding to students' ideas, dealing with structure investigations, guiding small group discussions, and making adaptations (Davis, 2002).

Metz (2009) examined teachers' views on the challenges of implementing a science reform curriculum as well as their learning as an outcome of interacting with the curriculum while concurrently engaging in a parallel professional development program. Data from case studies over five points across 32 months showed marked variability in what the teachers conceptualized as problematic in implementing science reform lessons.

Geier et al. (2008) investigated the impact of scaling up inquiry science curriculum with STEM-based professional development and learning technologies carefully embedded in a continuously redesigned process throughout the intervention. This was done in response to teacher evaluation of the professional development experiences, student achievement, and classroom observations. The findings demonstrated that "standards-based inquiry science curriculum can lead to standardized achievement test gains when the curriculum is highly specified, developed, and aligned with professional development and administrative support" (Geier, Blumenfeld Marx et al., 2008; p. 922). While these education researchers looked at various points of view about STEM-based professional development, policy, and the curriculum, they all uncovered new avenues for consensus building about curricular issues that matter (NRC, 2012b).

Professional Development and Policy: Cultural Pedagogical Issues

Focusing attention on pedagogy (Alonzo, Kobarg, & Seidel, 2012; Garza, 2012; Ladson-Billings, 1995; Shulman, 1987) and culturally relevant pedagogy (Gay, 2002; Nam, Roehrig, Kern, & Reynolds, 2013; Stafford & Bales, 2011) helped to shape the criteria for improving teaching and learning for all students. Both of these factors have proven to be critical in broadening participation for all students in a variety of ways. For example, Maurrasse, Kramer, Rukimbira, and Brewe (2008) explored ways to recruit STEM majors (e.g., chemistry, physics) into the teaching profession for placement in urban schools. As such, a critical component of the study was to direct attention to pedagogical content knowledge.

Researchers (Gay, 2002; Stafford & Bales, 2011) advocated for relevant pedagogy to be included in STEM-based professional development to help teachers become more aware of how to fully meet the academic needs of an increasingly diverse student body. Trumbull and Pacheco (2005) endorsed utilizing a students' culture as a venue for changing part of the total learning process. These researchers believed that real change will more likely occur when one's own values and cultures were visible when contrasted with others.

Giving attention to pedagogy and culture as part of STEM-based professional development, policy, and curriculum decision-making will likely build a stronger foundation for teaching science to a broader audience. The jury is still out on how many teachers and principals are actually ready to take on the issues of cultural responsive teaching and the inclusion of provisions of cultural competency in STEM-based professional development. While there is a growing field of research and curriculum products that focus on what cultural relevant pedagogy should be, increasingly studies are also exploring what it will actually take to place education research outcomes into practice (McAllister & Irvine, 2000).

This notion was of interest to Hazari (2010) who developed a pedagogical plan with sample lessons that detailed how to connect physics content to real-world contexts, counter stereotypes about physics, analysis underrepresentation in physics, and identify formative assessments. This pedagogical plan also focused more attention to affective factors (e.g., motivation, interest, identity threats) that sometime interfere with the achievement of students in STEM fields (Boykin & Noguera, 2011; Cohen & Garcia, 2008; Ford, 1992; Steele, 1997; Swarat et al., 2012).

Bertram and Crevensten (2012) planned STEM-based professional development for teachers of indigenous students to provide culturally relevant instruction where the vast majority of teachers are nonnative. These education researchers believe that long-term STEM-based professional development is needed as teachers enter preparation programs with inadequate background experiences in intercultural experience and with preconceived beliefs that undermine their ability to provide a high-quality science education to all students.

Stafford and Bales (2011) and Rivera Maulucci (2013) experimented with radical approaches about how preservice teachers might become more culturally competent. In the Stafford and Bales study, prospective teachers were allowed to "unpack beliefs about children unlike themselves" and grapple with ways to integrate culturally relevant pedagogy into disciplinary-based instructional strategies. The research participants were European American middle-class female preservice teachers with suburban and rural backgrounds who held negative and deficit attitudes and beliefs about cultures other than their own. The preservice teachers participated in a three-credit, field course in an urban context that involved a 50-h school-based experience in an urban school. These teachers also spent time in a pedagogy laboratory.

The results showed that the participants linked course content with pedagogical content knowledge, affirmed the value of case-based research, and connected the course content with laboratory activities and field observations. The outcomes demonstrated that participants gained a deeper understanding of their pedagogical content knowledge through the use of culturally relevant content and practices while the pedagogy laboratory helped participants to unpack their beliefs about children from other ethnic groups. The study highlighted an increase in the participants' confidence about using culturally relevant pedagogy despite prior beliefs about underrepresented students and personal beliefs about teaching.

Rivera Maulucci (2013) studied the emotions and positional identity in becoming a social justice chemistry teacher by an African American Caribbean preservice teacher. As such, Rivera Maulucci chronicled how this teacher struggled with oppression and cultural issues in urban schools and how she emerged as a social justice education. The results showed that emotions and positional identity help preservice teachers aim for achieving the great potentialities of our society by providing everyone with agency in the world and how the world works.

Metz (2009) studied teachers' conceptualization of students' inherent cognitive limitations and concluded that teachers may conceptualize the situation in ways that might help or hinder students' achievement. For example, if teachers believed that age is a determining factor in a student's ability to learn, in the teacher's mind of whether a student succeeds or fails without possible recourse is heavily based on age. On the other hand, if teachers believed their actions contributed to student comprehension, these teachers then looked for ways to improve their teaching, which is a powerful way to transform personal beliefs about students and science teaching.

Conclusion

Regardless of the reform activity, for over a half century, traditional professional development has been at the forefront of a myriad of reform efforts. As such, teachers have been both the target and the agent of change. However, evaluation and research show that traditional forms of professional development falls short of being effective in improving student achievement in science. With a growing body of research, meaningful descriptions of what professional development should be, evidence-based evaluations of large-scaled professional development, guidance from the *Science Framework* (2010), NGSS (2013), and the *Blueprint* (2010) for the first time in years, the nation has powerful tools for improving science learning for all.

Yet, new tools alone will not bring about the desired change without a comprehensive policy design strategy aimed at improving the performance of teachers, principals, and students. Such a strategy must send broad signals, set boundaries and requirements, allocate resources appropriately, establish accountability, render assistance, establish networks for flow of information, and distribute authority wisely (Elmore, 2004; Spillane, 2012). To do this, policymakers must learn to readily recognize and eliminate misconstrued and inappropriately placed policies that do little to further science education for a greater number of students.

Therefore, future efforts must be increasingly about learning, deeper learning, interactions, relationships, and the interplay between students and teachers (NRC, 2011; Porter et al., 2004; Willingham & Daniel, 2012) as they become progressively more aware of how learning unfolds over time. These lessons learned must be used to provide all students, particularly those in large urban centers and rural regions of the United States, with effective teachers each year who can see their potential, build on what they have, and prepare them for even greater successes in the future. Only when this scenario becomes ingrained in all schools will the United States have an education system that can be described as promoting equity and excellence for all.

To that end, more research on education is needed to inform future reforms from a list of factors that continues to grow rapidly. A few critical areas include learning more about learning and learning environments for teachers and students, stabilizing teacher assignment and teacher turnover, improving methods for determining teacher effectiveness, looking beyond best practices to ways that meet individual student needs, striking a balance between curriculum standards and customization for student differences, linking to twenty-first-century competencies with adult outcomes, and using policies that matter.

Unless relevant policies drive proven practices, efforts to provide emergent STEM-based highly effective teachers will be fraught with intractable challenges entrenched in political systems unjustly blocking paths to improvement. To avoid such a dismal future, going forward, the continuum from teacher preparation to STEM-based professional development must become an avenue for preparing highly effective science teachers with the capacity and resources to broaden the participation for a greater number of students in STEM and STEM careers.

References

Achieve. (2010). *A framework for K-12 science education: Practices, crosscutting concepts, and core ideas*. Washington, DC. www.achieve.org

Achieve. (2013). *Next generation science standards*. Washington, DC. www.achieve.org

Alonzo, A., Kobarg, M., & Seidel, T. (2012). Pedagogical content knowledge as reflected in teacher-student interactions: Analysis of two video cases. *Journal of Research in Science Teaching, 49*(10), 1211–1239. doi:10.1002/tea.21055.

Annetta, L. A., Frazier, W. M., Folta, E., Holmes, S., Lamb, R., & Cheng, M. (2013). Science teacher efficacy and extrinsic factors toward professional development using video games in a design-based research model: The next generation of STEM learning. *Journal of Science Education Technology, 22*, 47–61.

Bausmith, J. M., & Barry, C. (2011). Revisiting professional learning communities to increase college readiness: The importance of pedagogical content knowledge. *Educational Researcher, 40*(4), 175–178.

Bertram, K. B., & Crevensten, D. (2012). *Preparing responsive educators using place-based authentic research in earth systems*. www.nsf.gov/awards/abstracts/

Boykin, A. W., & Noguera, P. (2011). *Creating the opportunity to learn: Moving from research to practice to close the achievement gap*. Alexandria, VA: ASCD.

Business-Higher Education Forum. (2007). *An American imperative: Transforming the recruitment, retention, and renewal of our nation's mathematics and science teaching workforce*. Washington, DC. Retrieved from http://www.bhef.com/

Business-Higher Education Forum. (2010). *Increasing the number of STEM graduates: Insights from the U.S. STEM education & modeling project.* Washington, DC. Retrieved from http://www.bhef.com/

Bykerk-Kauffman, A. (2010). *Collaborative research: GARNET II: Self-regulated learning and the affective domain in physical geology.* www.nsf.gov/awards/abstracts/

Carnegie Foundation of New York. (2009). *The opportunity equation, transforming mathematics and science education for citizenship and the global economy.* New York. http://carnegie.org/

Church, E., Bland, P., & Church, B. (2010). Supporting quality staff development with best-practice-aligned policies. *Emporia State Research Studies, 46*(2), 44–47.

Cohen, G. L., & Garcia, J. (2008). Identity, belonging, and achievement: A model, interventions, implications. *Psychological Science, 17*(6), 365–369.

Corcoran, T. B. (2007). *Teaching matters: How state and local policymakers can improve the quality of teachers and teaching.* (CPRE Policy Brief No. RB-48). New Brunswick, NJ: Rutgers University, Consortium for Policy Research in Education.

Corcoran, S., Schwartz, A. E., & Weinstein, M. (2009). *The New York City aspiring principals program: A school-level evaluation.* New York: Institute for Education and Social Policy. http://steinhardt.nyu.edu/scmsAdmin/uploads/003/852/APP.pdf.

Cuban, L. (2012). For each to excel: Standards vs. customization: Finding the right balance. *Educational Leadership, 69*(6), 10–15.

Darling-Hammond, L., LaPointe, M., Meyerson, D., & Orr, M. T. (2007). *Preparing school leaders for a changing world: Executive summary.* Stanford, CA: Stanford University/Stanford Educational Leadership Institute.

Darling-Hammond, L., Wei, R.C., Andree, A., Richardson, N., & Orphanos, S. (2010). *Professional learning in the learning profession: A status report on teacher development in the U.S. and Abroad: Technical Report*, Dallas: National Staff Development Council and the School Redesign Network at Stanford University.

Davis, K. (2002). "Change is hard:" What science teachers are telling us about reform and teacher learning of innovative practices. *Science Education, 87*, 3–30.

Davis, E. (2011). *Contextual research – empirical – STEM teaching and learning: Investigating teachers' learning, practice, and efficacy using educative curriculum materials.* www.nsf.gov/awards/abstracts/

Davis, E., & Krajcik, J. S. (2005). Designing educative curriculum materials to promote teacher learning. *Educational Researcher, 34*(3), 3–14.

DeBoer, G. (1991). *A history of ideas in science education: Implications for practice.* New York: Teachers College Press.

DeBoer, G. (2000). Scientific literacy: Another look at its historical and contemporary meanings and its relationship to science education reform. *Journal of Research in Science Teaching, 37*(6), 582–601.

Drafo-Severson, E. (2007). Helping teachers learn: Principals as professional development leaders. *Teachers College Record, 109*(1), 70–125.

Drago-Severson, E. (2012). New opportunities for principal leadership: Shaping school climates for enhanced teacher development. *Teachers College Record, 114*(3), 1–44.

Duncan, R. G., & Rivet, A. E. (2013). Science learning progressions. *Science, 339*, 396–397.

Early, J. S., & Shagoury, R. (2012). The key to changing the teaching profession: What supports new urban teachers? *Educational Leadership, 67*(8), 1–4.

Education Development Center. (2009). *Quality measures for education leadership development systems and programs: Principal preparation programs quality self-assessment rubrics. Course content and pedagogy and clinical practice.* www.qualitymeasures.edc.org

Elmesky, R., Oliver, L., McNews-Birren, J., Van Duzor, A. G., Hogrebe, M., et al. (2012). Building capacity in understanding foundational biology concepts: A K-12 learning progression in genetics and protein expression informed by research on children's thinking and learning. Retrieved from http://www.artsci.wustl.edu/~relmesky/Papers/LP_Paper_final_uploaded.pdf

Elmore, R. (2000). *Building a new structure for school leadership.* Washington, DC: The Albert Shanker Institute, Consortium for Policy Research in Education.

Elmore, R. F. (2004). *School reform from the inside out: Policy, practice, and performance.* Cambridge, MA: Harvard Education Press.

Feldman, A. (2000). Decision-making in the practical domain: A model of practical conceptual change. *Science Education, 84,* 606–623.

Fields, E. T., Levy, A. J., Tzur, M. K., Martinez-Gudapakkam, A., & Jablonski, E. (2012). The science of professional development. *Phi Delta Kappan, 93*(8), 44–46.

Fleming, N. (2012, June). NAEP reveals shallow grasp of science. *Education Week, 6,* 19.

Ford, M. (1992). *Motivating humans: Goals, emotions, and personal agency beliefs.* Thousand Oaks, CA: Sage.

Fullan, M. G. (2007a). Change the term for teacher learning. *Journal of Staff Development, 28*(3), 35–36.

Fullan, M. G. (2007b). *The new meaning of educational change* (4th ed.). New York: Teachers College, Columbia University.

Fullan, M. G. (2010). System capacity building: What, why, and how. *Presentation at the symposium "Building capacity for whole school educational reform", Learning and teaching division,* Education Development Center, Newton, MA.

Furtak, E. (2012). www.nsf/gov/awards/abstracts/

Garet, M. S., Porter, A. C., Desimone, L., Birman, B. F., & Yoon, K. S. (2004). *Effective professional development in mathematics and science: Lessons from evaluation of the Eisenhower program.* Washington, DC: American Institutes of Research.

Garza, R. (2012). Initiating opportunities to enhance pre-service teachers' pedagogical knowledge: Perceptions about mentoring at-risk adolescents. *Journal of Urban learning, Teaching, and Research, 8,* 26–35.

Gay, G. (2002). Preparing for culturally responsive teaching. *Journal of Teacher Education, 53*(2), 106–116.

Geier, R., Blumenfeld, P. C., Marx, R. W., Krajcik, J. S., Fishman, B., Soloway, E., et al. (2008). Standardized test outcomes for students engaged in inquiry-based science curriculum in the context of urban reform. *Journal of Research in Science Teaching, 45*(8), 922–939. doi:10.1002/tea.20248.

Goldring, E., Huff, J., May, H., & Camburn, E. (2008). School context and individual characteristics: What influences principal practice? *Journal of Educational Administration, 46*(3), 332–352.

Goldring, E., Spillane, J.P., Huff, J., Barnes, C., & Supovitz, J. (2006, April). *Measuring the instructional leadership competence of school principals.* Paper presented at the 2006 meeting of the American Educational Research Association, New Orleans, LA.

Grobe, W. J., & McCall, D. (2004). Valid uses of student testing as part of authentic and comprehensive student assessment, school reports, and school system accountability. *Educational Horizons, 82,* 131–142.

Guskey, T. R. (2000). *Evaluating professional development.* Thousand Oaks, CA: Corwin Press.

Guskey, T. R. (2010, August). *Professional development: How best to spend your money.* Conference of the Near East South Asia Council for Overseas Schools, Kathmandu, Nepal.

Guskey, T. R., & Yoon, K. S. (2009). What works in professional development. *Phi Delta Kappan, 90*(7), 495–500.

Haney, J., & McArthur, J. (2002). Four case studies of prospective science teachers' beliefs concerning constructivist teaching practices. *Science Education, 86,* 783–802.

Hazari, Z. (2010). *Career: Changing the landscape: Towards the development of a physics identity in high school.* www.nsf.gov/awards/abstracts

Heller, J. I., Daehler, K. R., Wong, N., Shinohara, M., & Miratrix, L. W. (2012). Differential effects of three professional development models on teachers knowledge and student achievement in elementary science. *Journal of Research in Science Teaching, 49*(3), 333–362.

Ingersoll, R., & Merrill, L. (2010). Who's teaching our children? *Educational Leadership, 67*(8), 14–20.

Jackson, C. K., & Bruegmann, E. (2009). Teaching students and teaching each other: The importance of peer learning for teachers. *American Economic Journal: Applied Economics, 1*(4), 85–108.

Johnson, C. C., & Marx, S. (2009). Transformative professional development: A model for urban science education reform. *Journal of Science Teacher Education, 20,* 113–134.

King, C., Levinger, B., & Schoener, J. (2006). *Leadership development quality assessment process [LDQAP]*. Newton, MA: Educational Center.

Ladson-Billings, G. (1995). But that's just good teaching! The case for culturally relevant teaching. *Theory into Practice, 31*(4), 312–320.

Learning Forward. (2010). www.learningforward.org

Loucks-Horsley, S., Love, N., Stiles, K. E., Mundry, S., & Hewson, P. W. (2003). *Designing professional development for teachers of science and mathematics* (2nd ed.). Thousand Oaks, CA: Corwin Press.

Loveless, T. (2011). How well are American students learning? With sections on international tests, who's winning the real race to the top, and NAEP and the Common Core Standards. Brown Center on Education Policy at Brookings, 2(5). Washington, DC.

Lynch, S. J., Pyke, C., & Grafton, B. H. (2012). A retrospective view of a study of middle school science curriculum: Implementation, scale-up, and sustainability in a changing policy environment. *Journal of Research in Science Teaching, 49*(3), 305–332.

Maltese, A. V., & Tai, R. H. (2010). Eyeballs in the fridge. Sources of early interest in science. *International Journal of Science Education, 12*(5), 669–685.

Maskiewicz, A. C., & Winters, V. A. (2012). Understanding the co-construction of inquiry-based practices: A case study of a responsive teaching environment. *Journal of Research in Science Teaching, 49*(4), 429–464.

Maurrasse, F., Kramer, L., Rukimbira, P., & Brewe, E. (2008). *Get educators in mathematics and science (GEMS)*. www.nsf.gov/awards/abstracts/

McAllister, G., & Irvine, J. J. (2000). Cross cultural competency and multicultural teacher education. *Review of Educational Research, 70*(1), 3–24.

Metz, K. E. (2009). Elementary school teachers as "targets and agents of change": Teachers' learning in interaction with reform science curriculum. *Science Education, 93*, 915–954. doi:10.1002/sce.20309.

Mizell, H. (2008, July). *National Staff Development Council's State Affiliate Summer Conference*, Orlando, FL. http://www.learningforward.org/

Muijs, D., & Lindsay, G. (2008). Where are we at? An empirical study of levels and methods of evaluating continuing professional development. *British Educational Research Journal, 34*(2), 195–211. doi:10.1080/01411920701532194.

Nam, Y., Roehrig, G., Kern, A., & Reynolds, B. (2013). Perceptions and practices of culturally relevant science teaching in American Indian Classrooms. *International Journal of Science and Mathematics Education, 11*, 143–167.

National Academy of Sciences. (2010, September). News from the National Academies. *The U.S. must involve underrepresented minorities in science and engineering to maintain competitive edge*. Retrieved from http://www8.nationalacademies.org/onpinews/newsitem.aspx?RecordID=12984

National Center for Education Statistics. (2009). Institute of Education Sciences. The Nation's Report Card. Science 2009. *National Assessment of Educational Progress at Grade 8*. U. S. Department of Education, Washington, DC. Retrieved from http://nces.ed.gov/pubsearch/pubsinfo.asp?pubid=2011451

National Center for Education Statistics. (2011). Institute of Education Sciences. The Nation's Report Card. Science 2011. *National Assessment of Educational Progress at Grade 8*. U. S. Department of Education, Washington, DC. http://nces.ed.gov/pubsearch/pubsinfo.asp?pubid=2012465

National Research Council. (2007). *Taking science to school: Learning and teaching science in grades K-8*. Committee on Science Learning, Kindergarten Through Eighth Grade. Richard A. Duschl, Heidi A. Schweingruber, and Andrew W. Shouse, Editors. Board on Science Education, Center for Education. Division of Behavioral and Social Sciences and Education. Washington, DC: The National Academy Press.

National Research Council. (2011). *Successful K-12 STEM education: Identifying effective approaches in science, technology, engineering, and mathematics*. Committee on highly successful science programs for K-12 science education, Board on Science and on Testing and

Assessment, Division of Behavioral and Social Sciences and Education. Washington, DC: The National Academy Press.

National Research Council. (2012a). *A framework for K-12 science education: Practices, crosscutting concepts, and core Ideas*. Washington, DC: The National Academy Press.

National Research Council. (2012b). *Education for life and work: Developing transferable knowledge and skills in the 21st century*. Washington, DC: The National Academy Press.

National Research Council. (2013). *Monitoring progress towards successful K-12 STEM education: A nation advancing?* Committee on the framework for successful K- 12 STEM education. Board on Science Education and Board on testing and Assessment, Division of Behavioral and Social Science and Education. Washington, DC: The Academy Press.

Pea, C. (2012). Inquiry-based instruction: Does school environmental context matter? *Science Educator, 21*(1), 37–43.

Peled, Y., Kali, Y., & Dori, Y. J. (2011). School principals' influence on science teachers' technology implementation: A retrospective analysis. *International Journal of leadership in Education, 14*(2), 229–249.

Penuel, W. R., Fishman, B. J., Yamaguchi, R., & Gallagher, L. P. (2007). What makes professional; development effective? Strategies that foster curriculum implementation. *American Educational Research Journal, 44*(4), 921–959.

Penuel, W. R., McWilliams, H., McAuliffe, C., Benbow, A. E., Mably, C., & Hayden, M. M. (2009). Teaching for understanding in earth science: Comparing impacts on planning and instruction in three professional development designs for middle school science teachers. *Journal of Science Teacher Education, 20*, 415–436.

Porter, A., Birman, B., Garat, M. S., Desimone, L. M., & Yoon, K. S. (2004). *Effective professional development in mathematics and science: Lessons from evaluation of the Eisenhower program.* Washington, DC: American Institutes of Research.

Rivera Maulucci, M. S. (2013). Emotions and positional identity in becoming a social justice science teachers: Nicole's story. *Journal of Research in Science Teaching, 50*(4), 453–478. doi:10.1002/tea.21081.

Rogat, A., Anderson, C., Foster, J., Goldberg, F., Hicks, J., Kanter, D., et al. (2011). Developing learning progressions in support of the new science standards: A RAPID workshop series. *Consortium for Policy Research in Education*. New York: Teachers College, Columbia University.

Schneider, R. M., Krajcik, J., & Blumenfeld, P. (2005). Enacting reform-based science materials: The range of teacher enactments in reform classrooms. *Journal of Research in Science Teaching, 42*(3), 283–312.

Schwartz, M. S., Sadler, P. M., Sonnert, G., & Tai, R. H. (2008). Depth versus breadth: How content coverage in high school science courses relates to later success in college science coursework. *Science Education, 93*, 798–826.

Shaver, A., Cuevas, P., Lee, O., & Avalos, M. (2007). Teachers' perceptions of policy influences on science instructions with culturally and linguistically diverse elementary students. *Journal of Research in Science Teaching, 44*(5), 725–746.

Shulman, L. S. (1987). Knowledge and teaching: Foundations of the new reform. *Harvard Educational Review, 57*, 1–22.

Smith, C. L., Wiser, M., Anderson, C. W., & Krajcik, J. (2006). Implications of research on children's learning for standards and assessment: A proposed learning progression for matter and atomic-molecular theory. *Measurement: Interdisciplinary Research & Perspective, 4*(1–2), 1–98.

Spillane, J. P. (2000). *District leaders' perception of teacher learning*. CPRE occasional paper series. OP-05. Consortium for Policy Research in Education, University of Pennsylvania.

Spillane, J. P. (2012). The more things change the more they stay the same? *Education and Urban Society, 44*(2), 123–127.

Spillane, J. P., & Callahan, K. A. (2000). Implementing state standards for science education: What district policymakers make of the hoopla. *Journal of Research in Science Teaching, 37*(50), 401–425.

Stafford, F., & Bales, B. L. (2011). A new era in the preparation of teachers for urban schools: Linking multiculturalism, disciplinary-based content, and pedagogy. *Journal of Urban Learning, Teaching, and Research, 46*(5), 953–974.

Steele, C. (1997). A threat in the air: How stereotypes shape intellectual identity and performance. *American Psychologist, 52*, 613–629.

Stephens, R., & Richey, M. (2011). Accelerating STEM capacity: A complex adaptive system perspective. *Journal of Engineering Education, 100*(3), 417–423.

Stevens, S. Y., Delgado, C., & Krajcik, J. (2009). Developing a hypothetical multi-dimensional learning progression for the nature of matter. *Journal of Research in Science Teaching, 47*(6), 687–715.

Stylinski, C., Parker, C., & McAuliffe, C. (2011, April). *Examining real-world IT-Immersion teacher education experiences through the lens of two teacher roles.* Paper presented at the National Association for Research in Science Teaching, Orlando, FL.

Supovitz, J., Foley, E., & Mishook, J. (2012, July). In search of leading indicators in education. *Education Policy Analysis Archives, 20*, 19. http://epaa.asu.edu/ojs/article/view/952

Swarat, S., Ortony, A., & Revelle, W. (2012). Activity Matters: Understanding student interest in school science. *Journal of Research in Science Teaching, 49*(4), 515–537.

The President's Council of Advisors on Science and Technology. (2010). *Report to the President. Prepare and inspire: K-12 education in science, technology, engineering, and mathematics (STEM) for America's Future.* Retrieved from Washington, DC. www.whitehouse.gov/ostp/pcast

The President's Council of Advisors on Science and Technology. (2012). *Report to the President. Engage to excel: Producing one million additional college graduates with degrees in science, technology, engineering, and mathematics.* Washington, DC. www.whitehouse.gov/ostp/pcast

The Time Act. (2009). Retrieved from www.quickanded.com/2009/07/time-act-reintroduced.html

Trumbull, E., & Pacheco, M. (2005). *Leading with diversity: Cultural competencies for teacher preparation and professional development.* Providence, RI: The Education Alliance, Brown University.

U. S. Department of Education. (2010). *A blueprint for reform: Reauthorization of elementary and secondary education act.* Retrieved from http://www2.ed.gov/policy/elsec/leg/blueprint/blueprint.pdf

Wallace Foundation. (2011a). *The Wallace Foundation launches major "Principal Pipeline" initiative to help school districts build corps of effective school leaders.* Retrieved from www.wallacefoundation.org

Wallace Foundation. (2011b). *Reimagining the school day: More time for learning.* Washington, DC: A Wallace Foundation National Forum.

Wayne, A. J., Yoon, K. S., Zhu, P., Cronen, S., & Garet, M. S. (2008). Experimenting with teacher professional development: Motives and methods. *Educational Researcher, 37*(8), 469–479.

Wei, R. C., Darling-Hammond, L., & Adamson, F. (2009). *Professional development in the United States: Trends and challenges.* Dallas, TX. Learning Forward/National Staff Development Council. http://www.nsdc.org/news/NSDCstudytechnicalreport2010.p.df

Weinbaum, E. H., Weiss, M. J., & Beaver, J. K. (2012, September). Learning from NCLB: School responses to accountability pressures and student subgroup performance. *Consortium for Policy Research in Education, RB-54.* CPRE.org.

Wells, B. H., Sanchez, H. A., & Attridge, J. M. (2007, November). Modeling student interest in Science, Technology, Engineering and Mathematics. *Meeting the growing demand for engineers and their educators 2010–2020 International Summit IEEE, 50*; 1, 17, 9–11. doi: 10.1109/MGDETE.2007.4760362.

Willingham, D., & Daniel, D. (2012). For each to excel: Teaching to what students have in common. *Educational Leadership, 69*(5), 16–21.

Yin, R. (2006). *Cross-site evaluation of the urban systemic program. The final annual report: Baseline outcome analysis.* Washington, DC: The COSMOS Corporation.

Yoon, K.S., Duncan, T., Lee, S.W., Scarloss, B., & Shapley, K. (2007). *Reviewing the evidence on how teacher professional development affects student achievement.* Issues & Answers Report, REL 2007-No. 033. US Department of Education, Institute of Education Sciences, Washington, DC.

Policy Issues, Equity, Multicultural Science Education, and Local School District Support of In-Service Science Teachers

Bongani D. Bantwini

Introduction

Over the past few years we have witnessed increasing attention on science teachers, especially those that work with students from diverse cultural backgrounds, due to perpetually poor student performance and failure. Contributing to this attention are the disappointing student achievement scores from various states, national, and international assessment tests including the Trends in International Mathematics and Science Study (TIMSS) and Program for International Student Assessment (PISA), showing that students from low socioeconomic backgrounds perform below grade level (Jensen, 2009; Davis, 2003). This concern has also resulted in an ongoing discussion regarding how issues of race, ethnicity, gender, and economic background influence students' academic performance (Bergerson, 2006; Lacour & Tissington, 2011; Davis, 2003; Lee & Fradd, 1998). Much discussion has also focused on the quality and quantity of science teachers for students from the low socioeconomic backgrounds (Bantwini, 2010; Ingersoll, 2003; Goodrum, Hackling, & Rennie, 2001). Science teachers are also viewed as contributing to the undesirable or unacceptable students' performance with their science content knowledge, pedagogical content knowledge, general pedagogical approaches/knowledge, and disciplinary knowledge repeatedly being scrutinized. Fraser-Abder (2011) notes that ample evidence exists that shows that the problem is not the students, but instead how science is taught, which progressively diminishes students' interest in the subject and their confidence.

It has also been argued that teacher's knowledge, experiences, and beliefs greatly impact what takes place within the classroom and eventually influence student achievements (Ball, Thames, & Phelps, 2008; Davis, 2003; Haney, Lumpe,

---

B.D. Bantwini (✉)
Research Use and Impact Assessment Unit, Human Sciences Research Council,
134 Pretorius Street, Pretoria 0002, South Africa
e-mail: bongani.bantwini@gmail.com

Czerniak, & Egan, 2002). Davis (2003) contends that without the necessary subject matter knowledge, it is hard for teachers to learn strategies and techniques needed to respond to students thinking about the subject in ways that facilitate their learning. Sociocultural and political issues also play a critical role in educational complexities (Lee & Buxton, 2011; Parsons, 2008; Lee & Fradd, 1998). Lemke (2001) vehemently argues that changing your (one's) mind is not simply a matter of rational decision making, but a social process with social consequences. According to Lemke, it is not about what is right or what is true in the narrow rationalist sense; it is always also about how we feel, about who we are, about who we like, about who treats us with respect, about how we feel about ourselves, and others. In a community, he adds, individuals are not simply free to change their minds; the practical reality is that we are dependent on one another for our survival and all cultures reflect this fact by making the viability of beliefs contingent on their consequences for the community. Aggravating the focus on teachers is the belief that they are both the subjects and the agents of change (Sikes, 1992), and they ultimately decide the fate of national or state science standards implementation (U. S. Department of Education, 2010; Spillane & Callahan, 2000). Teachers are key components in the success of any new curriculum policy as they mediate between policy and practice.

This chapter intends to discuss policy issues, equity, multicultural science education, and the nature of support that local school districts should provide to inservice science teachers. The discussion is important for all stakeholders, especially science teachers and teacher educators since they are the key change agents who are usually the source of blame for the school failing. Thus, to achieve the purpose of this chapter, *firstly*, I will briefly discuss why this chapter focuses on local school districts and their officials/district offices. Empirical and theoretical studies conducted over the years will be used to argue this focus. *Secondly*, I will discuss why I view it imperative that school districts and their officials support science teachers, specifically for students from diverse and low socioeconomic backgrounds. Also to be argued is the belief that the effectiveness of science teachers somehow depends on the district offices and officials' support as many are confronted by various challenges including the following:

- Teacher science content knowledge limitations
- The rapidly changing student population, requiring change in our ways of teaching
- Interminable science curriculum reforms

Thirdly, using the sociocultural perspectives, I will discuss how school districts and their officials can effectively support science teachers. Spillane (2002) contends that reform failure at the district level is not solely a function of local actors' (*teachers*) inability or unwillingness to carry out policy proposals; rather it is in part a function of district office and its officials who are responsible for ensuring that teachers comprehend the new policies by providing them with a vision, interpretation, focus, and policy coordination (Corcoran, Fuhrman, & Belcher, 2001). District offices are major sources of capacity building for the teachers (Massell, 2000; Chinsamy, 2002), ensuring quality teaching and learning, effective assessment, increased learner performance, and achievement (Anderson, 2003; Iver, Abele, & Elizabeth, 2003; Rorrer, Skrla, & Scheurich, 2008). Important factors to address are as follows:

- Development of realistic policies
- Need to acknowledge the district office limitations
- Need to acknowledge the inequalities and existing injustices prevailing in schools/district
- Need for teaching and learning resources
- Promotion of viable working relationship by district offices

Lastly, I will conclude by highlighting some of the key issues advocated by this chapter and why it is imperative that we rethink more about the existing policies and the nature of support offered by school districts to science teachers.

Why Focus on School Districts and Their Officials?

Through the years we have had several empirical and literature review studies that focused on the role of school districts (Abele, Iver, & Farley, 2003; Anderson, 2003; Bantwini & Diko, 2011; Corcoran et al., 2001; Daly & Finnigan, 2011; Farley-Ripple, 2012). Common from these studies is the consensus regarding the significant role that school districts play or are supposed to play in supporting schools and teachers. These roles include mediation between schools and the government (Abele et al., 2003; Honig, 2008; Southern Regional Education Board, 2010; Spillane, 2000, 2002). School districts are viewed as legal entities required by state law to provide education to all students regardless of race, ethnicity, socioeconomic background, and disability within the attendance boundaries (Hightower, Knapp, Marsh, & McLaughlin, 2002). Some studies view districts as implementers of state policies (Honig, 2008; Marsh, 2002), as professional learning laboratories (Stein & D'Amico, 2002), as teacher educators for beginning teachers as they struggle with the daily decisions about what and how to teach (Grossman, Thompson, & Valencia, 2002), as boundary spanners in the context of collaborative education policy implementation (Honig, 2006), and as initiators of a variety of other policies that shape the way professional development is conducted (U. S. Department of Education, 2010; Youngs, 2001). Overall, school districts are key elements and authorized agents that administer and guide schools (Massell, 2000), vital institutional actors in educational reforms (Rorrer et al., 2008), and the major sources of capacity building for the schools (Massell, 2000). Using their influential role they ensure quality teaching and learning, effective assessment, increased learner performance, and achievement (Anderson, 2003; Iver et al., 2003).

Indisputably, the significant role of school districts goes beyond the US borders. In other countries such as South Africa, Roberts (2001) describes the primary function of school districts as twofold: to support the delivery of curriculum in schools and to monitor and enhance the quality of learning experiences offered to learners. In his observations, district offices have a particular role to play in working closely with local schools and ensuring that local educational needs are met. In supporting the primary function of the district, he notes that there are five possible areas of operation: policy implementation, leading and managing change, creating an enabling environment for schools to operate effectively, intervening in failing schools, and offer administrative and professional services to schools and teachers. Roberts believes that these different areas of operation should be aligned to support the district's primary purposes, teaching, and learning.

Evidently, school districts can play a critical role in supporting science teachers and addressing their various classroom needs. However, despite this potential and mandate to support science teachers, some research studies show that several districts are still falling short in this area, leaving teachers struggling to teach all students from diverse backgrounds (Archibald, Coggshall, Croft, & Goe, 2011; Bantwini & Diko, 2011). According to Bantwini (2010) teachers continuously complain about lack of support from their school districts and the amount of paper work required of them daily. These are the teachers who admittedly noted their lack of confidence in teaching science in their classroom. Bantwini (2010) suggests that it is very difficult for science teachers, especially those dealing with students from low socioeconomic backgrounds, to succeed in their work without effective support from their local district officials, as they also lack the necessary resources to address or combat the challenges confronting them. Similar views are shared by Glenn (2000) who contends that for teachers to deliver high-quality teaching, they must be empowered to do so. He also insists that policy makers and all the key stakeholders must be willing to stand up for teachers as the primary drivers of student achievement. Unfortunately, many teachers still report feeling more pressure than support from their districts (Rutledge, 2008; Bantwini, 2010).

Why Support InService Science Teachers?

Teacher Science Content Knowledge Limitations

Inservice science teachers are confronted by various challenges that impact their teaching and student learning (Moreno & Erdmann, 2010; Johnson, 2007). In their observation, Abell et al. (2007), and Ingersoll and Perda (2009) reveal that too many classrooms are taught by individuals who are not certified in the subject matter that they are teaching, and this problem is acute in areas of science and mathematics and typically occurs most often in low-income, high-poverty areas. The sentiment is also shared by the US Department of Education (2011) as they report that over half of all districts have difficulty recruiting highly qualified teachers in science education, and over 90 % of high-minority districts also concur in this challenge. Also, literature shows that there is a widely recognized lack of confidence from many science teachers which leads to many either avoiding teaching science or teaching it minimally (Kenny, 2012; Appleton, 2005; Tyler, 2007).

In the "Before it's too late" report, John Glenn, the National Commission Chairman on Mathematics and Science, teaching for the twenty-first century, emphasized that "the future well-being of a nation and people depends not just on how well they educate children generally, but on how well they educate them in mathematics and science specifically" (Glenn, 2000, p. 4). Justifying the imperative of effective teaching, he argues that high-quality teaching requires that teachers have a deep knowledge of subject matter and for this there is no substitute. According to Wheeler (2007) "a nation's ability to remain an economic and technological leader in a global marketplace relies on how well that nation educates its students in science, technology, engineering, and mathematics" (pp. 30–31). However, in their observation Goodrum et al. (2001) reveal that though the actual picture of science teaching and learning is one of great variability, on average the picture is disappointing. They report that the common trends in various countries show that science teachers, especially those that teach at primary level, lack confidence in teaching science. Furthermore, literature confirming that primary teachers may have a good range of pedagogical skills, but may lack specific science content knowledge and the confidence to apply their pedagogical skills in a science context thrives (Hudson, 2005; Kenny, 2012).

Teachers in diverse and multicultural classrooms have been cited as facing increasing challenges in providing an appropriate classroom environment and high standards of instruction that foster the academic achievement of all students, particularly students from low socioeconomic backgrounds (McAllister & Irvine, 2000). Using the South African case to present my argument, most of the Black science teachers acquired their teaching qualifications during the apartheid era. The nature of the preservice education they received did not sufficiently prepare them to teach the twenty-first century science learners. Research, reports, and other studies show that these teachers are confronted by a multitude of challenges that are also attributed to the current challenge of poor student performances in both science and mathematics (Centre for Development and Enterprise, 2007; Muwanga-Zake, 2003). Clearly, these inservice teachers cannot be sent back to universities to learn more about science teaching, rather school districts can intervene and assist them by developing various professional development programs aimed at instilling the science teaching confidence. The South African situation used above may sound awkward, but the issues they are confronted with are not unique but common in many countries including both the developing and underdeveloped countries. This issue also receives more attention in the Obama administration plan for teacher education reform and improvement, as the US Department of Education (2011) acknowledges the challenge of teacher science content knowledge and appeals that more should be done to support and reward excellent teaching at various stages in the education system.

Changing Population Requiring New Pedagogical Approaches

According to Lemke (2001), diversity is not just a matter of exceptionality or exotic and radial difference, but in some degree a condition of every community. Trend analysis reveals that the US K-12 student population is increasingly becoming diverse with one of every three students being from underrepresented groups (Atwater & Riley, 1993; Cox-Petersen, Melber, & Patchen, 2012; Kose & Lim, 2010). Noting the large and growing cultural gaps between children and teachers in school, Sleeter (2001) indicates that in 1994 39 % of teachers had students with limited English proficiency in their classrooms, with one quarter of those teachers having received training for working with them. However, according to the Migration Policy Institute (2011), the number of limited English proficient students in the United States grew by 80 % between 1990 and 2010. In Cox-Petersen et al. (2012) recent analysis, currently about 21 % of the students live with at least one foreign-born parent, whereas 19 % live with families who speak languages other than English at home. Lee and Fradd (1998) state that though commonalities can apply across groups, however, the differences in immigration history, socioeconomic status, acculturation within mainstream society, and family attitudes towards education all contribute to the increasing diversity among the student population.

The changes in student population underscore the significant need for a comprehensive knowledge base to provide effective science instruction. Teachers require knowledge to be able to teach students from diverse racial, ethnic, social class, and language background. It is imperative that school district professional development policies ensure that they have provision for teachers who all of sudden have to deal with the growing diversity in our schools. Learning science is viewed as crossing a border (language, methods, and different contents from what ethnic minority or lower socioeconomic status children experience in their homes), thus culturally relevant teaching helps the students to navigate across and beyond this border that has traditionally served as a barrier to student learning (Cox-Petersen et al., 2012; Aikenhead, 1996). Furthermore, it also helps the teachers to better support the teaching and learning of science for all students.

Villegas and Lucas (2002) observe that the typical response of teacher education programs to this growing diversity has been to add a course or two on multicultural education. However, these authors contend that though these courses may play a crucial role in teacher preparation for diversity, but the approach does not go far enough because the courses are often optional and students can complete their education program without receiving any preparation whatsoever in issues of diversity. This may mean that a majority of the teachers leave college without taking a single course that prepares them to deal with students from diverse backgrounds. In this case the school district has a responsibility to assist those teachers by providing the necessary professional development that will address the teachers' challenges or weaknesses. Assistance is an indispensable requirement as teachers are confronted by frequent and endless changes requiring them to be ready at all times. The following section provides more focus on the science education curriculum reforms.

Interminable Science Curriculum Reforms

Since the launch of Sputnik in 1957, the science education field has shown tremendous growth relative to new reform changes which includes use of new curriculums, teaching approaches, and new understanding of the student learning styles (Carter, 2005; Bybee & Fuchs, 2006; DeBoer, 1991). Moreover, there are also increasing demands that science teachers modify knowledge, understanding, beliefs, and practice in order to facilitate student learning and understanding of science concepts. Also evident is that these changes are at levels above the preparation that most teachers received, requiring teachers to keep up with them. This has proven to be very stressful, as this process also has time implications. On their own, the chances that teachers can be successful without hiccups are almost impossible and hence the proposal for school district support.

Literature reveals that a number of teachers complain about the lack of support from their district offices, causing them a stress as they have to find their own ways of understanding and ensuring successful implementation of these new reforms (Bantwini & Diko, 2011). Mutegi (2011) argues that the prevailing reform-guided approach in science education is not likely to meet the social needs of underrepresented student group. There are also reports from various countries that the majority of teachers rarely implement new science curricular reforms (Bantwini, 2010; Johnson, 2007) and as a result this hinders progress towards reform. Spillane and Callahan (2000) also argue that it is unlikely that teachers will implement the science standards absent of any support from the local policy environment. They contend that school districts play an important role in the implementation of state and federal policy. According to Rorrer and Skrla (2005), the ability of districts and school leaders to reconceptualize the purpose of the policy, adapt the policies to local needs, and integrate the policy into individual's context provides a starting point to achieve substantially improved opportunities and outcomes for all children in their district, particularly students from low-income homes and students of color.

How to Encourage and Support Science Teachers

Development of Realistic Science Education Policies

Jansen (2002) observes that all nation states develop education policies with symbolic purposes in mind. However, some reform policy developments are political symbols, as they are developed without a clear implementation strategy as well as understanding of those who will eventually implement Jansen (2002). He notes that education policy making can demonstrate the preoccupation of the state with settling policy struggles in the political domain rather than in the realm of practice. Jansen argues that such political investment in the production of policy is important to politicians in selling their advantage to the broader democratic alliance. Obviously, there are some crucial lessons that can be learned from this notion including that of developing realistic policies that will stress the practical considerations above the symbolic function. These are policies that will focus and address the real issues from the ground, policies that can be adapted to the local contextual needs, and policies that aim to empower people. Rorrer and Skrla (2005, p. 55) view the ability to adapt policies to local contextual needs as a valuable survival skill, a fact of administrative life and a requisite for an integrated, cohesive responsive to policy.

Over the decades various countries have developed a plethora of science education policies that later had to be abandoned or revised due to their failure to address the challenges they were intended to solve. The development of new curricula as Rogan and Grayson (2003) highlight is a common event in many countries. However, in many cases, these curricula are well designed and the aims they are intended to achieve are laudable. Rogan and Grayson argue that all too often the attention and energies of policy makers and politicians are focused on the "what" of desired educational change neglecting the "how." Similarly, most school district policies declare their support for teachers to improve the teaching and learning in classroom. However, in some the intention and the practical application hardly correspond. In this case, the policy can be viewed as "a blunt instrument ill-suited to forging fundamental change in instruction" (Spillane & Callahan, 2000).

Effective and efficient district support policies should acknowledge the existing realities, past and prevailing injustices, and school and classroom cultures as these factors possess the ability to nullify the reform efforts being initiated (Bantwini & Diko, 2011). Spillane and Callahan (2000) argue that policy and reformers rarely take account of implementers' prior knowledge in designing policy and propose that we need to reconsider the criteria brought into policy design. They assert that those who design standards and other reform proposals need to begin where local policy makers are rather than dwell solely on the brave new world for science education they want to create. Literature indicates that policy should support the environment for learning rather than rigid systems and programs which can lead to meaningless activities and out-of-date structures (Southern Regional Education Board, 2010; Darling-Hammond & McLaughlin, 1995). Spillane and Callahan (2000) observe that policy problems and their solutions are in great part constructed by implementers, albeit with the aid of policy makers' cues. They argue that it is not only the design of policy (which is its authority, consistency, clarity, and more) that influences what implementers do and do not do by way of implementing policy, but also implementers' knowledge and beliefs which they use to make sense of the policy message. Therefore, it is essential that policy development should consider the implementation strategies that take into account the local context including diversity that may exist within that context and psychological factors that influence learning and change (Rogan & Grayson, 2003).

District Office Limitations

According to Spillane and Thompson (1997), the factors that make up a district's capacity to support ambitious instructional reform are highly intertwined, and therefore, the capacity to support instructional reforms should best be understood as a complex and interactive configuration. One of the critical issues facing school districts is the deficit of personnel, hindering the few officials from effectively servicing schools and the teachers (Bantwini & Diko, 2011). The lack of sufficient personnel has negative impacts on the expected results, especially in the implementation of the ongoing curriculum reforms. Considering the district officials' and school/teacher ratio in many districts, it would be unrealistic to expect a profound amount of change in the current teaching and learning in schools. This typically points to policy development that does not correspond with reality. It is imperative as Davis (2003) suggests that much thought and effort be given to how teachers learn to teach, what teachers know, how their knowledge is acquired, how it changes over time, and what processes bring about change in individual teacher practices as well as deep and long lasting change in science classroom. This is crucial if new reforms are intended to be worthwhile and not political symbols.

Narsee (2006) asserts that the central dilemma for education districts could be their structural conditions. Narsee emphasizes that school districts operate at the intersection of dual, related dichotomies of support and pressure, centralization, and decentralization. However, she believes that it is only through conscious engagement with these dichotomies will districts be able to resolve the tensions between the policy, support, and management roles expected of them. The research of Walberg and Fowler Jr. (1987) suggests that bigger districts yield lower achievements. This was also evident in Bantwini and Diko (2011) findings which reveal that it is difficult for district officials to assist schools that are in dire need of help, especially when they themselves are still confronted by limitations in their personnel, an issue that receives more focus later in this chapter.

Inequalities and Existing Injustices Prevailing in Schools

Building instructional capacity extends beyond the focus on teaching and learning (Rutledge, 2008). According to Chisholm and Leyendecker (2008), local cultural and contextual realities and capacities still appear to be overlooked. These authors suggest that curriculum reforms probably work best when curriculum developers acknowledge existing realities, classroom cultures, and implementation requirements. Effective science teacher growth is possible when the district support policies acknowledge the existing realities, past and prevailing injustices, and school and classroom cultures (Bantwini & Diko, 2011). Lee and Buxton (2011) argue how nonmainstream students have been ignored and at worse been oppressed by schooling in general and science education in particular. They assert that these inequalities do not go unnoticed both by students and their families.

At college level, the teaching methods that preservice teachers are taught or exposed to are usually for a generalized context, since we hardly know where they are going to work after completion of the teaching degree. In this case, they graduate from college with generalized ideas of teaching at various contexts such as students from diverse background. Many graduates from college have taken very few, while others have not even taken a single course on multicultural or diversity courses. Many colleges do not mandate their teacher candidates to take multicultural or diversity courses. These teacher candidates end up teaching students from diverse multicultural and low socioeconomic backgrounds. Their lack of understanding regarding students from diverse backgrounds perpetuates the inequalities and injustices in their schools. One of the injustices is that later there is no help for teachers to master specific teaching strategies, approaches, or methods for these contexts when they are employed in their school districts.

On the district level teachers should receive support through ongoing professional development that addresses equity in teaching and promotes a discourse on diversity issues. Lack of preparation in culturally relevant teaching or pedagogical practices often affects the success of many beginning and veteran teachers (Bantwini & Diko, 2011) and correspondingly their students (U. S. Department of Education, 2010). Rorrer et al. (2008) view school districts as particularly well positioned to address issues of equity as they are uniquely equipped to address differences and discrepancies between schools. Sometimes these realities exist on both sides: the district office and the schools which require policies to be flexible enough to fit particular contexts and needs (King, 2004).

Need for Teaching and Learning Resources

In the Framework for K-12 Science Education, the National Research Council (NRC, 2012) indicates that equity in science education requires that all students are provided with equitable opportunities to learn science with access to quality space, equipment, and teachers to support and motivate that engagement and learning and adequate time spent on science. This is one of the focuses of the Next Generation Science Education Standards intended to be used in every state in the United States. The NRC (1996) emphasizes that educational system must act to sustain effective teaching and to use the routines, rewards, structures, and expectations of the system to endorse the vision of science teaching portrayed by the standards. It argues that teachers must be provided with resources, time, and opportunities to make change as described in the program and system standards and should work within a framework that encourages their efforts. The significance of providing science teachers with resources and materials in order to facilitate effective teaching and student learning is widely recognized (Lee & Buxton, 2011; Moreno & Erdmann, 2010: Johnson, 2007; Darling-Hammond, 2003; Goodrum et al., 2001; Glenn,

2000). However, despite this indisputable call for science materials, many schools are still struggling to acquire resources to facilitate teaching and learning in the science classroom (Moreno & Erdmann, 2010; Johnson, 2007; Johnson, Monk, & Hodges, 2000; Anderson, 1996). Moreno and Erdmann (2010) assert that most teachers still have insufficient access to quality continuing education, teaching resources, up-to-date content, and preparation in current laboratory techniques.

Science is a complex subject that when taught without the proper materials can perpetuate especially in students from low socioeconomic backgrounds. Districts need to provide more resources for science teachers to promote scientific understanding and literacy facilitating students' use of scientific evidence, engagement in scientific discourse, development of science knowledge, and excitement about science (Moreno & Erdmann, 2010). Enrollment statistical trends from many countries show that a large group of students do not pursue science because it is considered a difficult and abstract subject. To be attributed to this negative perception about science is the lack of resources that can make teaching and learning of it fun and exciting. Used properly, resources have the capability to make abstract concepts more relevant and meaningful to students. According to Anderson (2003) districts that believe that quality of student learning is highly dependent on the quality of instruction organize themselves and their resources to support instructionally focused professional learning for teachers.

Schools, as Fullan (1992) argues, cannot redesign themselves; districts play an important function in establishing the conditions for continuous and long-term improvements for schools as they control and coordinate all the development projects implemented in their schools. According to Lee and Fradd (1998) the involvement of policy makers is essential in establishing effective programs, securing resources, and promoting public awareness of the importance of science for all students. From a literature review, Rutledge (2008) who indicates high-performing districts had a stronger infrastructure focused on instruction than their low-performing counterparts, engaging in a range of activities supporting school and classroom instruction including professional development, a focus on student achievement, alignment of curriculum across schools, and systemic monitoring of instructional practices. School districts that value student quality teaching and learning invest in providing schools and teachers with necessary resources.

Promoting Viable Working Relationships with Schools

Leaders, according to Rorrer and Skrla (2005), who serve as policy mediators in responsible and positive ways within a strong accountability environment develop and nurture relationships and interaction to facilitate the reconceptualization and integration of accountability policies. These authors view the relationships and interactions between and among school and district personnel as serving as vital organizational linkages through which leaders communicate and build support for the achievement of all children and in part for the accountability policy. Additionally, they advise that relationships in the district and schools that forge successful and productive policy implementation require the leadership to be involved and to coordinate efforts across organizational levels.

The successful implementation of instructional reforms as Spillane and Callahan (2000) argue depends in some measure on the broader policy environment in which classrooms are nested, a complete territory of the school district. These authors argue that if teachers work in environment where they have few opportunities and no incentives to learn about the science standards, they are less likely to implement the reform ideas advanced by standard. They argue that school districts can and do influence these conditions, with district policy makers making decision about numerous instructional guidance instruments—including staff development, curriculum guides, and curricular materials. Spillane and Callahan argue that it is unlikely that most teachers will implement the science standards absent of any support from the local policy environment. According to Darling-Hammond and McLaughlin (1995) it is imperative to establish an environment of professional trust and encourage problem solving. Working conditions should be one target for policies aimed at retaining qualified teachers in high-need schools as they have a direct effect in what teachers do in the classroom, how well students achieve, and their experience of school (Ladd, 2009; Hanushek & Rivkin, 2007; Leithwood & McAdie, 2007; Darling-Hammond, 2003).

McLaughlin (1992) mentions that district officials need to initiate and facilitate discussion about how teachers feel about their work and how they see district policies and practices as supporting or inhibiting their practice or sense of professional worth. Bantwini and Diko (2011) state that as futile and time consuming this process might seem, it should be realized that the success of any new reforms depends on the good working relationships and synergy between the local district officials and the teachers. In his study, Bantwini (manuscript in preparation) found that contributing towards the myriad challenges that confronted district officials was their lack of good relationship with their teachers. The district officials believed that teachers harbored suspicions and discomfort with regard to professional interactions with the district offices. These suspicions were attributed to the historical relationship that previously prevailed between the two parties. Bantwini describes the nature of their work relationship as power-authority driven, top down in nature, creating "us and them" situation. Also, he contends that the issue of trust between teachers and district officials was a critical challenge, long overdue but ironically, not receiving the attention it deserves. He argues that it is an issue that cannot be overlooked any longer with the hope that desired change will just come. Certainly time is a crucial factor in the healing of the past injustices; however, initiation of the process to repair the broken relationship should occur.

Reina and Reina (1999) argue that when leaders create trusting working environment, people are safe to challenge the system and perform beyond expectations. Employees feel more freedom to express their creative ideas and are willing to take risks, admit mistakes, and learn from those mistakes (Reina & Reina). Citing Friedkin and Slater (1994), Rorrer and Skrla (2005) emphasize the necessity of developing trust, respect, and credibility to coordinate and integrate desired changes. The lack of good relationship and synergy only leads to complications that can only serve to hold back the desired outcomes. As Fullan (2007) argues "...change with any depth must be cultivated by building relationships while pushing forward..." (p. 211). Darling-Hammond and McLaughlin (1995) propose that professional development policy should support the environment for learning rather than rigid systems and programs, which can lead to meaningless activities and out-of-date structures.

The success of the teachers is not only cognitive related, but also contextually situated and intrinsic to the context within which and with which the individual interacts (Jurasaite-Harbison & Rex, 2010), and thus the desired teacher change should correspond with effective district support. According to Glenn (2000) teacher empowerment also means that we should accord them the respect that they deserve for their judgment about learning and rewarding their professionalism. Rorrer and Skrla (2005) note that in the district that they studied, leaders acknowledged that the relationship they established with their faculties, staffs, and constituents were used to gain commitment for changes in organizational structures. They state that these relationships also increased the capacity of districts and schools to create learning environment in which all children could be successful, a goal compatible with the stated purpose of accountability policies. Clearly, without effective working relationships between the district officials and the science teachers, chances of successful collaboration are limited.

Conclusion

Undeniably, the repeated emphasis on effective science teaching and learning indicates how crucial science education has become in our lives. However, it is also clear that science teachers on their own can hardly achieve the desired goals, to help student perform well or achieve the desirable outcomes. This chapter, premised on the notion that the school district play or can play a crucial role in the teaching and learning, proposes the imperative of school districts increasing their science teacher support and develop science education policies that are realistic and not just political symbols but promote equity among students. Realistic policies take into consideration the contextual needs and are collaboratively developed including individuals that will eventually implement them. Furthermore, it is imperative that each district develops a program directed to assisting their teachers with issues that they confront in their teaching area. This includes the fact that many preservice teachers graduate with very little content knowledge which in turn adversely impacts their confidence to teach science in their classrooms. Thus, district should provide more professional development focused on equity and diversity issues as they are also contributing factors in the science educational challenges. The major focus of many teacher professional developments is on science and pedagogical knowledge that shy away from multicultural issues, equity, and diversity. Clearly, avoiding addressing these issues will not benefit the teachers; rather it will impact students as their diverse needs are not being addressed. When students believe that their needs are not met, they will develop a mistrust which according to Lee and Buxton (2011) presents a serious barrier to science teaching and learning until it is dealt with explicitly in the classroom.

However, it will be unfair to expect teachers who do not have expertise in multicultural and diversity issues to be able to address those issues in their classrooms. The NRC (2012) in the Framework for K-12 Science Education notes the profound differences among demographic groups in their educational achievements and patterns of science learning, which are complex in nature. The differences among demographic groups require teachers to start confronting them by openly discussing and inviting personnel with expertise to assist them understand how they can address them in their classrooms. Clearly, learning to address multicultural and equity issues will require time and an ongoing teacher support. The process will also help many teachers to embrace the existing diversity and also strive to promote equity among their students. Such practice will help teachers recognize discontinuities between their students' worldviews and scientific views and avoid student labeling.

Colleges of Education have been noted to contributing to some of the challenges discussed above. However, continual blame for not adequately preparing future teachers can be unproductive. Thus, Colleges of Education should mandate their teacher candidates to take some multicultural or diversity courses before they graduate. To consolidate that knowledge, they can also conduct their field practicum in settings that will expose them to the realities of multiculturalism and diversity. The world is changing rapidly and it is imperative that all graduating teachers are fully equipped to address students' diverse needs. Lack of these skills usually, in a subtle manner, results to low learning expectations and biased stereotypical views about the interests or abilities of particular students or demographic groups (NRC, 2012).

Additionally, inservice teachers as agents of change need to understand that personal-professional growth is part of their responsibility that does not only depend on their district, as some tend to think that way. Bantwini (2012) argues that shifting responsibility to the Department of Education just proves to be highly unproductive and irresponsible. Teachers need to take initiative in setting professional development programs that respond to their needs, as they are more familiar with what these needs are than is the district office. This should be used as a platform to generate and contribute towards knowledge development.

School districts should also be realistic about their own limitations and challenges including having fewer district officials to assist teachers. It is highly unlikely that few district personnel can manage to assist every teacher in their district. Thus, provision of adequate district personnel to work with all the teachers will benefit the struggling schools. Moreover, I believe that if school districts can play their role, collaborate and synergize with science teachers including other stakeholders that will increase chances to yield the desired results in student performances and schools.

References

Abele, M., Iver, M., & Farley, E. (2003). *Bringing the district back in: The role of central office in improving instruction and student achievement.* Centre for Research on the Education of Students Placed at Risk. John Hopkins University.

Abell, S. K., Lannin, K. K., Marra, R. M., Ehlert, M. W., Cole, J. S., Lee, M. H., et al. (2007). Multi-site evaluation of science and mathematics teacher professional development programs: the project profile approach. *Studies in Educational Evaluation, 33*(1), 135–158.

Aikenhead, G. S. (1996). Science education: Border crossing into subculture of science. *Studies in Science Education, 27*(1), 1–52.

Anderson, R. D. (1996). *Study of curriculum reform. Volume I: Findings and conclusions. Studies of Educational Reform.* (ERIC Document Reproduction Service No. ED 397535).

Anderson, S. E. (2003). *The school district role in educational change: A review of the literature.* International Centre for Educational Change. Ontario Institute for Studies in Education. Toronto: Canada.

Appleton, K. (2005). *Elementary science teacher education: International perspectives on contemporary issues and practice.* London: Routledge.

Archibald, S., Coggshall, J., Croft, A., & Goe, L. (2011). *High quality professional development for all teachers: Effectively allocating resources.* Washington, DC: National Comprehensive Center for Teacher Quality.

Atwater, M. M., & Riley, J. P. (1993). Multicultural science education: Perspectives, definitions, and research agenda. *Science Education, 77*(6), 661–668.

Ball, D. L., Thames, M. H., & Phelps, G. (2008). Content knowledge for teaching: What makes it special. *Journal of Teacher Education, 59*(5), 389–407.

Bantwini, B. D. (2010). How teachers perceive the new curriculum reform: Lessons from a school district in the Eastern Cape Province, South Africa. *International Journal of Educational Development, 30*(1), 83–90.

Bantwini, B. D. (2012). Primary school science teachers' perspectives regarding their professional development: Implications for school districts in South Africa. *Professional Development in Education Journal, 38*(4), 517–532.

Bantwini, B. D. (Manuscript in preparation). *Analysis of district officials and teaches' working relationships: Implications for new curriculum reforms in South Africa.*

Bantwini, B. D., & Diko, N. (2011). Factors affecting South African district officials' capacity to provide effective teacher support. *Creative Education, Scientific Research, 2*(3), 103–112.

Bergerson, T. (2006). *Race, poverty and academic achievement.* Retrieved from: http://sboh.wa.gov/ESS/documents/Race&Poverty.pdf

Bybee, R. W., & Fuchs, B. (2006). Preparing the 21st century workforce: A new reform in science and technology education. *Journal of Research in Science Teaching, 43*(4), 349–352.

Carter, L. (2005). Globalization and Science Education: Rethinking Science Education Reform. *Journal of Research in Science Teaching, 42*(5), 561–580.

Centre for Development and Enterprise. (2007). *Doubling for growth: Addressing the math and science challenge in South Africa's schools.* Johannesburg, South Africa: The Centre for Development and Enterprise.

Chinsamy, B. (2002). *Successful school improvement and the educational district office in South Africa: Some emerging propositions.* Developed under the District Development Support Programme Project, USAID. Retrieved from: http://www.rti.org/pubs/Chinsamy_Full.pdf

Chisholm, L., & Leyendecker, R. (2008). Curriculum reform in post-1990 Sub-Saharan Africa. *International Journal of Educational Development, 28*, 195–205.

Corcoran, T., Fuhrman, S. H., & Belcher, C. L. (2001). The district role in instructional improvement. *Phi Delta Kappan, 83*(1), 78–84.

Cox-Petersen, A., Melber, L. M., & Patchen, T. (2012). *Teaching science to culturally and linguistically diverse elementary students.* Boston: Pearson Publishers.

Daly, A. J., & Finnigan, K. S. (2011). The ebb and flow of social network ties between district leaders under high-stakes accountability. *American Educational Research Journal, 48*(1), 39–79.

Darling-Hammond, L. (2003). Keeping good teachers: Why it matters, what leaders can do. *Educational Leadership, 60*(8), 6–13.

Darling-Hammond, L., & McLaughlin, M. W. (1995). Policies that support professional development in an era of reform. *Phi Delta Kappan, 76*(8), 597–604.

Davis, K. S. (2003). "Change is hard": What science teachers are telling us about reform and teacher learning in innovative practices. *Science Education, 87*, 3–30.

DeBoer, G. (1991). *A history of ideas in science education: Implications for practice*. New York: Teachers College Press.

Farley-Ripple, E. N. (2012). Research use in school district central office decision making: A case study. *Educational Management Administration and Leadership, 40*(6), 786–806.

Fraser-Abder, P. (2011). *Teaching budding scientists: Fostering scientific inquiry with diverse learners in grades 3–5*. Boston: Pearson.

Friedkin, N. E., & Slater, M. R. (1994). School leadership and performance: A social network approach. *Sociology of Education, 67*(2), 139–157.

Fullan, M. G. (1992). *The new meaning of educational change*. London: Cassel Educational Limited.

Fullan, M. (2007). *The new meaning of educational change* (4th ed.). New York: Columbia University/Teachers' College Press.

Glenn, J. (2000). *Before it is too late. A Report to the Nation from the National Commission on Mathematics and Science Teaching for the 21stCentury*. A report submitted to the US Secretary of Education, September 27, 2000.

Goodrum, D., Hackling, M., & Rennie, L. (2001). *The status and quality of teaching and learning of science in Australian schools*. A Research Report prepared for the Department of Education, Training and Youth Affairs. Retrieved from: http://www.detya.gov.au/schools/publications/index.htm

Grossman, P. L., Thompson, C., & Valencia, S. W. (2002). Focusing the concerns of new teachers: The district as teacher educator. In M. Knapp, M. McLaughlin, J. Marsh, & A. Hightower (Eds.), *School districts and instructional renewal: Opening the conversation* (pp. 129–142). New York: Teachers College Press.

Haney, J. J., Lumpe, A. T., Czerniak, C. M., & Egan, V. (2002). From beliefs to actions: The beliefs and actions of teachers implementing change. *Journal of Science Teacher Education, 13*(3), 171–187.

Hanushek, E. A., & Rivkin, S. G. (2007). Pay, working conditions, and teacher quality. *The Future of Children, 17*(1), 69–86.

Hightower, A. M., Knapp, M. S., Marsh, J. A., & McLaughlin, M. W. (2002). The district role in instructional renewal: Setting the stage for dialogue. In A. M. Hightower, M. S. Knapp, J. A. Marsh, & M. W. McLaughlin (Eds.), *School districts and instructional renewal* (pp. 1–6). New York: Teacher College, Columbia University.

Honig, M. I. (2006). Street-Level bureaucracy revisited: Frontline district central-office administrators as boundary spanners in education policy implementation. *Educational Evaluation and Policy Analysis, 28*(4), 357–383.

Honig, M. (2008). District central offices as learning organizations: How socio-cultural and organizational learning theories elaborate district-central-office administrators' participation in teaching and learning improvement efforts. *American Journal of Education, 114*, 627–664.

Hudson, P. (2005). Identifying mentoring practices for developing effective primary science teaching. *International Journal of Science Education, 27*(14), 1723–1739.

Ingersoll, R. (2003). *Is there really a teacher shortage? A research report co-sponsored by the consortium for policy research in education and the center for the Study of Teaching and Policy*. Available online: http://depts.washington.edu/ctpmail/PDFs/Shortage-RI-09-2003.pdf

Ingersoll, R. M., & Perda, D. (2009). *The mathematics and science teacher shortage: Facts and myth*. Retrieved from: https://www.csun.edu/science/courses/710/bibliography/math%20science%20shortage%20paper%20march%202009%20final.pdf

Iver, M., Abele, M., & Elizabeth, F. (2003). *Bringing the district back in: The role of the central office in improving instruction and student achievement*. Baltimore, MD: Center for Research on the Education of Students Placed at Risk.

Jansen, D. J. (2002). Political symbolism as policy craft: Explaining non-reform in South African education after apartheid. *Journal of Education Policy, 17*(2), 199–215.

Jensen, E. (2009). *Teaching with poverty in mind: what being poor does to kids' brains and what schools can do about it*. Alexandria, VA: ASCD.

Johnson, C. C. (2007). Technical, political and cultural barriers to science education reforms. *International Journal of Leadership in Education, 10*(2), 171–190.

Johnson, S., Monk, M., & Hodges, M. (2000). Teacher development and change in South Africa: A critique of the appropriateness of transfer of Northern/Western practices. *Compare, 30*(2), 179–192.

Jurasaite-Harbison, E., & Rex, L. A. (2010). School cultures as contexts for informal teacher learning. *Teaching and Teacher Education, 26*(1), 267–277.

Kenny, J. D. (2012). University-school partnerships: Pre-service and in-service teachers working together to teach primary science. *Australian Journal of Teacher Education, 37*(3), 56–82. Available online: http://ro.ecu.edu.au/ajte/vol37/iss3/6.

King, M. B. (2004). School- and district-level leadership for teacher workforce development: Enhancing teacher learning and capacity. In M. A. Smylie & D. Miretsky (Eds.), *Developing the teacher work-force: 103rd yearbook of the NSSE Part I* (pp. 303–325). Chicago: University of Chicago Press.

Kose, B. W., & Lim, E. Y. (2010). Transformative professional development: Relationship to teachers' beliefs, expertise and teaching. *International Journal of Leadership in Education: Theory and Practice, 13*(4), 393–419.

Lacour, M., & Tissington, L. (2011). The effects of poverty on academic achievement. *Educational Research and Reviews, 6*(7), 522–527.

Ladd, H. (2009). *Teachers' perceptions of their working conditions: How predictive of policy-relevant outcomes?* Retrieved from: http://www.urban.org/publications/1001440.html

Lee, O., & Buxton, C. (2011). Engaging culturally and linguistically diverse students in learning science. *Theory into Practice, 50*(4), 277–284.

Lee, O., & Fradd, S. H. (1998). Science for all, including students from non-English-language backgrounds. *Educational Researcher, 27*(4), 12–21.

Leithwood, K., & McAdie, P. (2007). Teacher working conditions that matter. *Education Canada, 47*(2), 42–45.

Lemke, J. L. (2001). Articulating communities: Sociocultural perspectives on science education. *Journal of Research in Science Teaching, 38*(3), 296–316.

Marsh, J. A. (2002). How districts relate to states, schools, and communities: A review of emerging literature. In A. M. Hightower, M. S. Knapp, J. A. Marsh, & M. W. McLaughlin (Eds.), *School districts and instructional renewal* (pp. 25–40). New York: Teacher College, Columbia University.

Massell, D. (2000). *The district role in building capacity: Four strategies* (Consortium for policy research in education). Philadelphia: Graduate School of Education, University of Pennsylvania.

McAllister, G., & Irvine, J. J. (2000). Cross cultural competency and multicultural teacher education. *Review of Educational Research, 70*(1), 3–24.

McLaughlin, M. W. (1992). How district communities do and do not foster teacher pride. *Educational Leadership, 50*(1), 33–35.

Migration Policy Institute. (2011). *Limited English proficient individuals in the United States: Number, share, growth, and Linguistic diversity*. Retrieved from: http://www.migrationinformation.org/integration/LEPdatabrief.pdf

Moreno, N. P., & Erdmann, D. B. (2010). Addressing science teachers needs. *Science, 327*(1), 1589–1590. Available online: http://www.sciencemag.org/content/327/5973/1589.full.pdf.

Mutegi, J. W. (2011). The inadequacies of "Science for All" and the necessity and nature of a socially transformative curriculum approach for African American science education. *Journal of Research in Science Teaching, 248*(3), 301–316.

Muwanga-Zake, J. W. F. (2003). *Is science education in a crisis? Some of the problems in South Africa*. Science in Africa. On-line Science Magazine, Issue 2, June 11, www.scienceinafrica.co.za, Science Education.

Narsee, H. (2006). *The common and contested meanings of education districts in South Africa*. A Thesis submitted to the faculty of education of the University of Pretoria in fulfilment of the requirements for the degree of Doctor of Education. Retrieved from: http://upetd.up.ac.za/thesis/available/etd-03232006-094442/unrestricted/00front.pdf

National Research Council. (1996). *National science education standards*. Washington, DC: National Academy Press.

National Research Council. (2012). *A framework for K-12 science education: Practices, crosscutting concepts, and core ideas*. Washington, DC: The National Academies Press.

Parsons, E. R. C. (2008). Positionality of African Americans and a theoretical accommodation of it: Rethinking science education research. *Science Education, 92*(6), 1127–1144.

Reina, D. S., & Reina, M. L. (1999). *Trust & betrayal in the workplace. Building effective relationships in your organization*. San Francisco: Berrett-Koehler Publishers.

Roberts, J. (2001). *District development – The new hope for educational reform*. Johannesburg: JET Education Services.

Rogan, J. M., & Grayson, D. F. (2003). Towards a theory of curriculum implementation with particular reference to science education in developing countries. *International Journal of Science Education, 25*(10), 1171–1204.

Rorrer, A. K., & Skrla, L. (2005). Leaders as policy mediators: The reconceptualization of accountability. *Theory into Practice, 44*(1), 53–62.

Rorrer, A. K., Skrla, L., & Scheurich, J. J. (2008). Districts as institutional actors in educational reform. *Educational Administration Quarterly, 44*(3), 307–358.

Rutledge, S. A. (2008). *Policy perspective on the role of districts in increasing instructional capacity*. Retrieved from: http://hookcenter.missouri.edu/wp-content/uploads/2009/06/Rutledge_AERA08.pdf

Sikes, P. J. (1992). Imposed change and the experienced teacher. In M. Fullan & A. Hargreaves (Eds.), *Teacher development and educational change* (pp. 36–55). Washington, DC: Famer Press.

Sleeter, C. E. (2001). Preparing teachers for culturally diverse schools: Research and the overwhelming presence of Whiteness. *Journal of Teacher Education, 52*, 94–106.

Southern Regional Education Board. (2010). *The three essentials: Improving schools requires district vision, district and state support, and principal leadership*. Retrieved from: http://www.wallacefoundation.org/knowledge-center/school-leadership/district-policy-and-practice/Documents/Three-Essentials-to-Improving-Schools.pdf

Spillane, J. P. (2000). Cognition and policy implementation: District policy makers and the reform of mathematics education. *Cognition and Instruction, 18*(2), 141–179.

Spillane, J. P. (2002). Local theories of teacher change: The pedagogy of district policies and programs. *Teacher College Record, 104*(3), 377–420.

Spillane, J. P., & Callahan, K. C. (2000). Implementing state standards for science education: What district policy makers make of the Hoopla. *Journal of Research in Science Teaching, 37*(5), 401–425.

Spillane, J. P., & Thompson, C. L. (1997). Reconstructing conceptions of local capacity: The local education agency's capacity for ambitious instructional reform. *Educational Evaluation and Policy Analysis, 19*(2), 185–203.

Stein, M. K., & D'Amico, L. (2002). The District as a Professional Learning Laboratory. In A. M. Hightower, M. S. Knapp, J. A. Marsh, & M. W. McLaughlin (Eds.), *School districts and instructional renewal* (pp. 61–75). New York: Teacher College, Columbia University.

Tyler, R. (2007). *Re-imaging science education: engaging students in science for Australia's future*. Australian Education Review. Retrieved from: http://www.acer.edu.au/documents/AER51_ReimaginingScieEduc.pdf

U. S. Department of Education. (2010). *A blueprint for reform: The reauthorization of the Elementary and Secondary Education Act*. Washington, DC: Author. Retrieved from http://www2.ed.gov/policy/elsec/leg/blueprint/blueprint.pdf

United States Department of Education. (2011). *Our future, our teachers: The Obama administration's plan for teacher education reform and improvement*. Retrieved from: http://www.ed.gov/sites/default/files/our-future-our-teachers.pdf

Villegas, A. M., & Lucas, T. (2002). Preparing culturally responsive teachers: Rethinking the curriculum. *Journal of Teacher Education, 53*(1), 20–32.

Walberg, H. J., & Fowler, J. T., Jr. (1987). Expenditure and size efficiencies of public school districts. *American Educational Research Association, 16*(7), 5–13.

Wheeler, G. F. (2007). Strategies for science education reform. *Educational Leadership, 64*(4), 30–34.

Youngs, P. (2001). District and state policy influences on professional development and school capacity. *Educational Policy, 15*(2), 278–301.

Policy Issues in Science Education: The Importance of Science Teacher Education, Equity, and Social Justice

Sheneka M. Williams and Mary M. Atwater

As the number of students of color grows significantly in the nation's schools, policymakers, administrators, and teachers alike must work in tandem to ensure that all students receive equitable learning opportunities. As such, science teachers and science teacher educators face increasing pressure to bridge the gap between their pedagogical content knowledge and students' learning outcomes. In the chapter "Equity and Diversity in Science and Engineering Education," the National Academy of Science (2011) provides two reasons for the differences among specific groups of students in their educational performance and patterns of science learning. One reason provided by the Academy includes inequities across schools, districts, and communities, and differences. This also includes differences in opportunities related to curricular and instructional materials and assessment/evaluation. Additionally, the Academy lists elementary science preparation, literacy, and mathematics understandings as pressing challenges to students' performance in science. While the onus for low student performance is often placed on teacher effectiveness, educators and policymakers should also consider how curriculum and policy decisions impact student learning and student outcomes. In the case of science teacher education programs, an emphasis on multicultural course offerings might provide teachers with a better understanding of students. For example, courses that delve into students' cultural and social capital should be foundational in teacher preparation courses. In turn, this might encourage students to exhibit a better

---

S.M. Williams (✉)
Department of Lifelong Education, Administration, and Policy,
College of Education, University of Georgia, 850 College Station Road,
324 River's Crossing, Athens, GA 30602, USA
e-mail: smwill@uga.edu

M.M. Atwater
Department of Mathematics and Science Education, College of Education,
University of Georgia, 376 Aderhold Hall, Athens, GA 30602-7126, USA
e-mail: matwaterchemi@bellsouth.net

appreciation for the subject matter. We posit that teacher preparation programs should encourage teachers to understand and value student differences and respond to those differences in their teaching styles. To be clear, we do not advocate teaching from a deficit perspective; however, we suggest that teachers should value the diverse perspectives and knowledge that all students bring to classrooms (Milner, 2010). Furthermore, we suggest that policy initiatives that seek to increase performance for low-income rural and urban students consider a framework that speaks to the differences that students bring to classrooms. Thus, the purpose of this chapter is to review science teacher education policy in conjunction with standards to which teachers teach. Moreover, this chapter sets forth a new policy agenda to improve science teacher practices and science performance among low-income rural and urban students of color.

Teaching Strategies for Meeting the Needs of Today's Students for Tomorrow's Future

Before prospective teachers enter a teacher preservice program, they come to the program with their own epistemologies or ways of seeing the world. For some teachers, this lens does not include low-income rural and urban students of color excelling in STEM subjects (Bryan & Atwater, 2002). This deficit way of depicting students' interests in science trickles down to the way some teachers teach. Jones and Carter (2007) suggest that teachers' epistemological beliefs tend to be relatively stable and resistant to change. Thus, many teachers rarely depict low-income rural and urban students or students from traditionally underrepresented groups (i.e., African American, Latino/a) as future scientists. While this culture is endemic in many of today's science education classrooms, it is also reified in the larger school community. As such, teachers' contexts often influence their practices. Given that, then, how do science teacher education programs break the stereotypic cycle that some teachers bring to the science teacher education classroom? Social justice researchers propose that the major goal of research is to develop action agenda to address the lives of marginalized, oppressed groups (Atwater, 1996; Barton, Ermer, Burkett, & Osborne, 2003; Cochrane-Smith & Fries, 2011; Darling-Hammond, 2006; Rodriguez, 1998; Seiler & Elmesky, 2005). More specifically, social justice researchers propose the following: (a) include the history of science in the curricula to demonstrate that science is a human endeavor and aids students in understanding that social and political powers are tied to science; (b) teach the history of science so that students can understand the many contributions of other cultures to science; and (c) teach the history of science so that students can learn about their cultural heritage and provide them role models for the their future endeavors. As a result, a social justice approach to teaching science education might provide students with tools and concepts to better understand their role in producing science knowledge.

Loughran (2007) suggests that science teachers, along with science teacher educators, traditionally utilize transmission approaches to teaching. Subsequently, the transmission approach to teaching proposes that telling students will promote science learning as opposed to engaging students in their own learning (Tishman, Jay, & Perkins, 1993). Students learn science by *doing* science. Therefore, in order for teachers to engage learners, teachers should relate scientific concepts and processes to students' background and heritage. This, we believe, will help students to view science as a more attainable subject. This approach may also lead to better academic performance gains among low-income students and underrepresented students of color in science education (Julyan & Duckworth, 1996; Parsons, 2003).

O'loughlin (1992) maintains constructivist teaching is fallacious because of its inability to come to terms with the essential issues of culture, power, and discourse in the classroom. He argues that a sociocultural approach to teaching and learning takes seriously the notion that learning is situated in the following contexts: (a) students bring their own subjectivities and cultural perspectives to bear in constructing understanding; (b) issues of power exist in the classroom that need to be addressed; and (c) education into scientific ways of knowing requires students to understand modes of classroom discourse. If students understand classroom discourses, then they will be able to negotiate these modes effectively. This will allow students to master and critique scientific ways of knowing without sacrificing their own personally and culturally constructed ways of knowing.

A Review of Science Teacher Education Policy and Standards

Science teacher education policy implemented during the 1960s and 1970s emphasized teacher competency and science mastery learning (Yager, 2000). During the 1980s, science teachers were viewed as "knowers"; therefore, teachers' practical knowledge dominated science teacher education literature (Abell, 2007). Teachers' pedagogical content knowledge (PCK), along with science content knowledge, was of interest to science teacher educators. During that time, science teacher educators focused on the following areas: (a) teachers' knowledge of goals for and general approaches to science teaching; (b) teachers' knowledge about the science curricula, including national, state, and district standards and specific science curricula; (c) teachers' knowledge about assessment of students; (d) teachers' knowledge about science instructional strategies, including representations, activities, and methods; and (e) teachers' knowledge of student science understanding (Abell). Although these standards were intended for teachers to use with all students, they were not designed for low-income rural and urban students. In an effort to ensure that low-income rural and urban students were taught science from the same standards, Shulman (1986) developed teacher knowledge bases that included the following: (a) content knowledge; (b) general pedagogical knowledge, with special reference to the broad principles and strategies of classroom management and organization that appear to transcend subject matter; (c) curriculum knowledge with particular grasp of the materials and programs that serve as "tools of the trade" for teachers; (d) pedagogical content knowledge (a special amalgam of content and pedagogy); (e) knowledge of learners and their characteristics; (f) knowledge of educational contexts, ranging from the workings of groups or classrooms, the governance and financing of school districts, and the character of communities and cultures; and (g) knowledge of educational ends, purposes, and values and their philosophical and historical grounds. These standards allowed teachers to reach inside the lived experiences of students, thus extending science education beyond "generic" learners. As a result, Shulman's model (1986) for understanding teacher knowledge became of great interest in science education.

During the 1990s *The National Science Education Standards* (1996) created standards related to science teaching, assessment in science education, science content, and science education programs. In the assessment standard, Standard D states, "Assessment practices must be fair" (p. 85). However, this standard focuses on bias and includes "Assessment tasks … must not assume the perspective or experience of a particular gender, racial, or ethnic group" (p. 85). This standard, which is developed on a color-blind ideology (Bonilla-Silva, 2006), poses problems for students of color and women in science interested in science education. In a traditional sense, assessment items are based upon a White dominant paradigm, and it is assumed that all races and genders of students should understand concepts through a White male epistemological lens (Linn & Harnish, 1981). This lens negates equity or social justice as it relates to the preparation of science teachers. As Milner and Williams (2008) note, "standardized" policies that do not take into account the multiple layers of needs and issues in particular contexts often result in inequities and inequalities that are difficult to control.

The turn of the twenty-first century brought with it reform of the 1965 *Elementary and Secondary Education Act*. This reform, known as the *No Child Left Behind Act* (NCLB) of 2001 called for all students to be proficient in all subjects by the year 2014. This federal education policy required disaggregated data of student subgroups. These data indicate that students from underrepresented groups (i.e., ELL, African American, and students with disabilities) lag behind their White counterparts in most subject areas, but particularly in mathematics and science. The exposure of such data reveal that one possible cause of the differences among student performance is the widened gulf between teacher subject matter knowledge to other forms of teacher knowledge, teacher beliefs and values, and classroom practice (Ferguson, 2003; Gess-Newsome, 1999; Norman, Ault, Bentz, & Meskimen, 2001; Parsons, 2005). It also suggests teacher classroom practices and students' cultural backgrounds are disconnected. As federal policymakers prepare to reauthorize and make legitimate changes to NCLB, science teacher educators must continue to ensure that required objectives and goals have a multicultural component as a means of meeting the needs of all students.

If science teachers envision science teaching as aligning with the national standards, then it is imperative that the standards include issues related to equity and social justice in the learning and teaching of science and the assessing and evaluating of students' science knowledge and skills. Currently, the nation's proposed common core of standards does not include a standard that focuses specifically on equity and social justice. Instead, the common core includes a standard that aims to meet equity and social justice objectives. The standard, *Connections in Teaching Science*, provides learning objectives for students that include the following: (a) the examination of science applications in their personal lives and interests and in the examination of local issues and (b) relating knowledge of other disciplines, particularly mathematics and social sciences, to concepts of science in applications to their personal lives. While these objectives provide students with an opportunity to apply science knowledge to their daily-lived experiences, it does not allow teachers to highlight such experiences as a teaching focus. For instance, the knowledge objective for teachers includes understanding how students can identify and utilize science concepts in their daily lives. In order to improve science performance for low-income students of color in rural and urban schools, then teachers must be committed to being change agents in the profession.

Unlike the common core of standards, the National Board of Professional Standards of Teaching includes standards related to equity and social justice. For the adolescent and young adult (high school), the standards are as follows: (a) VI—Promoting Diversity, Equity, and Fairness-Accomplished Adolescence and Young Adulthood/Science teachers ensure that all students, including those from groups that have historically not been encouraged to enter the world of science and that experience ongoing barriers; (b) XII—Connecting with Families and the Community-Accomplished Adolescence and Young Adulthood/Science teachers proactively work with families and communities to serve the best interests of each student; and (c) XI—Family and Community Outreach-Accomplished/Science teachers proactively work with families and communities to serve the interests of students. While these standards exist, most schoolteachers do not adhere to the standards because many do not seek National Board Certification. Again, this suggests a disconnect between the world of policymakers and practitioners in terms of teaching students through an equitable framework.

The National Council for Accreditation of Teacher Education (NCATE) does not include standards that are specifically related to teachers and social justice. Instead, NCATE standards are based on a belief that caring, competent, and qualified teachers should teach every student. Given that, NCATE standards indirectly prepare teachers for a diverse community of students. Within NCATE, the National Science Teacher Association (2003) designed standards for science teacher education programs. These standards call for candidates to show how they take into account student differences in their planning and teaching. However, even within these standards, there are no standards specifically for science teacher educators.

Although other national organizations struggle with including standards that address equity and social justice, The Association for the Education of Teachers in Science (AETS) in 1997 clearly defined a framework for the knowledge, skills, experiences, attitudes, and habits of mind essential for highly qualified science teacher educators at the beginning of their professional careers. These standards were established to guide the development and revision of graduate-level programs that prepare science teacher educators, criteria for the qualifications of a university-level science educator, and guidelines for the qualifications of individuals. Those who could be science teacher educators were higher education faculty members who have coursework in the science subject matter and/or science pedagogy, school-based mentor teachers, school personnel who conducted professional development activities, and other agency personnel who provided professional development to science teachers. The standards are intended for early career science teacher educators since AETS believed that a lifetime effort is required to develop into an excellent science teacher educator. Given that, the standards focused on (a) the knowledge of science; (b) the knowledge of science pedagogy; (c) the theoretical and practical background in curriculum development, instructional design, and assessment; (d) the knowledge of learning and cognition; (e) the knowledge and skills for research/scholarly activity; and (f) the knowledge, habits of mind, and skills necessary to work with prospective and practicing science teachers as they move through a developmental process. Even though these standards were developed 15 years ago (Lederman et al., 1997), it is still very surprising that it was not one standard related to science teacher educators' knowledge and skills to prepare science teachers to teach students of color in urban and rural settings. In addition, there was no mention of equity or social justice. It was not until 2004 after the Association for the Education of Science Teachers (AETS) changed its name to Association of Science Teacher Education (ASTE) that its *Position Statement for Science Teacher Preparation and Career-long Development* called for science teacher education programs to engage prospective teachers in substantive clinical experiences where they develop and implement lesson plans appropriate for students from diverse backgrounds, assess their success on student learning, and plan next steps to improve their teaching.

Recently, The Carnegie Corporation of New York, along with the Institute of Advanced Study, took a bold move by calling for the creation of a common set of K-12 standards in science. In order to accomplish this task, the Carnegie Corporation initiated a two-step process by first developing a framework and then developing a set of science standards for the twenty-first century. *A Framework for K-12 Science Education: Practices, Crosscutting Concepts, and Core ideas* has been published (National Research Council, 2012) and is built upon the *Science for All Americans* and *Benchmarks for Science Literacy* (1993) and the *National Science Education Standards* (1996). As the presidents of National Academy of Sciences and National Academy of Engineering (National Research Council, 2012) assert, "The frameworks highlight the power of integrating understanding ideas of science with engagement in the practices of science and is designed to build students' proficiency and appreciation of science over multiple years of school" (p. x). There are several goals for the frameworks, but the most pertinent goals for this chapter include (a) "all students are careful consumers of scientific and technological information related to their everyday lives" (p. 1) and (b) "to guide the development of a new standards that in turn guide the revisions to science-related curriculum, instruction, assessment, and professional development for educators" (National Academy of Science, 2011, p. 2). Since US schools serve students from a variety of cultural backgrounds, one would assume that this document would discuss cultural issues and equity in some detail. However, we find that there are basically two sections that address cultural issues and equity in this framework. The sections are summarized below:

- Most students can engage in and learn complex subject matter, such as science and engineering, when they connect to their personal interests and consequences.
- Many students lack essential material resources and instructional support for exemplary science instruction due to their socioeconomic class, race, ethnicity, gender, language, disability designation, or nationality.
- Many students are at risk for academic failure in elementary schools in certain geographic locations.
- If science is viewed as a culturally mediated way of thinking and knowing about natural phenomena, then students and teachers do not leave their cultural world views at the classroom door.
- Many traditional science classroom practices are ineffective with certain students whose family discourse practices differ from those found in schools.
- The ways that science learning is evaluated are problematic due to language issues, students' beliefs and attitudes toward certain kinds of tests, and test bias.

We, the authors of this chapter, find it problematic that in these sections that little to no research conducted by African American, Latino/a, or Native American science education researchers guides the framework. It is as the research findings of only European American education researchers seem to matter (Scheurich & Young, 1997).

With 13 recommendations for providing guidance to future standard developers, one focuses on diversity and equity—"In designing standards and performance expectations, issues related to diversity and equity need to be taken into account. In particular, performance expectations should provide students with multiple ways of demonstrating competence in science" (National Research Council, 2012, p. 307). The problem is that equity is an issue that should be infused in each of the standards since student learning is pivotal in this discussion of frameworks and science is a human endeavor. But the most problematic proclivity of this group is the terminology used to characterize people in such a way that their commitment to equity can be questioned. For instance, the term "African American" is hyphenated and the term "minority" is used. These writing practices go against the sixth edition of the *Publication Manual* of the American Psychological Association in its Reducing Bias section (2010). Thus, it makes us raise the question: How committed is the National Research Council to developing standards so that students from different racial, ethnic, language, ability, socioeconomic backgrounds will truly experience high-quality science learning and teaching?

As we think about answers to the above question, we suggest that the policy agenda thus far has not made an honest commitment to including low-income rural and urban students of color in science education curricula. That calls for a new policy agenda to be established that includes students from these groups.

Setting a New Science Teacher Education Policy Agenda

Based on the history of science teacher education policy and standards, and the current political pulse, science teacher educators and policymakers must work cooperatively to infuse multiculturalism with a focus on equity and social justice into the current science education policy agenda. This must be done by redefining the policy problem in science teacher education. We assert that one part of the problem includes low student performance among low-income rural and urban students and underrepresented students of color. Policymakers, on the other hand, do not understand the other part of the problem: lack of students' culture represented in standards and objectives. Before any policy problem can be a part of the agenda for change, then policymakers must see it as a problem, and "a problem is a problem only if something can be done about it" (Wildavsky, 1979, p. 4). Once policymakers realize the connection between student performance and teachers' understanding of student culture and background, then it will be accepted as part of the education reform agenda in science education. Additionally, "the more people affected by a problem, the more likely the item will receive priority on the legislative agenda, particularly if the effects are concentrated and serious, or extreme" (Cooper, Fusarelli, & Randall, 2004, p. 66). Once education policymakers connect science teacher education and student performance to the global economy and sustainability of this nation, then they will find it easier to make a case for including standards and objectives related to student culture.

After science teacher educators and policymakers manage to get students' culture on the education policy agenda for change in science education, then the policy implementation phase begins (Thompson, Wilder, & Atwater, 2001). Implementation is what happens when a policy is (or is not) carried out (Sabatier & Mazmanian, 1981), and it is in the implementation phase when most policy agenda items go awry. Oftentimes those who develop policy are loosely connected to those who implement policy. This, in turn, creates a divide that often results in poor implementation. In the case of science teacher education, it often results in mediocre classroom practice. Lackluster practice is not necessarily a characteristic of ineffective teachers, but it is often related to teachers not understanding what is being asked of them. This results from teachers not been asked to play a major role in the policy formulation phase. Thus, it is imperative that science teacher educators have a voice in setting the policy agenda as it relates to science teacher education and science education. Rarely, if ever, are teachers asked to partake in the policymaking process. However, they are expected to act as "street-level bureaucrats" and implement policies with fidelity. This, in many instances, creates a breakdown in the intended consequences of policy implementation.

Based on a review of teaching strategies and policies related to science teacher education, we propose an equity and social justice framework for science teacher education that includes the following elements: (a) equity in the development of science teacher education policy, (b) curriculum framework that encourages culturally relevant and culturally responsive teaching, and (c) equity in learning opportunities for marginalized students. If science teacher educators have a larger voice in the development of policy that impacts students from various backgrounds, then policy will be more inclusive and more equitable. For example, science teacher educators from Alabama should have a seat at the table as well those from Massachusetts. In that way, the lived experiences and realities of students will be represented in the development of science teacher education policy.

In addition to more equitable development in science teacher education policy, we suggest that curriculum development should center around culturally relevant and culturally responsive teaching. This, we believe, must be bolstered by curriculum in science teacher education programs and district professional learning opportunities.

Once policies and curriculum are developed to be more inclusive (Atwater & Suriel, 2010), then we also suggest the equitable learning opportunities are afforded to students. This exists beyond the local level. It also includes more access and representation in internships and fellowships at nationally and internationally acclaimed think tanks, foundations, and universities.

As we have identified a science teacher educator policy and practice framework for equity and social justice, we reiterate the role of science teacher educators and science teachers as change agents in the process. Since most US students do not perform well on international science tests (Fleishman, Hopstock, Pelczar, & Shelley, 2010), then few students will become scientifically literate. This is the case even though inclusive science instruction, science learning as a cultural accomplishment, scientific discourse, students' prior interest and identity, students' cultural funds of knowledge, making diversity visible, and multiple modes of expression are advocated. Hence, science teacher educators are expected to prepare teachers to be at least competent in (a) inclusive science instruction, which includes using students' informal or native language and familiar modes of interactions, building on students' prior interests and science identities, and leveraging students cultural funds of knowledge, (b) understanding that science learning is a cultural accomplishment, and (c) valuing multiple modes of expression, especially in terms of assessment/evaluation. If science teacher educators prepare science teachers in this way, then we will see long-term gains in science performance.

References

Abell, S. K. (2007). Research on science teacher knowledge. In S. K. Abell & N. G. Lederman (Eds.), *Handbook of research on science education* (pp. 1105–1066). Mahwah, NJ: Lawrence Erlbaum Associates.
American Association for the Advancement of Science. (1993). *Benchmarks for science literacy. Project 2061*. New York: Oxford University Press.
Association for Science Teacher Education. (2004). *Position statement science teacher preparation and career-long development*. Retrieved September 05, 2011 from http://theaste.org/aboutus/AETSPosnStatemt1.htm

Association for the Education of Teachers in Science. (1997). *Position statement: Professional knowledge standards for science teacher educators.* Retrieved September 05, 2011 from http://www.umd.umich.edu/casl/natsci/faculty/zitzewitz/curie/TeacherPrep/178.pdf

Atwater, M. M. (1996). Social constructivism: Infusion into multicultural science education research. *Journal of Research in Science Teaching, 33,* 821–838.

Atwater, M. M., & Suriel, R. L. (2010). Science curricular materials through the lens of social justice: Research findings. In T. K. Chapman & N. Hobbel (Eds.), *Social justice pedagogy across the curriculum: The practice of freedom* (pp. 273–282). New York: Routledge.

Barton, A. C., Ermer, J. L., Burkett, T. A., & Osborne, M. D. (2003). *Teaching for social justice.* New York: Teachers College.

Bonilla-Silva, E. (2006). *Racism with racists: Color-blind racism and the persistence of racial inequality in the United States.* Lanham, MD: Rowman & Littlefield.

Bryan, L. A., & Atwater, M. M. (2002). Teacher beliefs and cultural models: A challenge for science teacher preparation programs. *Science Education, 86*(6), 821–839.

Cochran-Smith, M., & Fries, K. (2011). Teacher Education for Diversity: Policy and politics. In A. F. Ball & C.A. Tyson (Eds.), *Studying diversity in teacher education* (339–362). Lanham, MD: Rowman & Littlefield Publishers.

Cooper, B. S., Fusarelli, L. D., & Randall, E. V. (2004). *Better schools, better policies.* Boston: Pearson Education.

Darling-Hammond, L. (2006). *Powerful teacher education: Lessons from exemplary programs.* San Francisco: Jossey-Bass.

Ferguson, R. (2003). Teachers' perceptions and expectations and the Black-White test score gap. *Urban Education, 38*(4), 460–507.

Fleishman, H. L., Hopstock, P. J., Pelczar, M. P., & Shelley, B. E. (2010). *Highlights form PISA 2009: Performance of U.S. 15-year-old students in reading, mathematics, and science literacy in an international context.* Washington DC: National Center for Education Statistics, Institute of Education Sciences, U. S. Department of Education. Retrieved from: http://www.eric.ed.gov/PDFS/ED513640.pdf.

Gess- Newsome, J. (1999). Secondary teachers' knowledge and beliefs about subject matter and their impact on instruction. In J. Gess-Newsome & N. G. Lederman (Eds.), *Examining pedagogical content knowledge: The construct and its implications for science education* (pp. 51–94). New York: Springer.

Jones, M. G., & Carter, G. (2007). Science teacher attitudes and beliefs. In S. K. Abell & N. G. Lederman (Eds.), *Handbook of research on science education* (pp. 1067–1104). Mahwah, NJ: Lawrence Erlbaum Associates.

Julyan, C., & Duckworth, E. (1996). A constructivist perspective on teaching and learning science. In C. T. Fosnot (Ed.), *Constructivism: Theory, perspectives and practice* (pp. 55–72). New York: Teachers College.

Lederman, N. G., Ramey-Gassert, L., Kuerbis, P., Loving, C., Roychoudhuray, A., & Spector, B. S. (1997). Professional knowledge standards for science teacher educators. *AETS Newsletter, 31*(3), 5–10.

Linn, R. L., & Harnish, D. (1981). Interactions between item content and group membership on achievement test items. *Journal of Educational Measurement, 18*(2), 109–118.

Loughran, J. J. (2007). Science teacher as learner. In S. K. Abell & N. G. Lederman (Eds.), *Handbook of research on science education* (pp. 1043–1066). Mahwah, NJ: Lawrence Erlbaum Associates.

Milner, H. R. (2010). *Start where you are but don't stay there. Understanding diversity, opportunity gaps, and teaching in today's classrooms.* Boston, MA: Harvard Education Press.

Milner, H. R., & Williams, S. M. (2008). Analyzing education policy and reform with attention to race and socio-economic status. *Journal of Public Management and Social Policy, 14*(2), 33–50.

National Academy of Science. (2011). *A framework for K-12 science education: Practices, cross-cutting concepts, and core ideas.* Retrieved September 6, 2011 from http://www.nap.edu/catalog.php?recored_id=1365

National Research Council. (1996). *National science education standards* (National Committee for Science Education Standards and Assessment). Washington, DC: National Academy Press.

National Research Council. (2012). *A framework for K-12 science education: Practices, crosscutting concepts, and core ideas*. Washington, DC: National Academies Press.

National Science Teacher Education. (2003). *Standards for science teacher educ*ation. Retrieved September 05, 2011 from http://www.nsta.org/pdfs/NSTAstandards2

Norman, O., Ault, C. R., Bentz, B., & Meskimen, L. (2001). The Black-White "achievement gap" as a perennial challenge of urban science education: A sociocultural and historical overview with implications for research and practice. *Journal of Research in Science Teaching, 38*(10), 1101–1114.

O'loughlin, M. (1992). Rethinking science education: Beyond Piagetian constructivism toward a sociocultural model of teaching and learning. *Journal of Research in Science Teaching, 29*(8), 791–820.

Parsons, E. C. (2003). Culturalizing instruction: Creating a more inclusive context for learning for African American students. *The High School Journal, 86*(4), 25–30.

Parsons, E. C. (2005). From caring as a relation to culturally relevant caring: A White Teacher's bridge to Black students. *Equity and Excellence in Education, 38*(1), 25–34.

Rodriguez, A. (1998). Strategies for counterresistance: Toward sociotransformative constructivism and learning to teach science for diversity and for understanding. *Journal of Research in Science Teaching, 35*(6), 589–622.

Sabatier, P. A., & Mazmanian, D. A. (1981). *Effective policy implementation*. Lexington, MA: Lexington Books.

Scheurich, J. J., & Young, M. D. (1997). Coloring epistemologies: Are our research epistemologies racially biased? *Educational Researcher, 26*(4), 4–16.

Seiler, G., & Elmesky, R. (2005). The who, what, where, and how of our urban ethnographic researcher. In K. Tobin, R. Elmesky, & G. Seiler (Eds.), *Improving urban science education: New roles for teachers, students, & researchers* (pp. 1–20). Lanham, MD: Rowman & Littlefield Publishers.

Shulman, L. S. (1986). Those who understand: Knowledge growth in teaching. *Educational Researcher, 15*(2), 4–14.

Thompson, N., Wilder, M., & Atwater, M. M. (2001). Critical multiculturalism and secondary teacher education programs. In D. Lavoie (Ed.), *Models for science teacher preparation: Bridging the gap between research and practice* (pp. 195–211). New York: Kluwer.

Tishman, S., Jay, E., & Perkins, D. N. (1993). Teaching thinking dispositions: From transmission to enculturation. *Theory into Practice, 32*(3), 147–153.

VandenBos, G. R., Appelbaum, M., Comas-Diaz, L., Cooper, H., Light, L., Ornstein, P., et al. (2010). *Publication manual of the American psychological association*. Washington, DC: American Psychological Association.

Wildavsky, A. (1979). *Speaking truth to power. The art and craft of policy analysis*. Boston: Little, Brown.

Yager, R. E. (2000). A vision of what science education should be like for the first 25 years of a new millennium. *School Science and Mathematics, 100*(6), 327–341.

Conclusion and Next Steps for Science Teacher Educators

Melody L. Russell, Malcolm B. Butler, and Mary M. Atwater

If there is no struggle there is no progress. Those who profess to favor freedom, and yet depreciate agitation, are men who want crops without plowing up the ground. They want rain without thunder and lightning. They want the ocean without the awful roar of its many waters. This struggle may be a moral one; or it may be a physical one; or it may be both moral and physical; but it must be a struggle. Power concedes nothing without demand. It never did and it never will (Douglass, 1849/1991).

What Do We Know: Central Ideas

The chapter authors of this coedited book provide valuable insight into the "struggle" every day to challenge the status quo and hegemonic policies in place that are barriers to school success for many students of color in the school systems across the United States. Many of the barriers that science teacher educators and advocates for equity and social justice in science teaching encounter are institutional and promote a "hidden curriculum" that marginalizes a large number of students from culturally

---

M.L. Russell, Ph.D. (✉)
Department of Curriculum and Teaching, College of Education, Auburn University, 5004 Haley Center, Auburn, AL 36849, USA
e-mail: russeml@auburn.edu

M.B. Butler
School of Teaching, Learning and Leadership, College of Education, University of Central Florida, PO Box 161250, Orlando, FL 32816-1250, USA
e-mail: Malcolm.Butler@ucf.edu

M.M. Atwater
Department of Mathematics & Science Education, College of Education, University of Georgia, 376 Aderhold Hall, Athens, GA 30602-7126, USA

and linguistically diverse backgrounds with the potential to experience success in science on both the secondary level and beyond. Frederick Douglass spoke prophetically in the aforementioned quote when he discusses that we will not progress without struggle.

One major impetus for this book was the need for a resource for science teacher educators that was both relevant and applicable to diverse classroom settings on the secondary level that many teachers experience daily. Moreover, it is no secret that oftentimes preservice teachers leave their teacher education programs ill prepared for the classroom since many of the preservice teachers come from backgrounds very different than the demographics of the schools they teach. It is essential that science teacher educators better prepare students with the accoutrement that will allow them more opportunities to understand their students and provide learning opportunities that help them realize their full potential, regardless of their background. This being said there are many questions and concerns that arise daily in the preservice classrooms, and until these questions are answered teachers will continue to go into the schools without the necessary understanding of pedagogy that will help students from underrepresented and traditionally marginalized groups realize their own potential.

The coeditors hope to provide a platform for not only more open dialogue and discourse on the critical areas of equity and social justice but transformation of curricula for preservice teachers towards facilitating their understanding of how culture, language, class, and so many other factors discussed in this book impact student outcomes in the science classroom. The authors address both rural and urban settings in their writing as poverty permeates throughout the essence of both of these environments serving as a way of further marginalizing students of color in the science classroom.

The chapters are written not only based on the authors own experiences as teacher educators or science educators but based on research in best practices in teaching for equity and social justice. The authors also provide a framework for science teacher educators to better prepare their preservice students with the necessary accoutrement to challenge the status quo and foster an environment that promotes equity and social justice for all of their science students.

More specifically, the contributing authors have clearly delineated practices and policies for supporting science teacher educators as these pedagogues are charged with the challenge of preparing science teachers to enhance and diversify the science outcomes for the twenty-first-century classrooms. Each contributing author also emphasized the critical role that equity, social justice, and multicultural education have on the teaching and meaningful learning and understanding of science for secondary students. Chapter authors for this book were asked to write from a lens relative to culture, equity, and social justice in the preparation of science teachers, as well as to highlight the need for integrating equity and social justice in science teacher education programs. Even though each chapter is not about a specific research study, contributing authors supported their ideas based on sound research on equity and multiculturalism in science education.

This being said, the chapter authors provided numerous strategies from their own research, expertise, or personal experiences, promoting equity in the science classroom, and provided a framework to help teachers become change agents in their schools and districts. Moreover, the first step towards promoting equity is in understanding and acknowledging that equity for all is seldom realized in the science classroom. In going a step further than the preserves of classrooms and teacher education, science education professional developers should find this book useful for their professional development activities and seminars involving beginning and veteran science teachers. Many issues that teachers face today revolve around "cultural dissonance," and as a result many questions arise, leading to much disequilibrium relative to discussions with teachers during these professional learning events. Hopefully, within these pages some assistance and support is provided to the facilitator.

By focusing on the central ideas and themes in the chapters, we discuss "what we know" and strategies for moving forward as science teacher educators in the twenty-first century. One key goal for this book is to hopefully transform the learning and teaching experiences of middle and high school science teachers through the framework that is delineated in the chapters of this book. The next few sections of this chapter explicate the fundamental positions that contributing authors have taken about science teacher education.

The Role of History, Culture, and Language in the Preparation of Science Teachers as Change Agents

It is paramount for science teacher educators to explicate the historical legacy of many students in today's classrooms and how this has impacted their participation in the STEM pipeline. Furthermore, it is essential that both preservice and science teachers understand how culture, language, and class play a critical role in determining "who will do science." Unfortunately there are a lot of students who do not see themselves as scientists, users of science, or decision makers of science policy practices. This book offers a framework for teachers to implement in order to better emphasize how equity, social justice, and multicultural education are all essential towards promoting the success of students in science and improving student outcomes.

Culture has been defined in many ways; Bullivant (1993) characterizes culture as a group's program for its continued existence and adaptations to its surroundings. Banks (2007) further delineates culture as "shared beliefs, symbols, and interpretations within a human group [and maintains] the essence of culture is not its artifacts, tools, or other tangible cultural elements but how the members of the group interpret, use, and perceive them" (p. 8). There are many different US microcultures defined by race, class, gender, and geographic location, and there is no one "*American culture*" as discussed by contributing authors Green, Butler, Walls, and Brand. Green maintains that sociohistorical events have influenced the development, self-identity, and self-esteem of African Americans so that many cannot envision themselves to be a "part of something that has continuously tried to disenfranchise them" – science. Butler continues the dialogue by sharing brief biological sketches of Black scientists during the Jim Crow era in the United States to assist science teacher educators to use these strategies to convince that African American students can be a part of the world of science. Walls maintains that another group – females of color – are part of a microcultural group that find themselves hampered by both race *and* gender and describes what an effective science classroom for females of color would look like. Finally, Brenda Brand suggests that "hidden" messages of inferiority force students from a microculture to overcome the social constraints imposed upon their identities in order to understand their value and their potential to achieve and make a contribution in science. These authors maintain that certain aspects of history and culture should be central themes in any science teacher education program but especially ones that are preparing teachers to teach in the twenty-first century.

Science teacher educators must also address in their teaching the critical role that language plays in the learning of science by students from culturally and linguistically diverse backgrounds. Halliday (1993) maintains that "Language is the essential condition of knowing, the process by which experience become knowledge" (p. 94). Lemke (1990) connects language to science by stating "Learning science means learning to talk science" (p. 1). If these two ideas are correct, then science teacher educators must assist their prospective science teachers and practicing science teachers to teach their students to talk science. Schooling then can be seen as a linguistic process, and language is often used to evaluate and differentiate students, hence setting up conditions for inequity, discrimination, and oppression.

While researchers like Fairclough (1989) and others explain the central role of language in teaching, most science teacher education programs either marginalize the role of language or do not include it at all but focus on content and pedagogical knowledge. Suriel points out in her chapter that learning, learning language, and learning through language are simultaneous processes. Suriel goes on to give examples of the many challenges that pose barriers to the success of Latino/a students in science classrooms, as well as strategies through scenarios for promoting equity for Latino/a students.

The Science Teacher Educators: Preparing Culturally Competent Science Teachers

Science teacher educators need to add to their own personal, professional, and experiential knowledge and skills for their own pedagogical problem solving and meeting the educational needs of prospective and working science teachers. Science teachers of students with diverse cultural and linguistic backgrounds must make their own appropriate pedagogical adjustments to meet the needs of their students each day. With this in mind contributing author, Russell, highlights the importance of motivation in the science classroom and shares effective strategies for enhancing student motivation and achievement in the STEM areas. Then Cone, Rascoe, Norman, Pringle and McLaughlin, Johnson and Atwater, and Hutchison focused on the role of curriculum and pedagogy in better preparing preservice science teachers for the science classrooms of today. One unique aspect of this book was how some of the authors (Norman and Hutchinson) examined science teaching through an international lens. More specifically, the underlying theme that was a common thread for their chapters addressed culturally relevant teaching, culturally competent teachers, and strategies for fostering this type of pedagogy in preservice programs.

It takes time to change any system, especially educational systems that are complex and entrenched. Kahle (2007) calls for science educators to address excellence and equity. Moreover, there are numerous policies that science teachers will encounter that impede their progress towards equity and social justice in their science teaching. It is critical that they recognize the policies and learn how to navigate in school systems to provide opportunities that encourage the participation of students from traditionally underrepresented and marginalized groups in the STEM pipeline. Hence, Bantwini, Pea, and Williams and Atwater further extend this perspective by closely examining the role that policy has on framing science teacher education and the US school systems, particularly.

Theoretical Frameworks in Science Teacher Education

Science teacher education programs, whether intentionally or not, will continue to prepare science teachers who lack cultural knowledge and practical applications of pedagogical knowledge to effectively teach through a lens of equity and social justice in science classrooms. In order to begin promoting change and transforming the teachers of today into change agents, theoretical frameworks must be used that provide a lens that challenges the inequities in science teaching that exist for many students from underserved and underrepresented groups. Several of the authors which include Butler, Suriel, Cone, and Green wrote the chapters through frameworks that challenge the culture of low expectations that exist in many of today's schools relative to traditionally underrepresented groups in the sciences. These authors' chapters were underscored by various theoretical frameworks that will hopefully facilitate the reader in interpretation and application of the chapters in their own classrooms. More specifically, the following theoretical frameworks were key to this book: a) sociocultural theory (Suriel), critical race theory (Butler), banking" ideology (Cone), and social Darwinism (Green). It must be understood that Green does not advocate for social Darwinism because he believes that the ideas in social Darwinism support the reality of institutional structures that already exist in the US society.

Where Do We Go From Here? Navigating the Road Less Traveled

It is a challenging task for science teacher educators to promote equity, social justice, and multicultural education in their curriculum and through their pedagogical strategies. The coeditors refer to taking on this challenge as the "road less traveled" because this is a difficult road to take since it requires those from the historically "dominant" culture to first examine their role in perpetuating the social inequities, "culture of poverty," and "bigotry of low expectations" that have long disenfranchised students from traditionally underserved and underrepresented groups for hundreds of years in our US school systems. It is critical that more science teacher educators "travel this road" with their students if they are to prepare these students for the increasing diversity in the classrooms of today, as well as promote equity in the STEM pipeline. The National Research Council (2012) suggests just such a perspective in its framework, suggesting that "concerns about equity should be at the forefront of any effort to improve the goals, structures, and practices that support learning and educational attainment for all [science] students" (p. 277).

Lastly, in discussing the central ideas and themes in this book, it is necessary to briefly discuss where we should be going as science teacher educators. It is imperative that we as science teacher educators move the field of science teacher education forward so that science teachers are able to successfully instruct students from diverse backgrounds to prepare them to compete in the job market and enhance their participation in STEM in the twenty-first century and beyond.

Hence, we propose the following questions on which teachers, researchers, and educators should continue to focus their efforts to find answers:

1. How do we assess high-quality teacher education for multicultural science education, equity, and social justice?
2. What do we find and what should we find relative to culturally relevant teaching when we follow science teachers into their classrooms instructing students?
3. What programs and pathways are successful at educating culturally competent science teachers, and what distinguishes these successful programs and pathways?
4. What can we learn from global and transnational teacher education work on culture, equity, and social justice to assist us in preparing and working with science teachers?

As we close this book we move forward in the continuing quest to bring about change in the "traditional" ways of thinking about science teaching and learning, while hopefully providing a framework through this book for a paradigm shift that promotes equity and social justice in science teaching in the twenty-first century. Taken holistically and summatively, all of the authors themselves are agents of change in the "call to action" for equity, social justice, and multiculturalism in science teacher education. We could not have asked for more nor expected any less.

References

Banks, J. A. (2007). Multicultural education: Characteristics and goals. In J. A. Banks & C. A. McGee Banks (Eds.), *Multicultural education: Issues and perspectives* (p. 26). Hoboken, NJ: Wiley.

Bullivant, B. (1993). Culture: Its nature and meaning for educators. In J. A. Banks & C. A. Banks (Eds.), *Multicultural education: Issues and perspectives* (pp. 29–47). Boston: Allyn & Bacon.

Douglass, F. (1849/1991). Letter to an abolitionist associate. In K. Bobo, J. Kendall, & S. Max (Eds.), *Organizing for social change: A mandate for activity in the 1990s*. Washington, DC: Seven Locks.

Fairclough, N. (1989). *Language and power*. New York: Longman.

Halliday, M. A. K. (1993). Toward a language-based theory of learning. *Linguistic and Education, 5*, 93–116.

Kahle, J. B. (2007). Systematic reform: Research, vision, and politics. In S. K. Abell & N. G. Lederman (Eds.), *Handbook of research on science education* (pp. 911–942). Mahweh, NJ: Lawrence Erlbaum Associates.

Lemke, J. (1990). *Talking science: Language, learning and values*. Westport, CT: Alex.

National Research Council. (2012). *A framework for K-12 science education: Practices, crosscutting concepts, and core ideas*. Washington, DC: National Academies Press.

Index

A
Access, 23, 42, 53, 56, 67, 69, 85, 96, 98, 106–109, 112, 134, 160, 161, 169, 170, 177, 179, 203, 215, 221, 222, 237, 262, 263, 281
Achievement, 5, 7, 11, 22, 23, 33–35, 49, 62, 64, 65, 68, 71, 73–75, 77, 82, 89, 104–112, 125, 133, 134, 160–162, 168, 177, 184, 187, 195, 197, 200, 204, 206, 209, 221, 223, 236–239, 241, 244–246, 253, 255–257, 261, 263, 266, 289
African American students, 5, 11, 26, 32, 65–67, 71, 78, 84–88, 91, 96, 106–107, 109, 144, 153, 154, 175, 177–183, 185–189, 288
African American teachers, 5, 175, 181–183, 185, 189

B
Black scientists, 4, 30–32, 34–36, 288
Black students, 1, 30, 98, 109, 177, 180, 186

C
Change, 3, 6, 14, 15, 26, 29, 30, 55, 70, 75, 77, 82, 85, 86, 88–90, 95, 97, 98, 110, 112–113, 121, 124, 127–129, 131, 134, 137, 139–140, 146, 150, 162–166, 185, 195, 199, 200, 204, 205, 234–243, 245, 246, 254, 256, 258–262, 264–266, 274, 276, 277, 280, 281, 287–290
Critical race theory (CRT), 4, 31, 32, 289
Cultural adaptation, 38
Cultural exchange, 141
Culturally competent science teachers, 287-290
Culturally relevant pedagogy, 5, 55, 81–98, 107, 244-246
Culturally responsive teaching (CRT), 61, 70–74, 83, 87, 137, 150, 155, 196, 245, 280, 281
Cultural pedagogy, 6, 233, 234, 243–246
Curriculum, 3, 6, 14, 16, 30, 51, 55, 61–63, 65, 68, 73, 75, 77, 82, 83, 85, 87, 89, 90, 92, 95, 97, 106, 107, 112, 113, 139, 152, 162, 163, 165, 169, 176, 189, 193–206, 222, 233, 234, 236, 240, 242–247, 254, 255, 258, 259, 261, 263, 264, 273, 275, 278, 280, 281, 285, 289, 290

D
Diversity, 2, 5, 54, 61, 83, 88, 91, 93–95, 109, 130, 151, 165, 182, 193–206, 215, 243, 258, 260, 262, 265, 266, 273, 277, 279, 281, 290

E
ELL. *See* English language learners (ELL)
Emergent Latino/a/bilinguals, 209, 210, 212, 213, 215–220, 223, 224
English language learners (ELL), 6, 88, 209–224, 276
Equity, 1–7, 11, 26, 41, 42, 46, 50, 56, 72, 82, 85–87, 89, 98, 103–113, 160–162, 169,

179, 184, 195–198, 202, 209, 215, 222, 253–266, 273–281, 285–290
Exclusion, 17, 46, 48, 68, 73, 182, 200, 202

F
Frameworks, 2, 4, 12–14, 30, 41, 54, 55, 82, 83, 104, 119–132, 167–162, 176, 186, 195, 199, 202, 211, 220, 234, 243, 246, 262, 266, 274, 277–281, 286, 287, 289, 290

G
Gender, 4, 17, 22, 33, 34, 42, 46–48, 50, 56, 64, 83, 98, 125, 130, 160, 195, 199, 221, 253, 276, 279, 287, 288
Globalization, 137–157

I
Immigrants, 2, 106, 138, 141, 142, 144, 146–152, 155, 156, 212, 221
Inservice teachers, 3, 5, 6, 53, 62, 72, 84, 88, 159, 166, 179, 196, 244, 257, 266
International students, 137–157, 253
International teachers, 141, 146–149, 151, 153

J
Jim Crow Era, 29–38, 288

L
Latino/a science learners, 240
Latino/a students, 209–211, 288

M
Marginalization, 5, 61, 62, 68, 72, 75, 160, 181, 182, 195, 197, 204
Middle school science curriculum, 193–206, 243
Misuse of science, 11–26
Motivation, 5, 37, 66, 87, 89, 103–113, 134, 160, 163, 175–190, 213, 239, 240, 245, 283
Multicultural science, 3, 6, 31, 54–56, 82, 195–199, 204–206, 253–266, 290

N
National science education standards (NSES), 14, 29, 52, 276, 278
Negotiating science content, 119–134

NSES. *See* National science education standards (NSES)
Nurturing, 5, 30, 175, 183–186, 189, 210, 220

P
Policy, 6, 23, 46, 55, 62, 74, 94, 112–113, 138, 142, 160, 161, 183, 189, 213, 223, 233–247, 254, 255, 258–265, 276, 280, 281, 285, 286, 289
Policy issues, 253–266, 273–281
Preparing change agents, 287–288
Preservice teachers, 3, 6, 7, 37, 50, 52–54, 83, 84, 87, 93, 95, 98, 105, 150, 161, 166, 187, 193–206, 245, 246, 262, 265, 286
Problem-based learning (PBL), 5, 111, 159–170

S
School district, 6, 107, 108, 113, 140, 143, 149, 183, 197, 198, 234, 243, 253–266, 276
Science and engineering practices, 2, 119, 120, 127, 128, 137, 169
Science attitude, 108, 223
Science core ideas, 119, 132
Science crossing cutting concepts, 121
Science teacher education, 1, 3–7, 37, 61–77, 81–97, 137, 138, 142, 146–147, 165–169, 198, 206, 273–281, 286–290
Science teacher education policy, 6, 274–281
Science teacher educators, 1–7, 12, 26, 29–32, 37, 41, 50, 52, 72, 77, 82, 90, 92–98, 105, 112, 113, 159, 168, 188, 197, 203–206, 273, 275–278, 280, 281, 285–290
Science, technology, engineering, and mathematics (STEM), 1, 3, 5, 6, 11, 12, 26, 41, 46, 49, 56, 77, 85, 93, 103–105, 107–110, 112, 113, 159, 166, 169, 175–190, 196, 197, 202, 204, 205, 233–247, 257
Scientists from underrepresented groups, 6, 61, 62, 193–206
Self-efficacy, 22, 68, 70, 77, 105, 109–110, 239
Social capital, 63, 188, 222, 273
Social Darwinism, 12–19, 289
Social justice, 1–7, 72–76, 81, 82, 85–87, 89, 94, 96–98, 103–113, 134, 137, 140,

Index   295

   197, 206, 209, 224, 246, 273–281, 285–287, 289, 290
Social justice teaching, 74–76
Sociocultural consciousness, 61–77
Socioculturally disadvantaged, 63–65, 68–71, 73, 76, 77
Sociocultural theory, 31, 210, 289
Standards, 2, 6, 14, 22, 26, 29, 38, 51, 52, 86, 88, 119, 127–132, 147, 149, 150, 152, 154, 157, 159, 160, 162, 167, 169, 201, 202, 206, 210, 216, 242–244, 247, 254, 257, 259, 260, 262, 264, 274–280
STEM. *See* Science, technology, engineering, and mathematics (STEM)
STEM-based professional development, 6, 233–247
Stereotypes, 4, 23, 46, 66, 71, 97, 107, 130, 161, 185, 188, 193–202, 204, 206, 245, 266, 274
Stigmatization, 5, 175–190
Structural disenfranchisement, 179

T
Teacher actions, 81, 84, 88, 89
Teacher beliefs, 81, 84, 85, 88, 97, 178, 276
Teacher education, 1, 3–7, 33, 37, 51, 61–77, 81–98, 137–140, 142, 146–150, 155, 156, 165–169, 186, 188, 189, 195, 197–199, 203–206, 215, 224, 257, 258, 273–281, 286–290
Teacher support, 6, 265, 266
Triangulation of cultures and languages, 209–224

U
Underrepresented, 2–6, 11, 25, 42, 50, 61, 62, 64–67, 77, 95, 97, 98, 103, 106–110, 112, 159, 164, 169, 175–190, 193–206, 246, 258, 259, 274–276, 280, 286, 289, 290
Urban students, 6, 11, 31, 159, 160, 162–164, 169, 170, 177, 274, 275, 279, 280

Printed in the USA
CPSIA information can be obtained
at www.ICGtesting.com
LVHW021240221223
766979LV00005B/321